MPEG VIDEO COMPRESSION STANDARD

Edited by

Joan L. Mitchell,
William B. Pennebaker,
Chad E. Fogg, and
Didier J. LeGall

Join Us on the Internet

WWW: http://www.thomson.com
EMAIL: findit@kiosk.thomson.com

thomson.com is the on-line portal for the products, services and resources available from International Thomson Publishing (ITP). This Internet kiosk gives users immediate access to more than 34 ITP publishers and over 20,000 products. Through *thomson.com* Internet users can search catalogs, examine subject-specific resource centers and subscribe to electronic discussion lists. You can purchase ITP products from your local bookseller, or directly through *thomson.com*.

Visit Chapman & Hall's Internet Resource Center for information on our new publications, links to useful sites on the World Wide Web and an opportunity to join our e-mail mailing list. Point your browser to: **http://www.chaphall.com/chaphall.html** or **http://www.chaphall.com/chaphall/electeng.html** for Electrical Engineering

A service of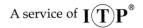

MPEG VIDEO COMPRESSION STANDARD

Edited by

Joan L. Mitchell,
William B. Pennebaker,
Chad E. Fogg, and
Didier J. LeGall

CHAPMAN & HALL

 INTERNATIONAL THOMSON PUBLISHING

New York · Albany · Bonn · Boston · Cincinnati · Detroit · London · Madrid · Melbourne
Mexico City · Pacific Grove · Paris · San Francisco · Singapore · Tokyo · Toronto · Washington

Cover design: Curtis Tow Graphics

Copyright © 1997 by Chapman & Hall, New York, NY

Printed in the United States of America

Chapman & Hall
115 Fifth Avenue
New York, NY 10003

Chapman & Hall
2-6 Boundary Row
London SE1 8HN
England

Thomas Nelson Australia
102 Dodds Street
South Melbourne, 3205
Victoria, Australia

Chapman & Hall GmbH
Postfach 100 263
D-69442 Weinheim
Germany

International Thomson Editores
Campos Eliseos 385, Piso 7
Col. Polanco
11560 Mexico D.F
Mexico

International Thomson Publishing–Japan
Hirakawacho-cho Kyowa Building, 3F
1-2-1 Hirakawacho-cho
Chiyoda-ku, 102 Tokyo
Japan

International Thomson Publishing Asia
221 Henderson Road #05-10
Henderson Building
Singapore 0315

1 2 3 4 5 6 7 8 9 10 XXX 01 00 99 98 97 96

Library of Congress Cataloging-in-Publication Data

MPEG video : compression standard / Joan L. Mitchell . . . [et al.].
    p. cm.
   Includes bibliographical references and index.
   ISBN 0-412-08771-5 (alk. paper)
    1. Digital video.   2. Video compression -- Standards.   3. Sound -
- Recording and reproducing -- Digital techniques -- Standards.
 4. Coding theory.   I. Mitchell, Joan L.
   TK6680.M64   1996
   621.388--dc20   96-31124
                                       CIP

British Library Cataloguing in Publication Data available

To order this or any other Chapman & Hall book, please contact **International Thomson Publishing, 7625 Empire Drive, Florence, KY 41042.** Phone: (606) 525-6600 or 1-800-842-3636. Fax: (606) 525-7778. e-mail: order@chaphall.com.

For a complete listing of Chapman & Hall titles, send your request to **Chapman & Hall, Dept. BC, 115 Fifth Avenue, New York, NY 10003.**

Dedication

To
   Nancy, Don, and Sandy
   Margaret, Betsy, Patty, and Erik
   Lori and John.

Series Preface

This book initiates a new digital multimedia standards series. The purpose of the series is to make information about digital multimedia standards readily available. Both tutorial and advanced topics will be covered in the series, often in one book. Our hope is that users will find the series helpful in deciding what standards to support and use while implementors will discover a wealth of technical details that help them implement those standards correctly.

In today's global economy standards are increasingly important. Yet until a standard is widely used, most of the benefits of standardization are not realized. We hope that standards committee chairpeople will organize and encourage a book in this series devoted to their new standard. This can be a forum to share and preserve some of the "why" and "how" that went into the development of the standard and, in the process, assist in the rapid adoption of the standard.

Already in production for this series are books titled *Digital Video: Introduction to MPEG-2* and *Data Compression in Systems*.

Foreword

The reduction of bitrate of digitised television signals has been an R&D topic in research centres and academia for more than 30 years. Many were the purposes for which research was funded, but all shared the basic rationale. If you had a digital or digitised delivery medium and could bring the more than 200 Mbit/s bitrate of PCM television down to a value that could be handled economically by signal processing technology then you would achieve the goal of substantially improving existing products/services or even creating new ones. The bitrate of up to 1.5 Mbit/s was the value around which three industries shared a common interest towards the end of the '80s: consumer electronics with compact disc, broadcasting with digital radio and telecommunications with Narrowband ISDN and ASDL. The merit of MPEG was its ability to identify the opportunity and to bring disparate industries to work together. The unprecedented degree of international collaboration involving hundreds of researchers produced a standard - MPEG-1 - of very high technical quality. Even today, after several years and countless implementations not a single error was found in the specification. MPEG-1 devices are now counted by the millions in all continents. MPEG-1 encoded content can be found everywhere and is the first implementation of delivery media independence of content. The beginning of the '90s provided the next opportunity with a general interest in a new standard - MPEG-2 - for applications spanning all industries: broadcasting over satellite, terrestrial and CATV networks, Broadband ISDN, digital discs and tapes etc. The availability of another MPEG standards promised - and delivered - as scheduled was not a small contribution to the multimedia revolution we are living today. The book by Drs. Le Gall, Fogg, Mitchell and Pennebaker is a recommended reading for those who want to have a thorough understanding of the Video parts of MPEG-1 and MPEG-2. The concepts that were developed in those standards and presented in the book are a necessary reading for anybody who needs to work on them or go beyond them.

Leonardo Chiariglione
Convenor, ISO/IEC JTC1/SC29/WG11 (MPEG)

Acknowledgments

Several hundred technical experts worked hard, attended meetings all over the world, and spent many, many hours of personal and professional time before reaching a consensus on the MPEG-1 standard. We want to specifically acknowledge their dedicated work even though they are too numerous to be listed by name.

JM wishes to thank a former manager, Cesar Gonzales, for permission to start work on this book. She wishes to thank her current IBM management and in particular, Fred Mintzer and Howard Sachar, for giving her permission to take a two-year leave of absence from IBM. Her participation in writing this book would have been much more difficult without this extra freedom. She also acknowledges the support of the Computer and Electrical Engineering Department and the Beckman Institute at the University of Illinois, Urbana, Ill., for their 1996 part-time Visiting Professor and Visiting Scientist positions which she held during the final stages of finishing this book. She also wants to thank Christiana Creighton; Sandy and David Creighton; Anne Gilligan; Sandy Grover and Dorothy Cullinan of the Cutleaf Maples Motel, Arlington, VT; Jackie and Bruce Kelman; Joan and Don Lucas; Sherri and Peter McCormick; Doris, Bill, and Carol Mitchell; Nancy and Don Mitchell; Margaret and Bill Pennebaker; Ruth Redden; Frances and Charles Riggs; and Norma and Don Vance for encouragement, meals, and lodging during the writing of this book.

WP wishes to thank his wife, Margaret, for her patience and active support during the preparation of this book and the implementation of a MPEG-1 video encoder and decoder. Acknowledgment is also given to the neighbors who took the subjective quality tests, results of which are in this book.

We also want to thank the vendors who supplied information for this book; Cliff Reader, who contributed the IPR chapter; our many editors at Van Nostrand Reinhold and later at Chapman and Hall; Nancy Mitchell, who reviewed this book for consistency at the easiest technical level; and all of the other people who reviewed this book and helped us correct our errors before publication.

Contents

List of Figures

List of Tables

Trademarks

VIDEO*FLOW* is a registered trademark of Array Microsystems, Inc. Commodore and the Commodore logo are trademarks of Commodore Electronics Limited. Amiga and Amiga CD32 are trademarks of Commodore-Amiga, Inc. "MPEG Are US" is a trademark of CompCore Multimedia, Inc. VideoRisc is a trademark of C-Cube Inc. DIGITAL is a registered trademark of Digital Equipment Corporation. ReMultiplexer is a trademark of DiviCom, Inc. Dolby is a trademark of the Dolby Corporation. FutureTel, PrimeView, TeleMux, and MPEGWORKS are trademarks of FutureTel, Inc. DigiCipher and VideoCipher are registered trademarks of General Instrument Corporation. DigiCipher II is a trademark of General Instrument Corporation. Imedia is a trademark of Imedia Corporation. IBM is a registered trademark and IBM Microelectronics is a trademark of the International Business Machines Corporation. CoreWare is a trademark of LSI Logic Corporation. MS-DOS and MS-Windows are registered trademarks of Microsoft Corporation. Eikona is a registered trademark of Siemens Ltd. BetaCam is a registered trademark of Sony Corporation. SPARCstation and SunVideo are trademarks of Sun Microsystems, Inc. XingCD is a registered trademark of Xing Technology Corp.

MPEG VIDEO COMPRESSION STANDARD

Edited by
Joan L. Mitchell,
William B. Pennebaker,
Chad E. Fogg, and
Didier J. LeGall

1

Introduction

The subject of this book is the international standard for moving picture video compression, IS 11172-2, widely known as MPEG-1 video. The name MPEG is an acronym for Moving Picture Experts Group, a group formed under the auspices of the International Organization for Standardization (ISO) and the International Electrotechnical Commission (IEC) . Originally an ad hoc subcommittee of ISO/IEC JTC1/SC2/WG8 where the MPEG work started, MPEG was later given formal status within ISO/IEC and is now the working group ISO/IEC JTC1/SC29/WG11.

Although IS 11172 is officially entitled "Information technology — Coding of moving pictures and associated audio for digital storage media up to about 1,5 Mbit/s —," [1] as indicated by the title, MPEG-1 is concerned with coding of digital audio and digital video. It is also concerned with synchronization of audio and video bitstreams, including multiple interleaved video sequences. These three topics are covered in three separate parts of the standard: Part 1 deals with systems aspects, Part 2 with video compression, and Part 3 with audio compression.[2]

In this book the emphasis will be almost entirely on the compression of picture sequences, i.e., on IS 11172 Part 2: *Video*. While brief overviews of the audio and systems aspects of MPEG-1 are provided, it is our expectation that future volumes in this series will do more justice to these important topics.

If the current standard is known as MPEG-1, what does the "1" signify? It turns out that there are three different MPEGs: MPEG-1, MPEG-2, and MPEG-4,[3] and data rate and applications are distinguishing factors

---

[1] Note the European convention of a comma for the decimal point. We will follow the U.S. convention and use a period, unless we are quoting from the standard.

[2] Part 4 covers compliance testing, and Part 5 is a technical report on software implementation.

[3] There was once an MPEG-3, but it was merged with MPEG-2.

between them. MPEG-1 is intended for intermediate data rates on the order of 1.5 Mbit/s, MPEG-2 is intended for higher data rates (10 Mbit/s or more), and MPEG-4 is intended for very low data rates (about 64 Kbit/s or less). MPEG-2 and MPEG-4 have potential uses in telecommunications, and for this reason, a third major international standards organization, the International Telegraph and Telephone Consultative Committee (CCITT), is participating in their development.[4] At this point in time, MPEG-1 and MPEG-2 video are complete (some later parts of MPEG-2 are still being written), and MPEG-4 is still in the algorithm development stage.

One version of MPEG-2 video, the *main profile*, is a relatively straight-forward extension of MPEG-1. Therefore, while this book is concerned primarily with MPEG-1, an overview of the main profile of MPEG-2 will also be given. We note, however, that there is a lot more to MPEG-2, including four other profiles that provide major extensions in flexibility and function. Another volume in this series provides a far more comprehensive treatment of MPEG-2 [HPN97].

1.1   Why compress? ○

A sequence of pictures with accompanying sound track can occupy a vast amount of storage space when represented in digital form. For example, suppose the pictures in a sequence are digitized as discrete grids or arrays with 360 pels (picture elements) per raster line and 288 lines/picture, a resolution that is fairly typical for MPEG-1. Assuming the picture sequence is in color, a three-color separation (equivalent to the three primaries, red, green, and blue) can be used for each picture. If each color component in the separation is sampled at a 360x288 resolution with 8-bit precision, each picture occupies approximately 311 Kbytes. If the moving pictures are sent uncompressed at 24 pictures/second, the raw data rate for the sequence is about 60 Mbit/s, and a one-minute video clip occupies 448 Mbytes.

For the audio track the data rates are not quite as formidable. If the sound track is in stereo and each of the two channels is sampled with 16-bit precision at a 44 KHz sampling rate, the data rate is about 1.4 Mbit/s.

The problem MPEG-1 addresses is how to store a moving picture sequence, including the audio information, in a compressed format suitable for a digital storage medium such as CD-ROM (compact disc - read only memory). The magnitude of the problem surfaces immediately when one reflects that this is exactly the same digital storage technology that is used for high-quality compact disc (CD) sound reproduction. A high-quality stereo sound track *plus* the video information must be stored in about the same

---

[4]The name of the CCITT was recently changed to International Telecommunications Union-Telecommunication Sector (ITU-T).

space that is normally used for the audio alone. Further, this must be done in a way that permits playback with precise timing control and synchronization of audio and video bitstreams.

1.2   Why standardize video compression? ○

When the MPEG standardization effort was initiated in 1988, several different industries were converging on digital video technology. The computer industry was looking to expand beyond its traditional text and graphics capabilities into interactive multimedia with audio, video, and still images. The consumer electronics industry saw digital video as a means for improving the capabilities and interactivity of video games and entertainment media. Compact discs were already being used for digital storage, and the storage capacities were sufficient for compressed digital video. The telecommunications industry was taking advantage of the maturing compression technology to standardize teleconferencing, and the CCITT Recommendation H.261 digital video standard for teleconferencing was already in development. The debate about broadcasting a digital versus a higher bandwidth analog signal for high-definition television (HDTV) was about to start and direct broadcast of digital video from satellites was also being considered.

Cable television companies were exploring the use of digital video for satellite uplinks and downlinks, and they were also considering delivery of digital video to the home. Together with the microprocessors that would be used for control of home converters, direct digital links into the home raised the possibility of new function through the cable networks. Other links into the home such as Videotex also had potential uses for moving sequences.

With the sharing of technology among these diverse industrial interests, standardization was a natural development. If standardization of digital video could be achieved, the potentially large sales volumes in the consumer market would lower the cost of VLSI chips for all of the industries. Thus, the cost reduction implicit in having a common digital video technology for all has been a key driving force for standardization. The convergence of these interests and the resulting standardization is lowering the barriers that are usually present to the deployment of new technology, and therefore reducing the risks.[5]

The MPEG-1 algorithm development was in its final stages, but had not completed the standards balloting process, when MPEG-2 was started.

---

[5]Not everyone would agree with this thesis. A number of proprietary teleconferencing systems have been marketed over the past decade, and a video compression technique known as Indio is finding use in multimedia. However, some of the individuals and corporations that were involved in these alternatives have also actively contributed to MPEG, and it remains to be seen how these other techniques will fare in relation to MPEG.

The major focus of this second project was higher quality and bandwidth applications, but there were some secondary issues such as the increased flexibility and multiple resolutions needed by some applications. Although the door was open to totally different technologies, it turned out that the best proposals for MPEG-2 were based on enhancements of MPEG-1. In the authors' opinion the rapid acceptance of MPEG-2 is in no small way due to the fact that it is a natural extension of MPEG-1. The similarities between MPEG-1 and the main profile of MPEG-2 also made it relatively easy for us to include that aspect of MPEG-2 in this book.

1.3   Vocabulary ○

A number of basic terms will be used throughout this book, and the following few paragraphs of definitions may be of help to the non-experts among our readers. More extensive discussions of these terms are found in [PM93] and [RJ92].

1.3.1   Sequences, pictures, and samples ○

An MPEG video sequence is made up of individual pictures occurring (usually) at fixed time increments, as shown in Figure 1.1. Because the pictures are in color, each picture must have three *components*. Color is expressed in terms of a *luminance* component and two *chrominance* components. The luminance provides a monochrome picture such as in Figure 1.4, whereas the two chrominance compcnents express the equivalent of color hue and saturation in the picture. Although these components are not the familiar red, green, and blue (RGB) primaries, they are a mathematically equivalent color representation that can be more efficiently compressed. They are readily converted to RGB by relationships that will be described in Chapter 4.

Each component of a picture is made up of a two-dimensional (2-D) grid or array of samples. Each horizontal line of samples in the 2-D grid is called a raster line, and each sample in a *raster line* is a digital representation of the intensity of the component at that point on the raster line. Note, however, that the luminance and chrominance components do not necessarily have the same sampling grid. Because the eye does not resolve rapid spatial changes in chrominance as readily as changes in luminance, the chrominance components are typically sampled at a lower spatial resolution.

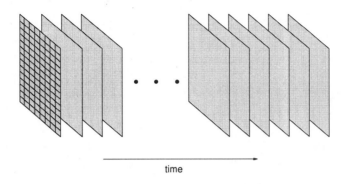

time

Figure 1.1: Illustration of an MPEG video sequence.

1.3.2 Frames and fields ○

In broadcast analog video standards such as NTSC, PAL, or SECAM[6] video
sequences are temporally subdivided into frames and raster lines in much the
same manner as in Figure 1.1. However, the signal within each raster line
is analog rather than digital. Further, as shown in Figure 1.2, each frame is
divided into two *interlaced* fields. Each field has half the raster lines of the
full frame and the fields are interleaved such that alternate raster lines in
the frame belong to alternate fields. The frame repetition rate is therefore
half the field rate.[7]

The relationship between analog video frames and fields and MPEG pictures is undefined for MPEG-1 and the assumption is made that appropriate
2-D pictures will be extracted from this field/frame format. The extensions
in MPEG-2 allow interlaced pictures, coding them either as individual fields
or full frames. Artifacts due to blanking are usually cropped before pre-
senting the pictures to an MPEG encoder. The typical TV uses overscan to
keep the blanking intervals hidden, and these areas are therefore normally
not visible. An MPEG decoder must in some clever way recreate these
cropped edge areas, and a poorly adjusted TV set might therefore exhibit
some curious artifacts near picture edges.

---

[6]NTSC, PAL and SECAM are three commonly-used analog video formats. They are
acronyms for National Television System Committee, phase alternating line, and sequentiel
couleur a memoire, respectively.

[7]A vertical blanking interval separates the active display of each field, and this blanking
interval typically inhibits the video signal for half of the top and bottom lines of field 1.

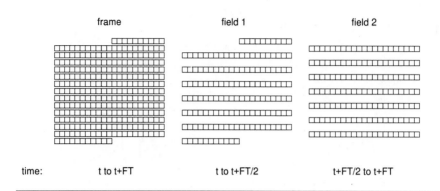

Figure 1.2: Analog video frame and field structure. FT is the time interval for a frame.

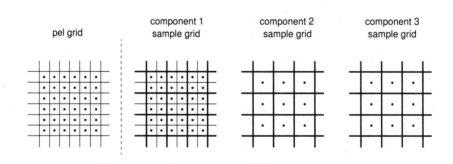

Figure 1.3: Component samples for MPEG-1 pels.

1.3.3   Pels and pixels ○

The component sample values at a particular point in a picture form a *pel* (picture element).[8] If all three components use the same sampling grid, each pel has three samples, one from each component. However, as noted above, the eye is insensitive to rapid spatial variations in chrominance information. For this reason, MPEG-1 subsamples the components that supply the color information, using a chrominance grid that is a factor of 2 lower in resolution in each dimension. Thus, a pel is defined to be the color representation at the highest sampling resolution, but not all samples that make up the pel are at that resolution. MPEG-1 pels are sketched in Figure 1.3.

Note that when lower sampling resolution is used for some components

---

[8] *Pixel* is also a commonly used abbreviation for picture element, but it is rarely used in the MPEG standard.

in a pel, the positioning of lower-resolution samples relative to the highest-resolution samples must be defined. In Figure 1.3 dots indicate the sample positions. The bold lines have been superimposed on the component 1 sample grid to illustrate the relative alignment of the component samples. In MPEG-1 the sample positioning is as in Figure 1.3; in MPEG-2 a different relative positioning is defined.

1.3.4 Compression vocabulary ○

The compression techniques used by MPEG are divided into two basic classes, *intra* and *nonintra*. Intra techniques compress a picture using information only from that picture; nonintra techniques also use information from one or two other pictures displaced in time.

The input to a data compressor is often termed the *source* data, whereas the output from a decompressor is called *reconstructed* data. Some compression techniques are designed such that reconstructed data and source data exactly match, and these techniques are termed *lossless*. With other techniques the reconstructed data are only a (hopefully) good approximation to the source data, and these techniques are termed *lossy*. The distinction between these two classes of compression techniques is extremely important, since purely lossless compression is normally far less efficient (in terms of bits/pel) than lossy compression. As will be discussed in Chapter 2, most data compression algorithms combine lossy and lossless elements. Very effective compression usually results when the lossy elements are designed to selectively ignore structure in the data that the eye cannot see (or the ear cannot hear).

1.3.5 Standards vocabulary ○

Standards documents separate material into two classes. *Normative* sections define the actual standard. *Informative* sections provide background material and descriptions, but do not in any way define the standard or any of the requirements and restrictions.

Standards documents also use a very particular vocabulary. Thus, the term *shall* has a particular meaning in "standardeze", and is used to signify that the particular rule or requirement is normative and the implementor has no choice but to do exactly as the standard states. *May* indicates that the implementor is allowed some latitude. The rule suggested in the standard can be followed, but if the implementor so desires, he or she can follow an alternative or perhaps not implement that part of the standard at all. *Must* indicates that no alternative exists.[9]

---

[9]However, the MPEG standard does not always follow these rules rigorously. For example, in Clause 2.3 of the standard, there is the following statement: "In particular,

As is typical of standards documents, the MPEG standard usually states precisely how something "shall" be done, but is often missing the explanations and the rationales for why it is done in that particular way. One of the principal goals of this book is to provide these explanations.

1.3.6   MPEG vocabulary ○

As is true for any standard, MPEG uses a specialized vocabulary that is carefully defined in that particular standard. Most definitions are consistent with conventional technical usage. However, in some cases the term is more narrowly defined or may even be uniquely defined for that standard. These definitions are normative and take precedence over any conventional usage.

For example, consider the MPEG terms *access unit* and *presentation unit*. According to the definitions in subclause 2.1 of the MPEG-1 standard, these two terms apply to both video and audio bitstreams, but with different definitions. For audio, the presentation unit is a decoded audio access unit, and the audio access unit is the smallest independently decodable segment of an audio bitstream. For video, the presentation unit is a picture and the video access unit is the coded representation of that picture. Most pictures in a typical sequence are not independently decodable.

When reading the MPEG standard (or any standard, for that matter), care must be taken to identify this specialized vocabulary. Further, the definitions in force in each part of the standard must be clearly understood, or misinterpretation is possible. Unless we state otherwise, our usage in this book will be consistent with the definitions for MPEG video.

1.4   A quick look at MPEG compression ○

Although there are many facets to data compression and image coding, perhaps the key aspect of moving picture compression is the similarity between pictures in a sequence such as in Figure 1.4. The similarity is illustrated more clearly by taking differences between the pictures, as is done in Figure 1.5. In this difference sequence, pel differences of zero are neutral gray, positive differences are proportionally lighter, and negative differences are proportionally darker.

The most visible differences in Figure 1.5(a) occur where objects, such as the ball, are in motion. As a general rule, less activity (smaller differences)

---

it [subclause 2.4.3] defines a correct and error-free input bitstream. Actual decoders *must* [our emphasis] include a means to look for start codes in order to begin decoding correctly, and to identify errors, erasures or insertions while decoding. The methods to identify these situations, and the actions to be taken, are not standardized." It is certainly possible to construct a decoder that does not have error recovery procedures, but it would be wiser to follow the MPEG committee requirement.

(a)

(b)

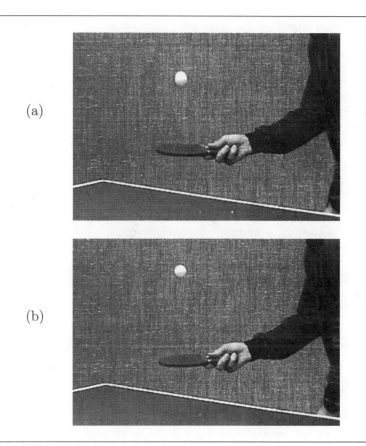

Figure 1.4: Pictures 30 (a) and 32 (b) of the tennis video sequence.

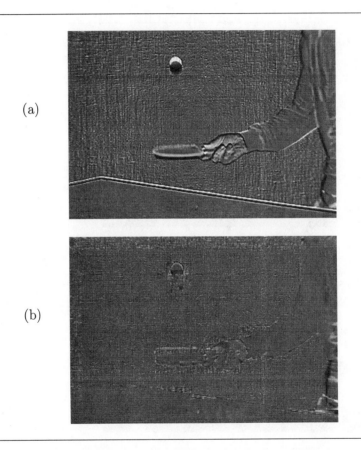

Figure 1.5: Picture differences: (a) simple differences. (b) motion-compensated differences.

in a picture leads to better compression. If there is no motion, as is the case in the center of the pictures, there is no difference, and the sequence can be coded as a single picture followed by a few bits for each following picture telling the decoder to simply repeat the picture. Not surprisingly, this is exactly what MPEG does in this situation.

Of course, coding a completely unchanging sequence is a very special case, and much more must be done to code a typical picture sequence. The sequence must start somewhere, and a technique is therefore needed for coding the first picture without reference to neighboring pictures. It turns out that this technique must also be used at regular intervals in a typical sequence. If pictures in a sequence are not occasionally coded independently of neighboring pictures, video editing becomes very difficult. Furthermore, because of bit errors during transmission or the small differences allowed between encoder and decoder image reconstructions, differences between the original source picture sequence and the sequence reconstructed by a decoder may build up to unacceptable levels.

The high compression needed by MPEG applications is achieved by coding most of the pictures as differences relative to neighboring pictures. MPEG accomplishes this in a variety of ways: Some parts of the picture — those where no significant changes occur — are simply copied. Other parts are imperfectly predicted from the adjacent picture and corrections must be coded. Still other parts may be best predicted by parts of the image that are displaced because of motion, and this requires the use of a technique called *motion compensation*. For example, in the peripheral areas of pictures in Figure 1.4, camera zoom caused a systematic displacement inward from (a) to (b). This is why the differences are more visible in the peripheral areas of Figure 1.5(b).

The differences are greatly reduced when the encoder searches picture (a) for a best match to (b), as is done in Figure 1.5(b). When parts of the picture are in motion, the best match for a given region between Figure 1.4(b) and (a) is shifted slightly because of the motion. If the position of (a) is shifted by that amount before taking the difference, Figure 1.5(b) results. Clearly, motion compensation eliminates much of the activity.

Some parts of a picture may have new content that does not exist in the reference picture — for example, areas uncovered when an object moves and new areas introduced by panning and zooming. This situation can be handled by coding these parts of the picture without reference to adjacent pictures. However, in many cases it can also be handled by predicting backward in time, assuming that an upcoming picture has already been coded and transmitted. An inviolable rule of data compression systems is that predictions can only be made from information available to the decoder.

A large asymmetry exists between MPEG encoders and MPEG decoders in terms of complexity and computational requirements, and this is partly due to motion compensation. The encoder has the task of finding the motion displacements (*vectors*), whereas the decoder merely uses the values transmitted to it by the encoder. In other places as well, the encoder must traverse a difficult decision tree to determine what to send to the decoder; the decoder simply does what it is told to do.

1.5   Compact pseudocode notation in MPEG ○

Those readers who have already had some exposure to MPEG will know that the standard uses a compact pseudocode notation in the normative sections of the standard. This is a very precise notation that is closely related to the C language; however, it can sometimes be quite cryptic and difficult to understand. In the authors' experience, those who have mastered this notation are very enthusiastic about it. However, there is a steep learning curve — especially for readers who do not have a background in C programming — that makes it difficult for many to understand MPEG. One of the goals of this book is to provide an alternative to this compact notation that allows readers without a C language background to understand MPEG video.

In the final analysis, however, no one who really needs to understand a standard can avoid reading the actual document. For this reason, we will also use pseudocode in some of our explanations, but with one significant improvement — we have added comments to almost every statement. Comments are informative, and therefore would have been inappropriate in the normative sections of the MPEG standard that define the pseudocode. We feel, however, that they add enormously to the readability of the code.

1.6   MPEG applications ○

Although MPEG started out as an anticipatory standard, its acceptance has been so rapid that in many cases the applications have been ahead of the approval process. For example, even before the standard was approved, software, multimedia boards, and chips implementing MPEG-1 were already available. Games, digital recorders, CD-ROM movies, digital TV, and multimedia are just a few of the current applications. Indeed, the headline, "Crooks Crack Digital Codes of Satellite TV," from the January 12, 1996 issue of the Wall Street Journal[TR96] hints at how financially important the digital satellite TV application has already become.

The committee made some effort to keep MPEG-1 stable and unchanged once technical consensus had been reached. This made possible these early implementations, but also prevented the standard from being modified to meet the higher bandwidth needs addressed by MPEG-2. This was one of the key reasons for the separate development of MPEG-2.

The acceptance of MPEG-2 has been even more rapid. High-definition television (HDTV) standards in the United States, Europe, and Japan are expected to use MPEG-2, and it will also be used in direct broadcast of high-quality digital video at current resolutions. Indeed, in at least one case deployment of decoders for this latter application has been delayed until the MPEG-2 standard is in final form [Rob94].

1.7   Organization of the book ○

The more technical sections of this book assume a basic knowledge of image coding concepts. However, we have included overview sections at the start of each chapter that provide background for non-technical readers. The chapters of the book are organized as follows:

Chapter 2 discusses the full MPEG-1 system, providing an overview of all three parts of the Standard. Although the primary emphasis in this book is on MPEG video compression, video data streams without audio have very limited value. In addition, without the synchronization and control provided by the MPEG systems layer, the compressed bitstreams are unusable. The basic structure of an MPEG-1 bitstream and the start code syntax are treated here. In addition, a high-level description is given of key aspects of the MPEG video compression algorithms such as the discrete cosine transform (DCT). MPEG-2 and MPEG-4 are briefly discussed.

Chapter 3 discusses the definition and mathematical structure of the 2-D DCT and describes the requirements for an MPEG-compliant DCT. Chapter 4 provides a review of selected aspects of visual perception and applications of this science to MPEG. Chapter 5 provides a review of the basic concepts and principles involved in MPEG compression of video sequences, ending with high-level block diagrams for an MPEG-1 encoder and decoder. Chapter 6 provides a tutorial on the MPEG pseudocode syntax and on the flowchart conventions used in this book. Chapter 7 provides an overview of the MPEG-1 system syntax.

Chapter 8 contains an in-depth discussion of MPEG-1 video bitstream syntax. Both commented pseudocode and flowcharts are presented. Chapter 9 provides a brief overview of the "main profile" of MPEG-2. Chapter 10 describes the syntax of MPEG-2 main profile. Chapter 11 provides an overview of motion compensation techniques, with an emphasis on the block matching techniques relevant to MPEG.

Chapter 12 covers pel reconstruction and includes an in-depth description of IDCT mismatch and the IDCT requirements for MPEG. Chapter 13 describes some of the techniques an encoder may use for estimating motion displacements. Chapter 14 reviews techniques for varying the DCT quantization in an MPEG encoder. Chapter 15 describes techniques for controlling bitrate.

Chapter 16, written by Cliff Reader, provides a brief analysis of patents that are relevant to MPEG. Chapter 17 provides a sampling of MPEG vendors and products. Chapter 18 gives a history of MPEG, and the final chapter of this book, Chapter 19, briefly describes other standardization efforts in the area of video compression.

1.8   Level of difficulty ○

The reader may have wondered why open circles have been placed after each section heading of this chapter. Since this book is intended for a rather diverse set of readers ranging from nontechnical to highly technical in background, each section is marked with a symbol indicating the approximate level of technical difficulty of the section. The three levels of difficulty are:

○   Sections suitable for all readers. For the most part, these sections avoid the use of mathematics and are not dependent on material from more difficult sections.

◑   Sections of intermediate technical difficulty. These sections build upon the material in the sections marked by ○ , and should be understandable to most readers with mathematical or engineering training. They should be of particular interest to MPEG implementors.

●   Sections that are either quite challenging or are primarily aimed at those readers who need to delve deeply into the details of MPEG theory or implementation.

1.9   An international collaboration ○

Several hundred technical experts typically attend the MPEG international committee meetings. Many more are involved at their national level. For example, the meetings of the American National Standards Institute (ANSI) committee, X3L3.1, often have more than 100 attendees. At the international level, simultaneous meetings of subgroups concerned with MPEG-1 video, MPEG-1 audio, MPEG-1 systems, MPEG-2 video, MPEG-2 audio, MPEG-2 systems, MPEG-4, testing, etc. are often held, and many companies therefore send more than one representative. With such broad participation, it is virtually impossible to acknowledge each individual contribution.

Many individuals contributed in very significant ways, and the MPEG standard would not have happened without their hard work and dedication and the support of their employers.

The authors of this book have had somewhat different exposures to MPEG during its development. CF and DL have been deeply involved in MPEG. DL chaired the MPEG video subgroup, and CF is a regular participant in MPEG and has been heavily involved in the MPEG software implementation that is Part 5 of the Standard. The other two authors, JM and WP, were occupied with parallel standardization efforts in JPEG and JBIG and did not contribute to the MPEG activities. In some ways these latter two authors felt a bit presumptuous in tackling this book. They started out, in fact, only as editors of this series, but concerted efforts to find alternate authors from the enormous pool of talent contributing to MPEG failed. With the exception of the MPEG video chair, and later, CF, all were too busy either completing the standard or implementing for their companies to be able to meet a tight publication schedule. In some ways, however, having two authors not intimately involved in the generation of the standard has been an advantage, as many things that might be obvious to someone who worked with the standard as it evolved were not so obvious to them.

1.10   An evolving standard ○

Complex standards such as MPEG are not easily arrived at, and require a lot of debate and discussion before consensus is reached. They evolve over time, partly because of a continuing process of refinement and the gradual elimination of competing concepts, and partly because of the influence of new applications that hope to use the standard. Early implementors therefore need to be fully aware that until the standard is fully finished, approved, and published, it may change. MPEG-1 has reached that status, but some parts of MPEG-2 could still be changed. Customers may prefer an early implementation of the Standard that will need upgrading later to maintain compliance rather than a proprietary algorithm, but should be made aware that changes are possible.

One source of information about various MPEG activities that should be current and up-to-date is the Internet World Wide Web MPEG page. The MPEG address is *http://www.mpeg.org* and this is a good place to get background information about the MPEG standards and committee activities. A WWW search should find many sources of MPEG movies, software, and hardware.

Because this standard, in common with most, is still going through a continuing process of refinement, implementors should obtain the latest version and not rely on the contents of either this book or early drafts of the

standard. In the United States, an up-to-date version can be obtained from
the American National Standards Institute (ANSI).

2

Overview of MPEG

ISO/IEC 11172, the document describing MPEG-1, currently contains five parts. Part 1 covers the MPEG system layer, Part 2 describes MPEG video, and Part 3 specifies MPEG audio. Part 4 is on compliance testing, and Part 5 is a software reference model for MPEG-1.

Although the main topic of this book is MPEG video, MPEG audio and MPEG systems are also crucial parts of a complete MPEG implementation. A video bitstream is fully decodable, but, by itself, is an incomplete specification. The MPEG system layer contains control information that enables parsing and precise control of playback of the bitstream. Furthermore, a video stream without audio has only limited use.

This chapter presents an overview of all three parts: systems, audio, and video. As an aid for people who have access to the official MPEG-1 video standard, we sometimes reference a particular clause in that document (clause is standardeze for section). These references have the format "IS[w:x.y.z]", where IS refers to ISO/IEC 11172, w is the part number (systems, video, audio, etc.) and x.y.z is the clause or sub-clause number. If w is 2, it may be omitted. Thus, IS[2:2.3] and IS[2.3] both refer to subclause 2.3 of the video part.

2.1 MPEG system layer ○

The MPEG system layer has the basic task of combining one or more audio and video compressed bitstreams into a single bitstream. It defines the data stream syntax that provides for timing control and the interleaving and synchronization of audio and video bitstreams.

From the systems perspective, an MPEG bitstream is made up of a system layer and compression layers. The system layer provides an envelope for the compression layers. The compression layers contain the data fed

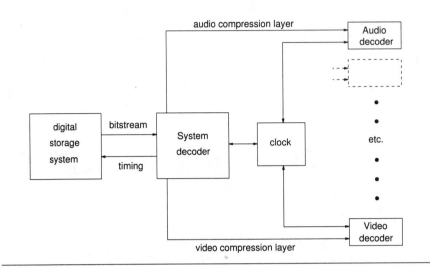

Figure 2.1: MPEG system structure.

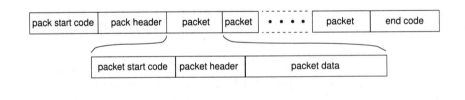

Figure 2.2: System layer pack and packet structure.

to the audio and video decoders, whereas the system layer provides the controls for demultiplexing the interleaved compression layers. A typical MPEG system block diagram is shown in Figure 2.1.

The MPEG bitstream consists of a sequence of packs that are in turn subdivided into packets, as sketched in Figure 2.2. Each pack consists of a unique 32-bit byte-aligned pack start code and header, followed by one or more packets of data. Each packet consists of a packet start code (another unique 32-bit byte-aligned code) and header, followed by packet data (compressed audio or video data). The system decoder parses this bitstream (the pack and packet start codes can be detected without decoding the data) and feeds the separated video and audio data to the appropriate decoders along with timing information.

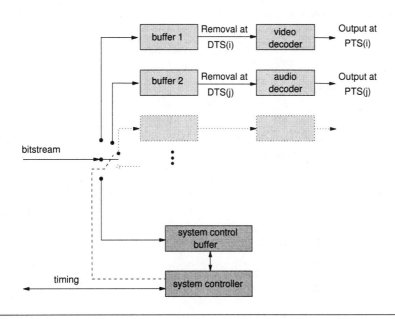

Figure 2.3: A system target decoder with two elementary bitstreams, one video and one audio. The decoders may need to internally reorder the decoded data.

2.1.1   System target decoder ○

The MPEG system uses an idealized decoder called the system target decoder (STD). This idealized decoder interprets the pack and packet headers, delivering the elementary bitstreams to the appropriate audio or video decoder. As shown in Figure 2.1, a number audio and video decoders may be present.

A major role of this idealized decoder is to prevent buffer overflow and underflow. The buffering requirements are described in terms of the decoder rather than the encoder; for this reason, buffer overflow occurs when the decoder does not remove data quickly enough, i.e., when the compression is too good. Conversely, buffer underflow occurs when the decoder removes data too quickly, i.e., when the encoding process produces too much data per picture.

In the STD the bits for an access unit (a picture or an audio access unit) are removed from the buffer instantaneously at a time dictated by a decoding time stamp (DTS) in the bitstream. The bitstream also contains another type of time stamp, the presentation time stamp (PTS). Buffer overflow and underflow is controlled by the DTS; synchronization between audio and video decoding is controlled by the PTS. At some points in a video sequence, the DTS and PTS are identical and only the PTS is used.

Data are fed to the respective buffers from the system decoder in Figure 2.1 at irregular intervals as needed to demultiplex the interleaved bitstreams. Conceptually, each of the interleaved bitstreams is allocated a separate buffer in the STD, with a size determined by the encoder and communicated to the decoder in the system layer. The situation is sketched in Figure 2.3 for a simple case where one audio and one video bitstream are interleaved.

On the basis of header information in the bitstream itself, the system control knows when to switch the bitstream to buffer 1 or buffer 2. Bits not intended for those buffers are sent instead to the system control buffer. The bits are removed in blocks called *access units*. The DTS time stamps (or PTS, if DTS is not needed) in the bitstream determine the times of transfer and presentation. In this idealized model the transfer, decoding and presentation occur instantaneously; in real systems some additional delays would occur.

2.1.2   System layer syntax ○

An MPEG packet of data has a length defined in the packet header. Packet lengths are typically structured to match to the requirements of the digital storage or transmission medium, and therefore do not necessarily conform to the audio and video access units created by the encoders. However, a packet may only contain one type of compressed data.

2.2   MPEG audio ◑

The MPEG audio coding standard defines three layers of increasing complexity and subjective quality.[1] It supports sampling rates of 32, 44.1, and 48 kHz. At 16 bits/sample the uncompressed audio would require about 1.5 Mbit/s. After compression, the bit rates for monophonic channels are between 32 and 192 Kbit/s; the bit rates for stereophonic channels are between 128 and 384 Kbit/s.

The MPEG audio coding techniques take good advantage of the psychoacoustic properties of human hearing. Just as there is a threshold for observing visible patterns, there is also a frequency-dependent threshold for perception of audio stimuli. There is also an effect known as simultaneous masking in which the presence of one audio signal can mask the perception

---

[1]The MPEG Standard uses the term layers in two different senses — as a descriptive term for the division of the bitstream into MPEG system, audio and video bitstreams, and to distinguish between the three audio algorithms.

of a smaller signal. A temporal masking effect also occurs in which audio signals immediately before and after a masking signal are less perceptible. All of these effects are used to good advantage in the MPEG audio coding models.

The MPEG audio system first segments the audio into windows 384 samples wide. Layers I and II use a filter bank to decompose each window into 32 subbands, each with a width of approximately 750 Hz (for a sampling rate of 48 kHz). As is typical for subband coding, each subband is decimated such that the sampling rate per subband is 1.5 kHz and there are 12 samples per window. A fast Fourier transform (FFT) of the audio input is used to compute a global masking threshold for each subband, and from this a uniform quantizer is chosen that provides the least audible distortion at the required bitrate. Layers I and II are quite similar, but layer II achieves higher performance by using a higher-resolution FFT, finer quantization, and a more efficient way of sending the scale factors for the subbands.

One of the problems introduced by quantization is pre-echoes. Pre-echoes can occur when a sharp percussive sound is preceded by silence. When the signal is reconstructed, the errors due to quantization tend to be distributed over the block of samples, thereby causing an audible distortion before the actual signal. In an 8 ms window, pre-echoes are not suppressed completely by temporal masking.

Pre-echo control is an important part of layer III of MPEG audio coding. Layer III adds a modified discrete cosine transform (DCT) decomposition of the subbands to get much finer frequency subdivision. Layer III also adds nonuniform quantization (larger signals can mask larger quantization errors), entropy coding, and dynamic window switching. Dynamic window switching provides better time resolution, and this allows better control of pre-echoes.

A recent review article by Pan [Pan94] gives more details on MPEG audio coding. The book by Bhaskaran and Konstantinides [BK95] has reviews of many recent standards, including MPEG audio.

2.3 MPEG video ○

MPEG video is specifically designed for compression of video sequences. A video sequence is simply a series of pictures taken at closely spaced intervals in time. Except for the special case of a scene change, these pictures tend to be quite similar from one to the next. Intuitively, a compression system ought to be able to take advantage of this similarity.

It should come as no surprise to the reader, therefore, that the compression techniques (*compression models*) used by MPEG take advantage of this similarity or predictability from one picture to the next in a sequence. Compression techniques that use information from other pictures in the sequence are usually called interframe techniques.

When a scene change occurs (and sometimes for other reasons), interframe compression does not work and the compression model should be changed. In this case the compression model should be structured to take advantage of the similarity of a given region of a picture to immediately adjacent areas in the same picture. Compression techniques that only use information from a single picture are usually called intraframe techniques.

These two compression techniques, interframe and intraframe, are at the heart of the MPEG video compression algorithm. However, because there might be confusion with the term *frame* as used in broadcast video systems, MPEG-1 uses the abbreviated terms, *inter* and *intra*. In addition, MPEG very often uses the term *non-intra* in place of *inter*. These definitions occur in IS[2.1.78, 2.1.79, and 2.1.95]. We will follow the MPEG-1 convention for the rest of this book.

2.3.1   MPEG video layers ○

The outermost layer of an MPEG video bitstream is the video sequence layer. Except for certain critical timing information in the MPEG systems layer, an MPEG video sequence bitstream is completely self-contained. It is independent of other video (and audio) bitstreams.

Each video sequence is divided into one or more groups of pictures, and each group of pictures is composed of one or more pictures of three different types, I-, P-, and B-, as illustrated in Figure 2.4.[2] I-pictures (intra-coded pictures) are coded independently, entirely without reference to other pictures. P- and B-pictures are compressed by coding the differences between the picture and reference I- or P-pictures, thereby exploiting the similarities from one picture to the next.

P-pictures (predictive-coded pictures) obtain predictions from temporally preceding I- or P-pictures in the sequence, whereas B-pictures (bidirectionally predictive-coded pictures) obtain predictions from the nearest preceding and/or upcoming I- or P-pictures in the sequence. Different regions of B-pictures may use different predictions, and may predict from preceding pictures, upcoming pictures, both, or neither. Similarly, P-pictures may also predict from preceding pictures or use no prediction. If no prediction is used, that region of the picture is coded by intra techniques.

In a closed group of pictures P- and B-pictures are predicted only from other pictures in the group of pictures; in an open group of pictures the prediction may be from pictures outside of the group of pictures.

---

[2]A fourth type, the D-picture, is defined but is rarely used. The D-picture is a low-resolution representation that may not be used in combination with I-, P-, or B-pictures.

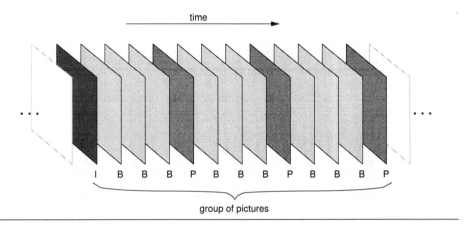

Figure 2.4: A typical group of pictures in display order.

2.3.2 Display and coding order ○

Since MPEG sometimes uses information from future pictures in the sequence, the coding order, the order in which compressed pictures are found in the bitstream, is not the same as the display order, the order in which pictures are presented to a viewer. The coding order is the order in which the pictures should be decoded by the decoder. The group of pictures illustrated in Figure 2.4 is shown in coding order in Figure 2.5.

2.3.3 Macroblock ○

The basic building block of an MPEG picture is the macroblock, sketched in Figure 2.6. The macroblock consists of a 16×16 sample array of luminance (grayscale) samples together with one 8×8 block of samples for each of two chrominance (color) components. The 16×16 sample array of luminance samples is actually composed of four 8×8 blocks of samples, and these 8×8 blocks are the units of data that are fed to the compression models. The reason for this will become clearer in the discussion of the coding models.

2.3.4 Slice ○

The MPEG picture is not simply an array of macroblocks, however. Rather, as sketched in Figure 2.7, it is composed of slices, where each slice is a contiguous sequence of macroblocks in raster scan order, starting at a specific address or position in the picture specified in the slice header. Each small

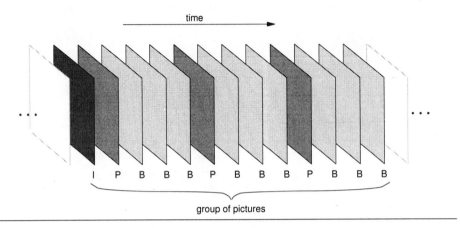

Figure 2.5: A typical group of pictures in coding order.

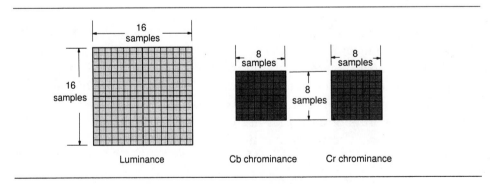

Figure 2.6: The MPEG macroblock.

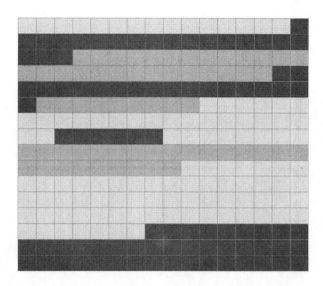

Figure 2.7: An illustration of a possible slice structure in an MPEG-1 picture.

block in the figure represents a macroblock, and contiguous macroblocks in a given slice have the same shade of gray. Slices can continue from one macroblock row to the next in MPEG-1. This slice structure, among other things, allows for great flexibility in signaling changes in some of the coding parameters. This is needed both to optimize quality for a given bitrate and to control that bitrate.

2.3.5   The discrete cosine transform in MPEG ○

At the heart of both intra and inter coding in MPEG is the discrete cosine transform (DCT). The DCT has certain properties that simplify coding models and make the coding efficient in terms of perceptual quality measures.

Basically, the DCT is a method of decomposing a block of data into a weighted sum of spatial frequencies. Figure 2.8 illustrates the spatial frequency patterns that are used for an 8×8 DCT (a DCT for an 8×8 block of pels). Each of these spatial frequency patterns has a corresponding *coefficient*, the amplitude needed to represent the contribution of that spatial frequency pattern in the block of data being analyzed. In other words, each spatial frequency pattern is multiplied by its coefficient and the resulting 64

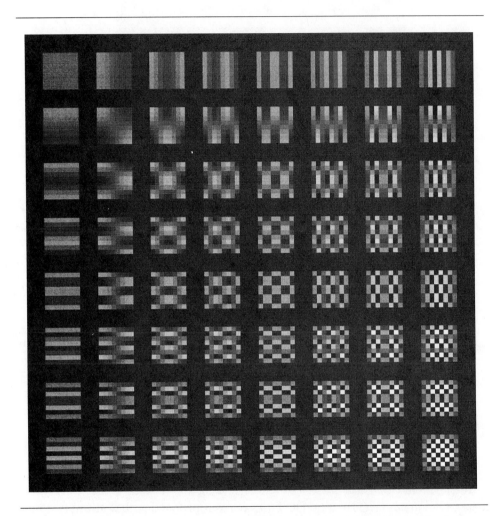

Figure 2.8: The 64 2-D cosine functions of the 8×8 DCT. Each block is an 8×8 array of samples. Zero amplitudes are neutral gray, negative amplitudes have darker intensities and positive amplitudes have lighter intensities.

8×8 amplitude arrays are summed, each pel separately, to reconstruct the 8×8 block.

If only the low-frequency DCT coefficients are nonzero, the data in the block vary slowly with position. If high frequencies are present, the block intensity changes rapidly from pel to pel. This qualitative behavior can be seen in Figure 2.8, where the 64 two-dimensional (2-D) patterns that make up the 8×8 DCT are seen to range from absolutely flat — the term that gives the average or "DC value" of the 8×8 block of pels — to a checkerboard with very large intensity changes from one pel to the next.[3] Note that at each axis of the plot, the variation is one-dimensional. As will be shown in Chapter 3, the complex 2-D behavior of Figure 2.8 is actually the product of the one-dimensional (1-D) oscillations seen along each axis. Qualitatively, this can be seen by closely inspecting Figure 2.8.

2.3.6   Quantization ○

When the DCT is computed for a block of pels, it is desirable to represent the coefficients for high spatial frequencies with less precision. This is done by a process called *quantization*. A DCT coefficient is quantized by dividing it by a nonzero positive integer called a quantization value and rounding the quotient — the quantized DCT coefficient — to the nearest integer. The bigger the quantization value is, the lower the precision is of the quantized DCT coefficient. Lower-precision coefficients can be transmitted to a decoder with fewer bits. The use of large quantization values for high spatial frequencies allows the encoder to selectively discard high spatial frequency activity that the human eye cannot readily perceive.

The DCT and visually-weighted quantization of the DCT are key parts of the MPEG coding system. In Chapter 3 we will discuss the DCT in some detail; in Chapter 4 we will explore properties of the human visual system that relate to quantization and other aspects of MPEG coding.

As noted above, a macroblock is composed of four 8×8 blocks of luminance (monochrome) samples and two 8×8 blocks of chrominance samples. The chrominance samples represent color in terms of the presence or absence of red and blue for a given luminance intensity. These 8×8 blocks of data are the unit processed by the DCT. A lower resolution is used for the chrominance blocks because the human eye resolves high spatial frequencies in luminance better than in chrominance. This will be covered in detail in Chapter 4.

The DCT turns out to have several advantages from the point of view of data compression. First, it turns out that for intra coding, the DCT

---

[3]The term *DC* stands for direct current, and comes from the use of the DCT to analyze electrical circuits. Similarly, currents varying with time are called *AC* (alternating current) terms; in our case, AC applies to spatially varying terms.

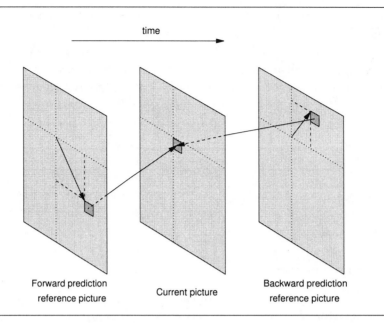

Figure 2.9: Motion compensated prediction and reconstruction.

coefficients are almost completely decorrelated — that is, they are independent of one another, and therefore can be coded independently. This makes it possible to design a relatively simple algorithm (called a coding model) for coding them. Decorrelation is of great theoretical and practical interest in terms of construction of the coding model. However, the coding performance is actually influenced much more profoundly by the visually-weighted quantization.

In nonintra coding — coding the difference between the current picture and a picture already transmitted — the DCT does not greatly improve the decorrelation, since the difference signal obtained by subtracting the prediction from a similar (i.e., correlated) picture is already fairly well decorrelated. However, quantization is still a powerful compression for controlling bitrate, even if decorrelation is not improved very much by the DCT. In practice, the DCT works quite well for both inter and intra coding.

2.3.7 Motion compensation ○

If there is motion in the sequence, a better prediction is often obtained by coding differences relative to areas that are shifted with respect to the area being coded, a process known as motion compensation. The process of determining the motion vectors in the encoder is called motion estimation.

The motion vectors describing the direction and amount of motion of

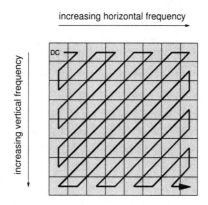

Figure 2.10: Zigzag scanning order of coefficients.

the macroblocks are transmitted to the decoder as part of the bitstream. The decoder then knows which area of the reference picture was used for each prediction, and sums the decoded difference with this motion compensated prediction to obtain the output. The encoder must follow the same procedure when the reconstructed picture will be used for predicting other pictures. The encoder's reconstruction process is sometimes called "local decoding". An example of motion compensation is sketched in Figure 2.9. The motion vector and corresponding vertical and horizontal displacements are shown for forward and backward motion compensation. The vectors are the same for every pel in the macroblock, and the vector precision is either to full pel or half-pel accuracy.

2.3.8   MPEG coding models ○

The quantized DCT coefficients are coded *losslessly*, such that the decoder can reconstruct precisely the same values. For MPEG, an approximately optimal coding technique based on Huffman coding was used to generate the tables of variable length codes needed for this task. Variable length codes are needed to achieve good coding efficiency, as very short codes must be used for the highly probable events.

The coefficients are arranged according to a 1-D sequence known as the zigzag scanning order. This zigzag ordering is shown in Figure 2.10, where we see that this scan approximately orders the coefficients in ascending spatial frequency. Since visually-weighted quantization strongly deemphasizes higher spatial frequencies, only a few lower-frequency coefficients are nonzero in a typical transformation. As a rough rule-of-thumb, the more zero coefficients, the better the compression. Consequently, it is very important for the coding model to use *symbols* (particular combinations of DCT coefficient information) that permit efficient coding of DCTs with mostly zero coefficients. One of these symbols is an end-of-block (EOB), that tells the decoder that all remaining DCT coefficients for the block are zero. More details about the coding model will be covered in Chapter 5.

The DCT is used to code both nonintra and intra picture information, but the properties are actually quite different in the two coding environments. Different quantization tables are used for nonintra compression, and the rounding is done differently. The data being transformed are also quite different, and therefore, different sets of symbols and slightly different code tables are used.

Many other aspects of MPEG, such as the methods for simply skipping over unchanged macroblocks and for coding of motion vectors, will be covered in Chapter 8 on MPEG video syntax. While it is inappropriate to get into such detail in this overview, these details are important in determining the overall coding performance that can be achieved by MPEG.

2.3.9   Encoder decision strategies ○

It is important to recognize that MPEG is potentially a highly asymmetric system. Decoders, since they merely follow the directions encoded in the bitstream, are relatively simple. Encoders, however, are much more complex than decoders and must have much more intelligence. Among other things, encoders must identify areas in motion, determine optimal motion vectors, control bitrate, control data buffering such that underflow and overflow do not occur, determine where to change quantization, determine when a given block can simply be repeated, determine when to code by intra and inter techniques, and vary all of these parameters and decisions dynamically so as to maximize quality for a given rate. The decision strategy that must be implemented in the encoder is probably the most complex and least published aspect of MPEG.

2.4   MPEG-2 ○

MPEG-2 is the follow-on to MPEG-1, and is intended primarily for higher bit rates, larger picture sizes, and interlaced frames. MPEG-2 builds upon the MPEG-1 capabilities. In fact, MPEG-2 decoders are required to be able to decode at least some MPEG-1 data streams.

Five different profiles have been defined for MPEG-2 video, providing particular sets of capabilities ranging from function close to MPEG-1 to the very advanced video compression techniques needed for HDTV. One of the profiles, the main profile (MP) of MPEG-2, is, however, a relatively straight-forward extension of MPEG-1. The main profile of MPEG-2 video is covered in this book. Briefly, MPEG-2 main profile introduces the concept of video fields and interlaced frames. Variable quantization is extended, DC precision can be customized, and another coding table is defined. Higher resolutions, bit rates, and frame rates are expected. Parameter ranges such as picture width, height, bit rate, and buffer sizes are extended. Copyright information can be recorded. Display characteristics can be transmitted including an offset for viewing a subset of the picture in a window.

MPEG-2 systems (Part 1) includes all of the features of MPEG-1 with enhancements that allow it to operate in more error-prone environments. In addition, more video and audio streams can be handled and they are no longer required to have a common time base. MPEG-2 systems still uses the pack and packet concepts, but combines them in both backwards compatible and noncompatible ways. Some of the noncompatible ways support asynchronous transmission mode (ATM) applications of MPEG-2.

MPEG-2 audio (Part 3) is backwards compatible with MPEG-1. Even though three more channels are allowed plus a sub-woofer to provide an enhanced surround sound environment, an MPEG-1 audio decoder can extract and decode two audio channels. Also, an MPEG-2 audio decoder can decode the two stereo channels of MPEG-1 audio. The requirement to be backwards compatible necessitated some coding efficiency compromises. A nonbackwards compatible (NBC) audio (Part 7) is also being developed.

MPEG-1 video was aimed at the digital storage media (DSM) application and thus fixed the video format (see Figure 1.3), assumed low errors rates, and ignored interlaced TV. MPEG-2 was intended to be a generic standard and therefore attempted to be much broader. MPEG-2 video (Part 2) maintains all of the MPEG-1 video syntax, but uses extensions to add additional flexibility and functions. "Scalable" extensions are added to provide video data streams with multiple resolutions for potential normal TV and HDTV coexistence. Other scalable extensions allow the data stream to be partitioned into two pieces. Others also offer some temporal flexibility so that not all frames have to be reconstructed.

2.5   MPEG-4 ○

Before MPEG-2 was finished a new project, MPEG-4, was started.[4] It was originally targeted primarily at very low bit rates, but over time its focus changed beyond just high compression. High priority was placed on content-based interactivity and "universal access", including error-prone wireless networks [ISO95]. Content-based manipulation, scalability, and editing are

---

[4]MPEG-3 was started and targeted at HDTV, but abandoned when MPEG-2 was demonstrated to meet that need.

also expected [Rea96] [5] MPEG-4 will provide a syntactic description language (MSDL). This MSDL is expected to describe how to parse and process the elementary data streams (which could be MPEG-1, MPEG-2, or even synthetic two-dimentionsal or three-dimensional objects). Missing pieces will be downloaded from an open set of tools and algorithms.

---

[5]A special issue on MPEG-4 is scheduled for the December 1996 issue of the IEEE Transactions on Circuits and Systems for Video Technology.

3

The Discrete Cosine Transform

In this chapter we present a review of the discrete cosine transform (DCT). This is an essential part of the MPEG standard, and must be understood if the reader is to make much sense of the chapters that follow. The treatment will, as much as possible, be nonmathematical and intuitive. We will first review the DCT from an intuitive graphical perspective; we then will briefly present the mathematics. Readers interested in more advanced topics such as fast DCTs may want to read [PM93] and [RY90].

The important topic of IDCT mismatch will not be covered in any detail until Chapter 12; at that point the reader will be familiar with the MPEG Standard pseudocode formalism used there.

3.1   The one-dimensional cosine transform ○

A video picture normally has relatively complex variations in signal amplitude as a function of distance across the screen, as illustrated in Figure 3.1. It is possible, however, to express this complex variation as a sum of simple oscillatory sine or cosine waveforms that have the general behavior shown in Figure 3.2(b). The sine or cosine waveforms must have the right spatial frequencies and amplitudes in order for the sum to exactly match the signal variations. When the waveforms are cosine functions, the summation is a *cosine transform*, the main topic of this chapter. When the waveforms are sine functions, the summation is another transform called the *sine transform*. The waveforms that make up these transforms are called *basis functions*.

Symmetry is a powerful tool for understanding some of the properties of these transforms. Figure 3.2(a) illustrates the basic properties of functions with *even* and *odd* symmetry. Given a function, $f(x)$, an even function,

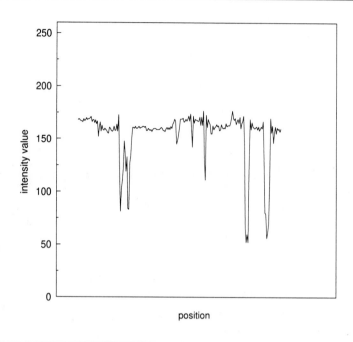

Figure 3.1: An example of a raster line of TV data.

$f(x)$ obeys the relationship $f(x) = f(-x)$. An odd function obeys the relationship $f(x) = -f(-x)$. Therefore, an even function has mirror symmetry across the vertical axis; an odd function must be negated when reflected across this axis. By inspection, if two even functions are added, the result will also have even symmetry. Similarly, sums of functions with odd symmetry have odd symmetry.

Denoting the cosine function as $\cos(\theta)$, where θ is the argument, the plot of $cos(\theta)$ as a function of θ in Figure 3.2(b) shows that the cosine function is an even function — that is, it has mirror symmetry across the vertical axis. The sine function in Figure 3.2(b) has odd symmetry. Therefore, a cosine transform has even symmetry and a sine transform has odd symmetry.

Since the cosine transform has even symmetry, the data being represented by the cosine transform must have even symmetry. This has nothing to do with the data *within* the interval, but does affect the properties of the transform *outside* of the interval. When points outside of the interval are calculated from the transform, they must have even symmetry.

The cosine basis functions are oscillatory, and the transform must therefore also be oscillatory. The period of the oscillation is set by the period of the lowest nonzero frequency cosine term. Since the symmetry and repeating pattern properties are outside of the interval being represented, they

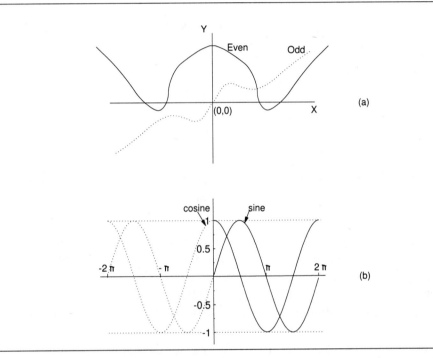

Figure 3.2: (a) Even and odd symmetry. (b) Cosine and sine waveforms. The cosine function has even symmetry, whereas the sine function has odd symmetry.

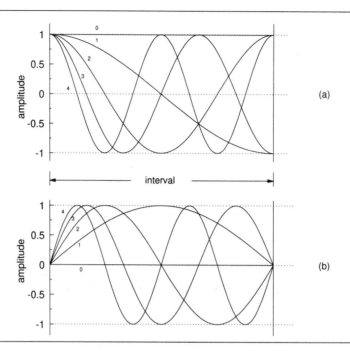

Figure 3.3: (a) First few cosine basis functions in a cosine transform, (b) first few sine basis functions in a sine transform.

don't affect the match between transform and data within the interval. The reader should recognize, however, that values computed from the transform for points outside of the interval will not match actual data outside of the interval. Only if the actual data just happen to have even symmetry and the right oscillatory behavior will they match the transform calculation, and that would be a coincidence.

The lowest frequency in the cosine transform is zero, reflecting the fact that $\cos(0) = 1$. This term, when multiplied by a scale factor, gives the average value of the data points in the interval. The lowest nonzero frequency in the cosine transform has a period equal to twice the interval of data. The plot of the first few cosine terms in the cosine transform are shown in Figure 3.3(a), and they show that the cosine functions always end at either a full or half-cycle. For the moment, we will defer the mathematical rationale for why this is so.

The plots of the first few sine basis functions in a sine transform in Figure 3.3(b) show that the sine basis functions have properties similar to the cosine basis functions. Note, however, that all of the sine basis functions are zero at the beginning and end of the interval. This means that the sine transform will also be zero at the beginning and end of the interval. Furthermore, a zero frequency sine function is always zero, and, therefore,

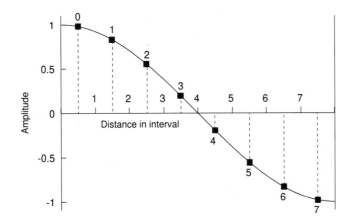

Figure 3.4: A discrete cosine basis function: the sampled cosine waveform.

does not contribute to the sine transform. There is no sine term that is constant across the interval, and the average value in the interval must be constructed from combinations of nonzero frequency sine functions. This, as will be seen, is a very awkward property of the sine transform.

3.2 1-D discrete cosine transform (DCT) ○

Both the cosine waveform and the TV raster are shown as continuous analog data in Figures 3.1 and 3.2. This implies an arbitrarily large number of data points and a correspondingly large number of cosine waveforms in the cosine transform. In practice, digital data systems only take samples of the data at discrete intervals, the interval size being set by the number of samples needed for a given picture resolution. When the smoothly varying cosine functions of Figure 3.2 are sampled at this same interval as shown in Figure 3.4, they become *discrete* cosine functions.

If the input signal frequency gets too high, the behavior shown in Figure 3.5 occurs. The upper bound on frequencies that can be reproduced in a sampled system is given by the Nyquist limit of two samples per cycle. Because the sampling is too sparse, frequencies above the Nyquist limit often appear as lower frequency oscillations, an effect called *aliasing*. This is an undesirable artifact, and aliasing is usually suppressed by low-pass filtering the signal before sampling it. One common way of doing this is to average the signal over the width of a sample interval.[1]

---

[1]There are a number of more sophisticated filters that do a much better job of sup-

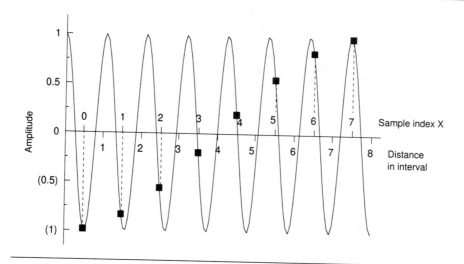

Figure 3.5: Sampling an input frequency above the Nyquist limit; an example of aliasing.

The lowest frequency in a cosine transform is zero, and as noted above, this represents the average value or DC value in the interval. Conversely, the nonzero frequency terms are called AC values. This terminology comes from the use of cosine transforms to decompose electrical currents into direct currents and alternating currents.

When the input data and cosine waveforms are sampled, the cosine transform becomes a *discrete cosine transform (DCT)*. The discrete 8-point cosine basis functions are plotted in Figure 3.6, and an 8-point sample sequence and corresponding numerical factors for scaling the basis functions are shown in Figure 3.7. These scaling factors are called *DCT coefficients*. In Figure 3.8 a reconstruction sequence is shown in which the basis functions are scaled and added one-by-one to reconstruct the source data in Figure 3.7(a)

In general, in order to fit a set of N arbitrary points, N independent variables are needed. Therefore, to fit N samples of data, a set of N independent cosine waveforms is required, and the resulting DCT is called an N-point transform. The cosine functions must be independent in the sense that any given cosine waveform in the set cannot be produced by scaling and summation of the other cosine waveforms in the set. In mathematical terms, this property means that the cosine waveforms (the basis functions) are *orthogonal*.

More formally, the expression of a set of eight samples, $f(x)$, as a sum

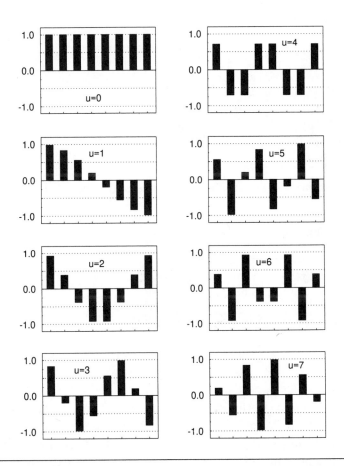

Figure 3.6: The basis functions for an 8-point discrete cosine transform.

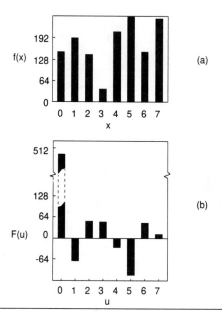

Figure 3.7: Example of a DCT: (a) Source data, (b) DCT coefficients.

of eight cosine basis functions is as follows:

$$f(x) = \sum_{\mu=0}^{7} \frac{C(\mu)}{2} F(\mu) \cos\left[(2x+1)\mu\pi/16\right] \tag{3.1}$$

where the constant, $C(\mu)$ is given by:

$$C(\mu) = 1/\sqrt{2} \quad \text{if } \mu = 0$$
$$C(\mu) = 1 \qquad \text{if } \mu > 0$$

In this equation, x is the displacement along the row of samples (one unit per sample). μ is an index that determines the spatial frequency of the cosine waveform, $F(\mu)$ is the DCT coefficient for that spatial frequency, and summation over integer values of μ from 0 to 7 is expressed by $\Sigma_{\mu=0}^{7}$. $\cos(2x+1)\mu\pi/16$ is the cosine waveform or basis function. The constants in the cosine function argument are such that the highest frequency is just below the Nyquist limit.[2] The cosine basis functions always finish the interval at a full or half-cycle, as these are the only cosine frequencies that satisfy the orthogonality criterion. The additive constant in $(2x+1)$ simply

---

[2]The Nyquist limit is at $\mu = 8$, whereas the lowest nonzero frequency has a period of 16 samples (twice the 8-point interval size).

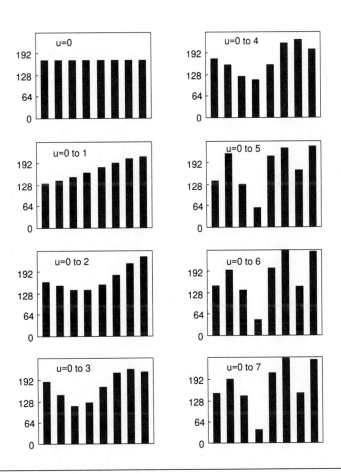

Figure 3.8: Reconstructing the source data of Figure 3.7(a).

shifts the sample points so that they are symmetric around the center of the N-point interval.[3]

The DCT coefficients are just as valid a representation of the sampled data as the original data itself. Assuming that the calculations are done with sufficient precision, the two representations are, in fact, interchangeable. The decomposition of the samples into a set of DCT coefficients is called the *Forward DCT* or *FDCT*. The reverse process in which scaled cosine waveforms are summed to recreate the original sampled data is called the *Inverse DCT* or *IDCT*. The coefficients, $F(\mu)$, are computed by the FDCT; the scaling in the IDCT is done by multiplying the coefficient value for a waveform by the value of the waveform at each sampling point. The 1-D IDCT is shown above in Equation 3.1; the corresponding equation for the 1-D FDCT is:

$$F(\mu) = \frac{C(\mu)}{2} \sum_{x=0}^{7} f(x) \cos\left[(2x+1)\mu\pi/16\right] \qquad (3.2)$$

Since MPEG uses only the 8-point DCT, we have restricted the discussion to that particular case. It is important to emphasize that the DCT is valid only within the interval containing the eight points. These eight points are sometimes called the *region of support*, and when two regions of support are adjacent, only ideal, perfect, DCT representations provide continuity from one region to the next. If approximations are made anywhere in the FDCT or IDCT, discontinuities between regions can be created.

3.3   2-D DCT ◗

The pictures compressed by MPEG are two dimensional arrays of samples. Not surprisingly, therefore, MPEG uses a two-dimensional (2-D) eight point by eight point (8×8) form of the DCT. The 2-D DCT is a separable transform, which means that it is composed of separate horizontal and vertical 1-D DCTs.

Suppose that a 2-D array of samples, $f(x,y)$, is to be transformed into a 2-D DCT. Horizontal 1-D DCTs can be run first, transforming the array into a set of 1-D DCTs — one horizontal DCT for each row of the array. The dependence on x is transformed into weighted sums of cosine functions of x, but if there is any vertical variation in the array, the weightings must be a function of y.

$$g(\mu,y) = \frac{C(\mu)}{2} \sum_{x=0}^{7} f(x,y) \cos\left[(2x+1)\mu\pi/16\right] \qquad (3.3)$$

---

[3]For other possible definitions of the DCT, see [RY90].

Vertical 1-D DCTs are then run to transform the y dependence of each column of the array $g(\mu, y)$ into weighted sums of cosine functions of y with frequency index ν.

$$F(\mu, \nu) = \frac{C(\nu)}{2} \sum_{y=0}^{7} g(\mu, y) \cos\left[(2y+1)\nu\pi/16\right] \qquad (3.4)$$

Note that because of the separability of the process, the order in which the horizontal and vertical transforms are taken does not affect the result.

When Equations 3.3 and 3.4 are combined, the 2-D DCT becomes:

$$F(\mu, \nu) = \frac{C(\mu)}{2}\frac{C(\nu)}{2} \sum_{y=0}^{7}\sum_{x=0}^{7} f(x, y) \cos\left[(2x+1)\mu\pi/16\right] \cos\left[(2y+1)\nu\pi/16\right]$$

$$(3.5)$$

where μ and ν are the horizontal and vertical frequency indices, respectively, and the constants, $C(\mu)$ and $C(\nu)$, are given by:

$$C(\mu) = 1/\sqrt{2} \quad \text{if } \mu = 0$$
$$C(\mu) = 1 \qquad \text{if } \mu > 0$$

The products of the cosine terms are the 2-D DCT basis functions, shown as 8×8 grayscale arrays in Figure 2.8 in Section 2.3.5. In this figure the horizontal frequency index increases from left to right, and the vertical frequency index increases from top to bottom. The top row is therefore the product of a constant and a horizontal 1-D DCT with frequency μ increasing from left to right; the left column is the product of a constant and a vertical 1-D DCT with frequency ν increasing from top to bottom. The block average (DC term) is the constant intensity block in the upper left corner.

As discussed in the preceding section, the 1-D DCT computation can be reversed by the IDCT to recreate the original samples. Similarly, a 2-D FDCT can be reverse by a 2-D IDCT. The simplest way to do this is to reverse the FDCT calculation, first calculating a 1-D IDCT for each column in the array, and then a 1-D IDCT for each row of the result. The 2-D IDCT is given by:

$$f(x, y) = \sum_{\mu=0}^{7}\sum_{\nu=0}^{7} \frac{C(\mu)}{2}\frac{C(\nu)}{2} F(\mu, \nu) \cos\left[(2x+1)\mu\pi/16\right] \cos\left[(2y+1)\nu\pi/16\right]$$

$$(3.6)$$

The products of the cosine terms are the two-dimensional basis functions for the 2-D DCT, and $F(\mu, \nu)$ is the 2-D DCT coefficient. The definitions of $C(\mu)$ and $C(\nu)$ are the same as in Equation 3.5.

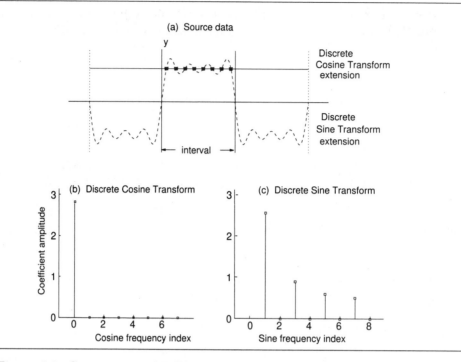

Figure 3.9: Comparison of DCT and DST for an absolutely flat line of data: (a) Source data, (b) DCT coefficients, (c) DST coefficients. The DCT is simply a flat line, whereas the DST is a square wave.

3.4   Why use the DCT? ○

There are two good reasons for using the DCT in a data compression system: First, DCT coefficients have been shown to be relatively uncorrelated [ANR74], and this makes it possible to construct relatively simple algorithms for compressing the coefficient values. A more detailed review of this is given in Chapter 5. Second, the DCT is a process for (approximately) decomposing the data into underlying spatial frequencies. This is very important in terms of compression, as it allows the precision of the DCT coefficients to be reduced in a manner consistent with the properties of the human visual system. This will be discussed in much more detail in Chapter 4.

3.4.1   Decorrelation and energy compaction ○

The reason for the decorrelation of the DCT coefficients can be qualitatively understood from the following argument: Suppose a perfectly flat line of data is transformed using a 1-D DCT. The data are perfectly correlated and since there is no oscillatory behavior, only the DC coefficient is present in the transform. In this instance the DCT concentrates or compacts all of the "energy" of the image into just one coefficient, and the rest are zero.

Alternatively, consider what happens if a *discrete sine transform* (DST) is used. Here, there is no DC term and the flat line must be built up from a fundamental and a large number of harmonics, as sketched in Figure 3.9. The energy is spread out among a large number of coefficients rather than being compacted. Further, since the source data are perfectly correlated, these DST coefficients must also be highly correlated. Finally, because of the artificial creation of many frequencies, visually-weighted quantization of a sine transform would have very complex interdependencies.

All of these points argue strongly for the use of the cosine transform instead of the sine transform. Indeed, the nearly ideal decorrelation and compaction achieved with the DCT were the original justification for its use[ANR74]. Note that a similar qualitative argument can be made for slowly varying data.

3.4.2   Efficient DCT algorithms ○

The formal equations for the DCT suggest a complexity that might argue against the use of this transform. For example, each 1-D DCT requires 64 multiplications and there are eight DCTs in each direction. If the number of operations is counted, the DCT equations would require 1024 multiplications and 896 additions.

However, fast algorithms exist that greatly reduce the complexity. For scaled DCTs in which quantization is folded into the DCT calculation, the best 2-D algorithm requires only 54 multiplications and 468 additions and shifts [FL90]. In addition, many VLSI chips are now available (see Chapter 17) with either MPEG encoding or decoding, so DCT complexity has not proven to be an obstacle for hardware. Algorithms for efficient computation of the DCT are described in detail in [PM93] and in [RY90].

3.5   Precision of the DCT ○

In the discussion of the DCT above, no consideration was given to the precision of the calculations. In fact, the MPEG FDCT and IDCT are defined with integer precision and the source data are integers. For intra coding the source data are 8-bit precision numbers from 0 to 255, whereas for inter coding the pel differences are signed nine-bit precision numbers.

The precision of the DCT coefficient can be obtained directly from the equations above. If the source data have eight-bit precision, the two summations over eight terms each potentially increases the precision to 14 bits, and another bit is needed for the cosine sign. However, the scaling by the constants reduces the precision by three bits to 12 bits (including the sign bit). This is the precision defined for the integer DCT, before any quantization is performed.

3.6   Quantization of the DCT ○

Quantization is basically a process for reducing the precision of the DCT coefficients. Precision reduction is extremely important, since lower precision almost always implies a lower bit rate in the compressed data stream.

The quantization process involves division of the integer DCT coefficient values by integer quantizing values. The result is an integer and fraction, and the fractional part must be rounded according to rules defined by MPEG. It is the quantized DCT values that are transmitted to the decoder.

The quantizing values are chosen so as to minimize perceived distortion in the reconstructed pictures, using principles based on the human visual system. This will be described in more detail in Chapter 4.

In order to reconstruct the DCT, the decoder must *dequantize* the quantized DCT coefficients, scaling them by the quantization value to reproduce the DCT coefficients computed by the encoder. Since some precision was lost in quantizing, the reconstructed DCT coefficients are necessarily approximations to the values before quantization. This is a major source of distortion in the encoding and decoding process, and can, under some circumstances, give rise to very visible artifacts in decoded video.

In this chapter the quantization values are simply assumed to be known. In fact, the proper choice of quantization value for a given basis function should be set according to the spatial frequency of the underlying DCT basis function and the response of the human visual system to that frequency. In Chapter 4 the rules by which one develops visually-weighted quantizing values will be discussed in detail. MPEG-1 defines default quantization tables, but custom tables appropriate for a particular set of viewing conditions are also permitted.

3.6.1   Rounding conventions for quantization ○

The quantization rules are different, depending on whether intra or inter coding is being performed. The quantization is basically a division by the quantizing value. However, in intra coding the quantization fractional values are rounded to the nearest integer (values exactly at 0.5 are rounded to the larger magnitude), whereas in nonintra coding fractional values are always rounded down to the smaller magnitude. A plot of the quantized values is shown for the two cases in Figure 3.10. The wide interval around zero for nonintra coding is called a *dead zone*.

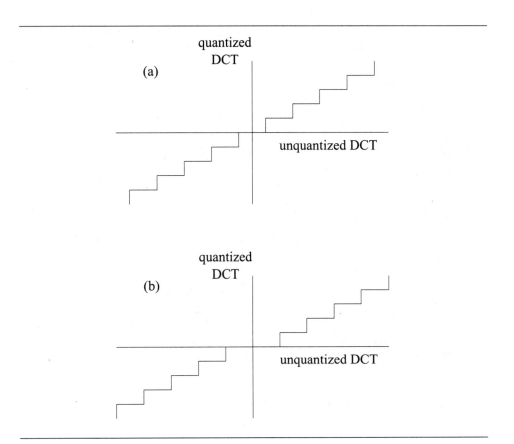

Figure 3.10: Rounding of quantized DCT coefficients for (a) intra coding and (b) nonintra coding. The nonintra rounding produces a dead zone around zero.

3.6.2 Scaling of quantizer tables ○

MPEG defines rules for changing the quantization of the AC coefficients from place to place in the image. For intra coding, the DC quantizer is always 8, whereas the AC quantization values can be set by downloading the values at certain points during the coding.

In addition, a multiplicative quantization scale factor, quantizer_scale, can be changed at the start of coding of each macroblock.[4] Each time this scale factor is changed, the new value must be coded in the bitstream and there is coding overhead in doing this. However, the bitrate is very sensitive to the quantizer_scale, and variable quantization is an important tool for improving picture quality and controlling bitrate.

The quantizer_scale has integer values in the range {1,..,31}, inclusive. The quantization scaling is a bit subtle, however, in that additional numerical scaling factors must be introduced in the calculation in order to maintain precision during rounding. The scaling in the FDCT and IDCT are done slightly differently.

3.6.2.1 Quantization of the FDCT ◑

For intra coding, the quantized DCT coefficient, $QDCT$, is calculated from the unquantized coefficient, DCT, by the following integer calculations:

$$QDCT = \frac{(16 \times DCT) + (Sign(DCT) \times \texttt{quantizer\_scale} \times Q)}{2 \times \texttt{quantizer\_scale} \times Q} \quad (3.7)$$

where Q is the quantization table value for the coefficient and the rounding term contains a function $Sign()$ with the following properties:

$$Sign(DCT) = \begin{cases} +1, & \text{when } DCT > 0 \\ 0, & \text{when } DCT = 0 \\ -1, & \text{when } DCT < 0 \end{cases} \quad (3.8)$$

The rounding term in Equation 3.7 produces the intra-coding quantizer characteristic shown in Figure 3.10.

A similar equation holds for nonintra coding, but the rounding is always to the smaller integer magnitude. Therefore, there is no rounding term in the equation.

$$QDCT = \frac{16 \times DCT}{2 \times \texttt{quantizer\_scale} \times Q} \quad (3.9)$$

The factor of 2 can be eliminated from this equation, but this form is easier to compare to Equation 3.7.

---

[4]Boldface font signifies that the term is used in the syntax or pseudocode.

3.6.2.2 Dequantization of the IDCT ◑

The procedures for dequantization in the IDCT calculation are the inverse of the FDCT quantization procedures. For the intra IDCT,

$$DCT = \frac{(2 \times QDCT) \times \texttt{quantizer\_scale} \times Q}{16} \qquad (3.10)$$

Note that since the intra quantization does not have a dead zone, there is no need for a rounding term before the division by 16.

Similarly, for the nonintra IDCT,

$$DCT = \frac{((2 \times QDCT) + Sign(QDCT)) \times \texttt{quantizer\_scale} \times Q}{16} \qquad (3.11)$$

In this case, however, a rounding term is needed to compensate for the quantizer dead zone.

3.7 Mismatch of IDCTs ○

The MPEG Standard does not precisely define how the IDCT is calculated, and this can lead to an increasing mismatch between the pictures encoders and decoders with different IDCT implementations used for predictions in nonintra coding and decoding. With each new stage in a chain of inter coded pictures, the differences between encoder and decoder pictures tend to increase, and this accumulating difference is called *IDCT mismatch*.

Even in the best implementations, mismatch will cause an accumulation of differences between encoder and decoder (unless they are identical). For this reason, intra coding of each area of the picture must be done periodically. The process is commonly called *forced updating*. MPEG mandates that forced updating be done at least once for every 132 P-pictures in the sequence. The forced updating doesn't need to be done in a single picture, however. In practice, this mandated forced updating is rarely encountered. I-pictures are typically coded at 10 to 15 picture increments to allow for ease of editing and fast forward/reverse playback.

The mismatch problem is aggravated by a fundamental problem in conversion of the ideal IDCT to an integer representation, and it has been the topic of much discussion and research in MPEG. Further discussion of IDCT mismatch is deferred to Chapter 12.

4

Aspects of Visual Perception

One problem in any video compression system is the large amount of data that must be handled. If every bit of data is sent exactly as captured from a video source, a huge bandwidth would be needed. Instead, data must be selectively discarded — an important part of any lossy data compression system — as part of the process of reducing the amount of data until it fits the available transmission bandwidth.

In this chapter we selectively review some basic aspects of the *human visual system* (HVS), in order to understand how to reduce the amount of data in moving picture sequences. Our goal for this chapter should be to describe how to get the best possible visual quality when a sequence is played back by a decoder. Unfortunately, we can't quite meet that goal. In practice, optimization of MPEG encoding and decoding systems is a very complex topic, and most MPEG implementors regard their optimization techniques as proprietary. What we can do, however, is describe some of the basic ground rules for this optimization.

Readers who have no familiarity with this area of technology will find the excellent introductory text by Cornsweet [Cor70] helpful, although somewhat dated. A recent book on vision by Wandell [Wan95] is also a good introduction to many aspects of MPEG. Poynton has written two very readable introductory monographs on color [Poy95a, Poy95b] that can be downloaded from a World Wide Web site on the Internet. The book edited by Benson [Ben85] contains a wealth of material relevant to this chapter from which we have drawn extensively. Rogowitz [Rog83b] has published a review of the human visual system as it affects display design; her treatment is general and of interest for MPEG.

4.1   Color representations ○

All of us, as children, learned about color wheels and the red-green-blue (RGB) primary color system. We also learned that mixing colors on the wheel produced other colors; e.g. mixing yellow and cyan (we probably called it blue) produced green. Since this artistry was done on paper, we were using a *subtractive* color system. In a subtractive system pigments absorb certain bands of light, leaving the rest to be reflected from the white underlying paper. The primaries for the subtractive color system are cyan, magenta and yellow (CMY); mixing yellow and cyan produces green, mixing cyan and magenta produces blue, and mixing magenta and yellow produces red.

For display systems that emit light, the *additive* red-green-blue primary system is used. For an additive system the emitted light from the primary sources is summed to produce a given color. Thus, mixing red and green produces yellow, mixing green and blue produces cyan, and mixing red and blue produces magenta.

In practice, actual implementations of the RGB and CMY systems are never ideal. Light emitting phosphors cannot reproduce pure spectral colors, and when, as young artists, we mixed all of the colors together, we got a muddy dark brown rather than black. For this reason, a fourth ink, blacK is needed in the CMYK system used for color printing. If we ignore the extra complications introduced by nonideal inks and phosphors, however, both RGB and CMY are capable of producing a subjectively pleasing full range of colors. Note, however, that in subtractive systems, the color of the illumination also affects the perceived colors.

This section presents a review of color representations, with particular emphasis on representations appropriate for MPEG.

4.1.1   Trichromatic theory ○

The use of only three colors to represent a color image is based on a fundamental property of the human visual system. According to the trichromatic theory of color vision, color vision is produced by the three classes of cone cells, the color photoreceptors in the eye. These receptors have differing sensitivities to the various wavelengths in the visible spectrum, as can be seen in Figure 4.1[RF85]. In this figure the curves are normalized to unity, making them appear more similar than they really are. The peak in sensitivity for the B cones (green-absorbing) is about 5% higher than the peak for A cones (red-absorbing), and both have peak sensitivities about 30 times greater than the C cone (blue-absorbing) sensitivity.

A given wavelength stimulates the three classes of receptors in a particular ratio, thereby producing the sensation of a particular color. For a given

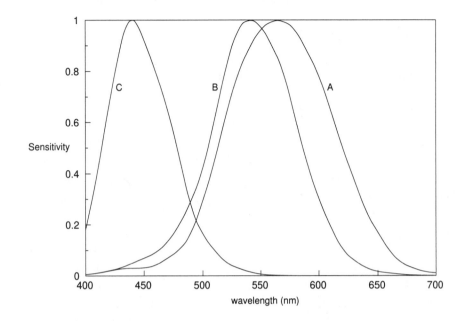

Figure 4.1: Sensitivities of the three types of cones in the human eye.

light intensity and background illumination, any stimulus that produces the same ratio will produce the same color.

If a three-color system is defined consisting of red, green, and blue primaries denoted respectively by R, G, and B, a match to an arbitrary color, C, can be obtained by a weighted sum of the three primaries [RF85]:

$$C = rR + gG + bB \tag{4.1}$$

where r, b, and c are called the tristimulus values for the primaries RGB and the color stimulus C. This tells us that a given color can be matched by the linear superposition (addition) of three primaries. Note, however, that for a given set of primaries and color being matched, tristimulus values may be positive or negative. A negative tristimulus value means that the primary must be added to the color being matched rather than to the RGB mixture.

Now consider color mixtures. If two colors, C_1 and C_2, are separately matched by different mixtures of the RGB primaries,

$$C_1 = r_1 R + g_1 G + b_1 B \tag{4.2}$$

$$C_2 = r_2 R + g_2 G + b_2 B \tag{4.3}$$

the match when the two colors are mixed is found experimentally to be:

$$C_1 + C_2 = (r_1 + r_2)R + (g_1 + g_2)G + (b_1 + b_2)B \tag{4.4}$$

That is, the mixing of colors is additive and linear. These are known as Grassman's laws [RF85], and are basic empirical rules for three-color representation of images.

4.1.2 Luminance and chrominance ○

One of the drawbacks to RGB representations is that the eye is most sensitive to green, less sensitive to red, and least sensitive to blue light. In addition, the eye's ability to resolve spatial detail is best in the greens at the center of the visible spectrum, and poorest in the blue. These effects are of interest in a compression system, as there is no reason to accurately reproduce details that the eye cannot see. If we don't send those details, we can compress the picture more effectively.

It is easier to take advantage of these effects when the picture is described in terms of luminance and chrominance. Luminance is closely related to the perception of brightness, whereas chrominance is related to the perception of color hue and saturation. By definition, luminance is proportional to the light energy emitted per unit projected area[1] of the source, but the energy in each band of wavelengths of the input is scaled by the corresponding sensitivity of the eye in that band. Therefore, luminance is a measure of the physical energy of the light source reaching the eye, but the incident energy is weighted according to the spectral sensitivity of the eye.

A number of different color representations use luminance/chrominance separations. However, before getting to color representation, we must first review how the eye perceives luminance.

4.1.3 Brightness perception ○

Perception of brightness is governed by a number of factors. According to Weber's law, for a wide range of luminance, if a light source area of luminance Y is adjacent to an area of luminance $Y + \Delta Y$, the difference is just detectable when:

$$\frac{\Delta Y}{Y} \approx 0.02 \qquad (4.5)$$

This implies a logarithmic relationship for the response of the visual system.

Weber's law can be used to explain effects such as brightness constancy, in which the perceived brightness of an object stays relatively constant, independent of illumination. According to Cornsweet [Cor70], the approximately logarithmic response is believed to occur in the first stage of the visual system, and this means that the stimulation of the neural elements is proportional to $\log Y$. Therefore, the perceived difference between areas

---

[1]The projected area is the effective area when the source is viewed at an angle.

illuminated by Y_1 and Y_2 is

$$\log Y_1 - \log Y_2 = \log Y_1/Y_2 \qquad (4.6)$$

and the perceived brightness change is determined by the ratio of the two brightnesses. Thus, as long as the reflectivity of the object does not change, the object appears essentially the same for a wide range of illuminations. Cornsweet's book provides some convincing illustrations of brightness constancy.

Brightness constancy breaks down for large changes in brightness, and indeed, partly because of this, Weber's law does not precisely describe the perceptual response for typical display usage.

In viewing displays, perceived brightness of a display field is dependent on a number of variables, including light reflected or emitted from regions surrounding the field of interest and the size of the stimulus change. According to studies by the Commission Internationale de l'Eclairage (CIE), for conditions more typical in display usage, the perceptual response to luminance, called Lightness ($L*$), is related to the luminance, Y, by the following empirical relationship:

$$L^* = \begin{cases} 116(Y/Y_n)^{1/3} - 16 & \text{if } Y/Y_n > 0.008856 \\ 903.3(Y/Y_n) & \text{otherwise} \end{cases} \qquad (4.7)$$

where Y_n is the white reference luminance. The relationship is plotted in Figure 4.2, in order to illustrate the nonlinearity of the perceived brightness.

This relationship for perceived brightness is used in the CIELUV and CIELAB perceptually uniform color spaces. A just detectable difference in L^* is $\Delta L^* \approx 1$, independent of L^*.

Using a representation that has uniform perception of brightness is important in compression systems such as MPEG. If the perception of brightness change is independent of brightness, no dependence on brightness levels is needed in the compression models.

4.1.4 Gamma correction ○

The typical video compression system, such as sketched in Figure 4.3 consists of four basic elements: a video camera, a gamma correction unit, a compression/decompression unit (probably physically separated), and a CRT-based display. These interact with an observer's eye and, presumably, the observer's brain.

As discussed in the preceding section, the observer's visual system has a nonlinear response to the light emitted from the display phosphors. However, the observer is not the only nonlinear part of the system. The CRT also

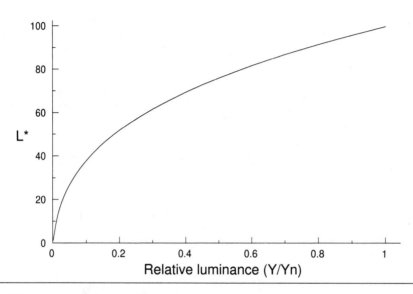

Figure 4.2: Relationship between perceived brightness and luminance.

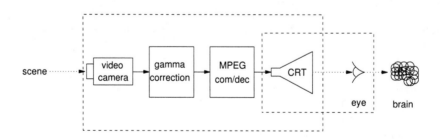

Figure 4.3: Diagram of a video compression system and observer.

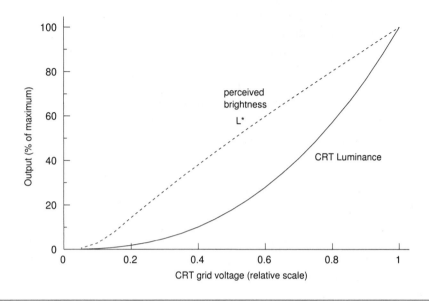

Figure 4.4: Cancellation of nonlinearities in observer's eye and CRT.

has a very nonlinear response to the input to the control grid, the output luminance varying according to a power law:

$$Y \propto V^{\gamma} \tag{4.8}$$

where V is the input signal relative to the cutoff voltage and γ is a numerical constant between 2 and 3 [RF85].

This would appear to be an awkward problem, but in fact, it is quite useful. The nonlinearity of the CRT is almost the inverse of the nonlinearity in the luminance response of an observer, thus providing a nearly linear relationship between the signal applied to the CRT grid and the perceived brightness. The effect of this approximate cancellation is shown in Figure 4.4, where the nonlinear CRT luminance produces a nearly linear perceived brightness. The smaller dashed rectangle in Figure 4.3 therefore encloses an almost linear combination of elements.

This fortuitous cancellation means that the compression system algorithms can be linear - that is, they do not have to compensate for signal amplitude. However, in order for the displayed image to appear the same as when the scene is viewed directly, the net effect of capture and display — the part enclosed by the larger dashed rectangle — should also be linear. Therefore, the display "gamma" should be reversed by applying a power law "gamma correction" to the input data. In fact, most video cameras have a gamma correction of 0.45, which would exactly compensate for a CRT gamma of 2.2.[2]

It is important to recognize that most video compression systems, MPEG included, use gamma-corrected video in order to have an internal representation for brightness that is nearly linear, perceptually. It is also important to recognize that the system elements enclosed in the two rectangles in Figure 4.3 are only approximately linear. Calibration and linearization of displays and cameras may be needed to get the best quality from MPEG.

4.1.5   Luma and gamma-corrected luminance ○

In video systems, gamma-corrected red, green, and blue (R', G', and B') are defined on a relative scale from 0 to 1, chosen such that shades of gray are produced when $R'=G'=B'$. Then, white is given by $R'=G'=B'=1$ and black by $R'=G'=B'=0$. White is actually defined in terms of standard "illuminants" that describe the spectral composition of the scene illumination. Modern television receiver phosphors produce a D_{65} white that is close to normal daylight illumination.[3]

R', G', and B' all contribute to the perceived brightness, but must be weighted according to the relative sensitive of the eye to each primary color. Thus,

$$Y' = 0.299R' + 0.587G' + 0.114B' \qquad (4.9)$$

where Y' is the *luma*, a quantity that is closely related to gamma-corrected luminance.

In common with most texts and papers on video compression, in the sections on compression we will sometimes use the term *luminance* when we should actually use *luma*. Since the weighting factors sum to 1 and R', G', and B' are defined to have values on a relative scale from 0 to 1, Y' will also have the same range.

4.1.6   YUV and YCbCr color systems and chroma ○

We now need to describe the hue and saturation of the colors in the picture. While it is possible to characterize hue and saturation directly, it is more convenient from a computational point of view to define color differences, U' and V':

$$U' = B' - Y' \qquad (4.10)$$

$$V' = R' - Y' \qquad (4.11)$$

To see how this YUV color representation relates to RGB, refer to Figure 4.5. This shows a plot of RGB color space in which the three values,

---

[2]In practice, most monitors have a gamma close to 2.5 [RF85].

[3]D_{65} nominally refers to a color temperature of 6500 K for the illumination.

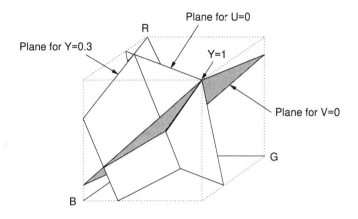

Figure 4.5: Relationship between the RGB and YUV color coordinate systems.

R', G', and B' are three independent coordinates forming a cube. Three planes, $Y' = 0.3$, $U' = 0$, and $V' = 0$, are sketched in this figure. Grayscale values occur on the line formed by the intersection of the $U' = 0$ and $V' = 0$ planes, as this is where $R' = G' = B' = Y'$. Values of U' and V' either greater than or less than zero describe colors of varying degrees of hue and saturation. It is important to recognize that U' and V' are not perceptually uniform color coordinates, even though they are derived from gamma-corrected RGB. Chrominance values computed from gamma-corrected RGB are called chroma.

The positive and negative values of U' and V' can be inconvenient. To eliminate negative values and give all three components roughly the same dynamic range, U' and V' are scaled and zero-shifted in the transformation to the Cb and Cr color coordinates used in MPEG-1. The luminance is the same for the two systems, and the chrominance transformation between the YUV and YCbCr color representations is:

$$Cb = (U'/2) + 0.5 \qquad (4.12)$$

$$Cr = (V'/1.6) + 0.5 \qquad (4.13)$$

If an 8-bit representation is used, R', G', and B' values are usually interpreted as integers from 0 to 255. The zero shift of 0.5 in the equations above should then be multiplied by 256. MPEG-1 uses the YCbCr representation.

Another color coordinate system that has been in use in the television industry for many years is YIQ. The luminance is given by Equation 4.9, whereas I and Q are given by [Lou85]:

$$Q' = 0.41(B' - Y') + 0.48(R' - Y') \qquad (4.14)$$

$$I' = -0.27(B' - Y') + 0.74(R' - Y') \qquad (4.15)$$

Therefore, given the definitions in Equations 4.14 and 4.15, the I' and Q' components are linear transformation of the U' and V' components. Again, the definitions are for gamma-corrected RGB. The YIQ system is allowed in MPEG-2, but not in MPEG-1.

Ignoring the effects of finite precision of the data, the conversion between different systems for representing color is a lossless mathematical transformation. Therefore, we neither lose nor gain by making a transformation to a luminance/chrominance representation. However, the luminance/chrominance representation puts us in a domain where we can take better advantage of the eye's ability to resolve spatial detail. Indeed, the representation of color in terms of hue and saturation is important in models for the perception of color.

4.1.7   Opponent model of human vision ◑

Some success has been achieved in terms of modeling the way the eye perceives color in terms of two opponent systems, one involving blue versus yellow and the other, red versus green (see [Rog83b]). If we call the three color sensing receptors A, B and C (with absorption spectra shown in Figure 4.1), the color opponent theory hypothesizes the interconnectivity sketched in Figure 4.6. In this model the huge number of interconnections in the visual system are modeled as simple functional blocks with two basic types of synaptic action, excitation and inhibition (excitation is additive in neural stimulation, whereas inhibition is subtractive).[4]

According to this model, the light absorbed in the three types of cones first undergoes a logarithmic transformation. Then, by sums and differences, three opponent systems are developed, blue-yellow, green-red, and black-white. The summation of red and green provides the luminance channel and the other two are chrominance channels.

In Section 4.1.3 we showed that the logarithmic transformation in the first stages of the visual system leads to the effect of brightness constancy in which relative contrast is constant, independent of illumination. The same mechanism appears to be responsible for color constancy, in which a given hue tends to remain perceptually constant, independent of the hue of the illumination. As with brightness constancy, color constancy holds only for a reasonable range of conditions [Cor70], but color constancy is why the responses of the blue-yellow and green-red opponent systems are relatively independent of luminance.

---

[4]The logarithmic response itself is attributed to inhibitory feedback in the first stages of the visual system [Cor70].

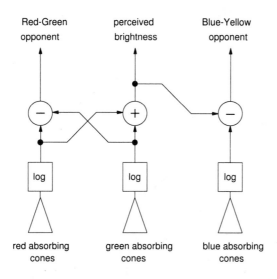

Figure 4.6: Opponent color model.

4.1.8   Chromaticity coordinates ◐

In Section 4.1.1 we discussed the additivity and linearity of color matching. Additivity and linearity require that for any two stimuli, $\{R_1, G_1, B_1\}$ and $\{R_2, G_2, B_2\}$, the color resulting from the two present simultaneously is given by the sum of the two stimuli.

Suppose that for a set of three primaries, RGB, the representation of color is normalized as follows:

$$r = R/(R + G + B) \qquad (4.16)$$

$$g = G/(R + G + B) \qquad (4.17)$$

$$b = B/(R + G + B) \qquad (4.18)$$

Since $r + g + b = 1$, only two of these three variables are needed to specify any color within the range allowed by the primaries. This is illustrated by Figure 4.7, in which only r and g are used. ($b=1$ is given by the absence of r and g.) r and g are the chromaticity coordinates for these particular RGB primaries.

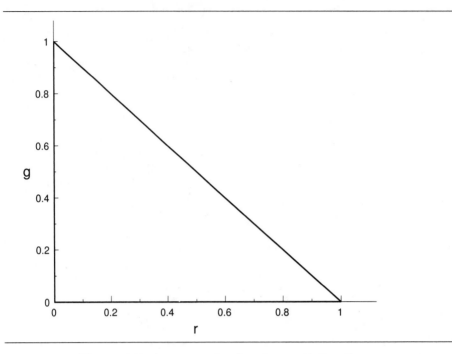

Figure 4.7: An example of a chromaticity diagram.

4.1.9   Center of gravity ◗

The additive and linear properties of color matching lead to the rule known as the *center of gravity law*. Given two stimuli, $\{R_1,G_1,B_1\}$ and $\{R_2,G_2,B_2\}$, the chromaticity resulting from the two present simultaneously is given by the sum of the two stimuli. Define

$$T_n = R_n + G_n + B_n \tag{4.19}$$

The red chromaticities are given by

$$r_1 = R_1/T_1 \tag{4.20}$$

for stimulus 1,

$$r_2 = R_2/T_2 \tag{4.21}$$

for stimulus 2, and

$$r = (R_1 + R_2)/(T_1 + T_2) \tag{4.22}$$

for both together. This last relationship can be recast to give the equations for the center of gravity law:

$$r = \frac{r_1 T_1 + r_2 T_2}{T_1 + T_2} \tag{4.23}$$

$$g = \frac{g_1 T_1 + g_2 T_2}{T_1 + T_2} \tag{4.24}$$

$$b = \frac{b_1 T_1 + b_2 T_2}{T_1 + T_2} \tag{4.25}$$

where the latter two equations are derived by exactly the same procedure used for r.

The center of gravity law shows that the color resulting from a mixture of any two colors lies on a straight line between those two colors in the chromaticity diagram, the position on the line being determined by the center of gravity of weights T_1 and T_2. In general, any point within the triangle in Figure 4.7 represents a mixture of the three primaries, whereas points outside of the triangle cannot be realized with values of r and g in the physically allowed range from 0 to 1.

One set of straight lines on this diagram is of particular interest. Since the luminance (true luminance in this case) is a weighted sum of R, B, and G,

$$L = w_r R + w_g G + w_b B \tag{4.26}$$

where the weights, w_r, w_g, and w_b are determined by the particular set of primaries.[5]

If we divide this equation by $R+G+B$, we get the following:

$$L/T = w_r r + w_g g + w_b b \tag{4.27}$$

and since $b = 1 - (r + g)$,

$$g = \left[\frac{L/T - w_b}{(w_g - w_b)} \right] - \left[\frac{w_r - w_b}{(w_g - w_b)} \right] r \tag{4.28}$$

This equation gives a set of straight lines with negative slope and an intercept determined by the luminance weighting coefficients and the luminance. The particular case where $L = 0$ is known as the *alychne* [RF85]. While color at zero luminance makes no sense physically, the alychne is important in the CIE primary system.

4.1.10 Coordinate transformations ◑

The linearity and additivity of color matching guarantees that a linear transformation always exists between one set of primaries and another. The transformations take the form

$$R' = \alpha_{11} R + \alpha_{21} G + \alpha_{31} B \tag{4.29}$$

---

[5]The weights in Equation 4.9 are for a particular set of NTSC phosphors and gamma-corrected RGB.

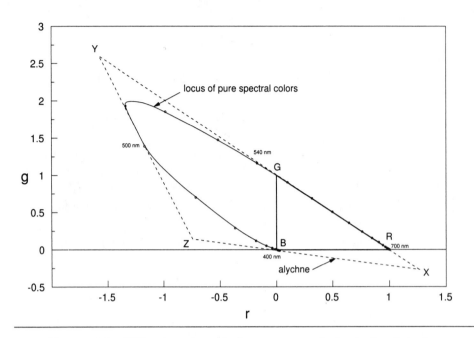

Figure 4.8: CIE primaries relative to a set of physical primaries.

$$G' = \alpha_{12}R + \alpha_{22}G + \alpha_{32}B \tag{4.30}$$

$$B' = \alpha_{13}R + \alpha_{23}G + \alpha_{33}B \tag{4.31}$$

These linear transformations are such that a straight line in one coordinate system remains a straight line in another. Furthermore, the center of gravity law proof is independent of the primary system, and therefore applies to any system which is a linear transform of another.

For most transformations, nonphysical weighting of some of the primaries will be required, reflecting the fact that some of the primaries in at least one of the systems must lie outside the chromaticity triangle of the other system.

4.1.11   CIE primaries ●

Although it is clear that nonphysical weightings must occur in transforming from one primary system to another, it is also possible (conceptually) to have nonphysical primaries. These nonphysical primaries cannot be produced by real physical light sources. They are, however, useful mathematical constructs that, by definition, follow the color matching laws of additivity and linearity.

The CIE has defined two primary systems for color matching, RGB and XYZ. RGB is based on a set of three pure spectral colors of nominally red,

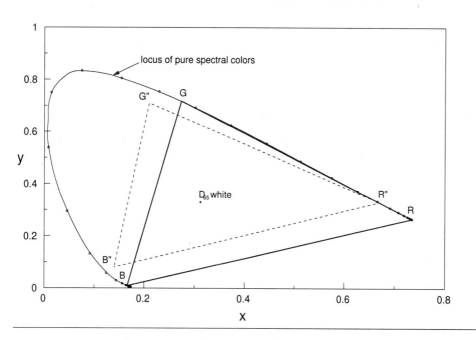

Figure 4.9: CIE chromaticity diagram.

green, and blue hues; XYZ is a nonphysical set of primaries defined such that chromaticity values are always positive.

The chromaticity diagram for the RGB set of pure spectral primaries at 700 nm (red), 546 nm (green), and 436 nm (blue) is shown in Figure 4.8.[6] This figure shows the locus of points for the full visible range of pure spectral colors, and illustrates that for a given set of physical primaries, not all colors can be produced. In fact, the center of gravity law tells us that only those colors within the right triangle with vertices labeled *RGB* are accessible. To match a color outside of the triangle requires at least one of the RGB chromaticities to be negative, as should be clear from Figure 4.8.

To get the full range of physical colors from sums of three primaries, the primaries must form a triangle that encloses the full locus of spectral colors. This, of course, requires points outside of the range accessed by physical light sources, and therefore implies nonphysical primaries at the triangle vertices.

One such triangle, labeled XYZ, is shown in Figure 4.8. This triangle is for the CIE XYZ primaries (which are at the vertices of the triangle). These primaries are chosen such that the Y primary spectral curve matches the eye's spectral sensitivity to luminance and the X and Z primaries lie on the

---

[6]The data for the CIE RGB and XYZ chromaticities are tabulated in [Fin85].

alychne (the line for zero luminance). In addition, the units of the primaries are chosen so as to be equal for an equi-energy stimulus[RF85].

Once the CIE primaries are defined, chromaticity coordinates can then be defined analogously to r, g, and b.

$$x = X/(X + Y + Z) \tag{4.32}$$

$$y = Y/(X + Y + Z) \tag{4.33}$$

As before, only two chromaticity coordinates are required to specify the color. Figure 4.9 illustrates the resulting CIE chromaticity diagram with these coordinates. The CIE spectral primaries are the vertices of the solid triangle, whereas the standard NTSC phosphors form the triangle labeled $R''G''B''$. As expected, the entire visible spectrum is included within positive values of x and y.

The triangle formed by phosphor coordinates such as $X''Y''Z''$ is called a *Maxwell triangle*. This triangle delineates the range of colors accessible with this set of phosphors.

4.1.12   CIELUV color space●

Neither the UV color representations nor the CIE chromaticity diagrams discussed in the previous sections provide perceptually uniform scales for color discrimination. The CIE had studied this problem extensively, and has defined two color spaces, CIELUV and CIELAB, that are reasonably perceptually uniform. CIELUV is recommended for additive display[LRR92], and thus is usually more relevant to MPEG.

In Section 4.1.3 the nonlinear transformation from luminance to the Lightness ($L*$) of these perceptually uniform spaces was discussed (see Equation 4.7). Comparably nonlinear equations hold for the CIELUV color space chrominance components[RF85]:

$$u* = 13L * (u' - u'_n) \tag{4.34}$$

$$v* = 13L * (v' - v'_n) \tag{4.35}$$

where $L*$ is the Lightness defined in Equation 4.9, u' and v' are defined by:

$$u' = \frac{4x}{-2x + 12y + 3} \tag{4.36}$$

$$v' = \frac{9y}{-2x + 12y + 3} \tag{4.37}$$

and where u'_n and v'_n are u' and v' for the reference white.[7] These equations attempt to fit the experimental data on perception of color differences, and

---

[7]u'_n and v'_n have no simple relationship to the gamma-corrected color differences, U' and V'.

as such, provide an approximately perceptually uniform color space. The scale is such that a color change of one unit (as with $L*$) represents a just detectable difference (approximately). A Euclidian distance is claimed to provide a composite measure of perceptual change in this color space,

$$\Delta E_{uv}* = \left[(\Delta L*)^2 + (\Delta u*)^2 + (\Delta v*)^2\right]^{1/2} \qquad (4.38)$$

and a just noticeable difference is $\Delta E_{uv} \approx 1$.

This perceptually uniform chrominance representation is important in the design of display systems needed by MPEG, but is not directly part of the MPEG standard. While MPEG incorporates many things that are much more complex (e.g. motion compensation), we will see that the portion of available bandwidth required to compress the gamma-corrected chrominance with good quality is already so small that using perceptually uniform chrominance would be of limited value. Having a perceptually uniform luminance representation is much more important, but we have already seen that gamma-corrected luminance provides a pretty good approximation to a linear system.

4.2   Resolving spatial detail ○

Color differences are a very efficient way to represent the color information in a picture, in that the eye resolves much more spatial detail in the luminance component than in the chrominance components. This can be seen qualitatively from the plot in Figure 4.10 of the contrast sensitivity functions as a function of angular frequency [NB67, Mul85]. Contrast is defined as the peak-to-peak amplitude of a sinusoidal variation in luminance, divided by the average luminance; the contrast sensitivity is defined as the inverse of the contrast for a just detectable change. Therefore, changes too small to be seen are in the region above the curve, whereas visible changes fall below the curve. The angular frequency increases on a logarithmic scale from left to right. [8]

The three curves in Figure 4.10 show the thresholds for luminance stimuli at zero chrominance (gray) [NB67] and for opposed-color chrominance stimuli [Mul85] at constant luminance. The responses to luminance and chrominance both roll off sharply at high angular frequencies, but the roll-offs start at much lower frequencies for the two chrominances. Since the upper range of frequencies in the plot corresponds to the upper range of

---

[8]Plots such as in Figure 4.10 are always done as a function of angular frequency, as this determines how rapidly the pattern focused on the retina of the eye changes. The angular frequency perceived by an observer is directly proportional to the spatial frequency of a pattern shown on a display, and inversely proportional to the distance between display and observer.

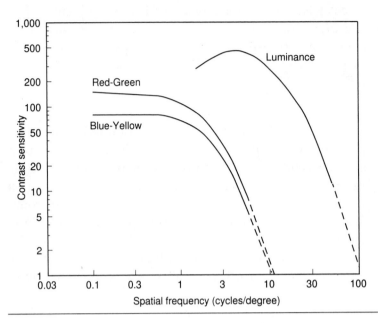

Figure 4.10: Contrast sensitivity for luminance and chrominance (after [NB67] and [Mul85]).

spatial frequencies in a typical display, the need for high spatial resolution in the chrominance samples is greatly reduced.

4.2.1 MPEG chrominance sampling ○

Because of the reduced sensitivity of the eye to high spatial frequency chrominance, most image compression systems use lower spatial resolution to represent the chrominance. In MPEG the horizontal and vertical resolution of the two chrominance components is reduced by a factor of 2.[9] In Chapter 1 we described a moving picture sequence with a sampling of 360 samples horizontally and 288 samples vertically for each picture, and a raw data rate of about 60 Mbits/s. Reducing the chrominance resolution by a factor of 2 in both directions cuts the data rate in half to about 30 Mbits/s. While this is a big effect, a further compression by 30:1 is still needed to get to the nominal 1 Mbit/s rate of MPEG-1. In large part, this additional compression is achieved by exploiting other properties of the human visual system.

   Although higher spatial resolution would be desirable (indeed, it is one reason for MPEG-2), the compressed data rate for most compression al-

---

[9]MPEG-2 allows other sampling ratios as well.

gorithms (including MPEG) is approximately proportional to the spatial resolution (the number of samples along a line or column). An increase in spatial resolution therefore quickly causes the data rate to exceed the available bandwidth. While an increase in spatial resolution can be achieved by increasing the amount of data reduction in the coding model, this usually leads to objectionable visible artifacts in the decoded pictures.

4.2.2   Varying precision ○

Simply discarding high-frequency spatial details by changing the sampling is not always a good idea. Sharp edges, for example, produce large amplitude high frequency variations that are very important to represent. Rather than discarding these terms, a better strategy is to reduce the precision with which they are represented.

To do this, we have to recast the data in a different format, transforming the data from a spatial to a frequency representation in which the data is decomposed into a set of sinusoidal waveforms. Since the frequencies of the waveforms are determined by the sampling of the image, all we need to characterize each waveform's contribution is the waveform amplitude. The visibility of a particular waveform is determined by the contrast sensitivity at that frequency, and this determines to what extent the amplitude can be scaled to reduce the precision.

The situation is not quite this ideal, however. In Chapter 3 we describe the discrete cosine transform (DCT), which is the particular method used by MPEG for decomposing the data into sinusoidal waveforms. The 8×8 DCT used by MPEG does not provide an ideal frequency decomposition, in that the cosine waveforms for a block end abruptly at the edges of the block. However, the truncated cosine waveforms are close enough to true sinusoids that we can regard them as such for the purposes of this discussion.

4.2.3   Frequency weighting and quantization ○

Once we have a frequency representation, precision is minimized by weighting each frequency term in accord with the contrast sensitivity for that frequency. This process, called quantization, produces an integer value for the waveform amplitude. The amplitude is divided by a quantization value and rounded to an integer, following rounding rules specified in the MPEG standard. The integer value, which may be positive or negative, is coded and transmitted to the decoder.

We will discuss coding in much more detail in later chapters, but a good rule-of-thumb is that the lower the precision, the fewer the bits required to code the value. Therefore, the more aggressively we can scale the values, the better our coding rate will be.

However, we can't scale a waveform by too large a value. In order to reproduce the waveform in the decoder, the integer value must be rescaled (dequantized) by multiplying it by the appropriate quantization value (which the decoder knows). The result is necessarily only an approximation to the initial amplitude. If, for example, the distortions produced by making this approximation are to be invisible, the amplitude discontinuities should be below the threshold of visibility. In theory, we can make a reasonable guess as to what this threshold should be from the contrast sensitivity at the waveform frequency, but this is only a rough guide. In practice, a much more sophisticated approach is needed to accurately determine these quantization values [AP92].

Exactly what value we use for quantizing is a function of both the spatial frequency of the waveform and the color coordinate. As should be apparent from Figure 4.10, the quantizing values for luminance should be very different from the quantizing values for chrominance. The two chrominance components are not quite identical, but are reasonably similar. However, the Cr and Cb coordinates used by MPEG are not perceptually uniform, and a finer quantization is therefore required. Therefore, MPEG makes a further approximation and uses the same quantization for all three components. Even with this approximation, chrominance usually requires less than 20% of the bandwidth.

There are, of course, some subtle aspects to the quantization process. First, as noted above, the 8×8 DCT is not an ideal frequency decomposition. Second, there is no guarantee that activity at one frequency will not affect the perception of activity at another frequency in the same block. Third, the threshold for perception may be different for a change in amplitude as opposed to the absence or presence of the waveform. Finally, in many cases compression targets can only be reached with quantization values well above the threshold for visibility.

4.2.4   Activity masking and facilitation○

Another property of the human visual system that can be used to advantage in compression systems is *activity masking*. In portions of a picture where strong edges and other rapid intensity changes occur, the eye has difficulty discerning small variations in intensity. Conversely, in very smooth areas, the same small changes may be quite visible. Since in image quality measurements the eye is drawn to the areas that show artifacts, a single array of quantization values for a DCT must be appropriate for the areas where quantization artifacts are most visible. It would be better to vary the quantization, depending on the activity in different regions.

However, some care must be exercised in using variable quantization. Vassilev [Vas73] has explored contrast sensitivity variations near edges, and

Rogowitz [Rog83a] has studied masking of a stimulus by an adjacent high contrast mask as a function of spatial frequency and temporal offset.[10] These experiments suggest that contrast sensitivity in the presence of adjacent masking stimuli is a complex phenomenon that may be difficult to exploit in data compression systems — at least in the spatial domain. The discussion in Wandell [Wan95] reinforces this point, and shows that in addition to masking, facilitation can occur. *Facilitation* is the enhancement of the visibility of one signal in the presence of another.

When stimulus and mask are overlapped, the threshold for detection of a given spatial frequency is increased if the mask has a similar frequency and relatively large amplitude (see [Rog83a]). However, the discussion in Wandell [Wan95] shows that the masking effects seen at large mask amplitudes may in fact change to facilitation (enhanced visiblity) at low amplitudes. Wandell also discusses experiments showing that masking is quite orientation dependent, the effect disappearing at an angle of 67 deg. (However, two superimposed masks at opposing 67 deg angles has a powerful masking effect.) This suggests that quantization thresholds can be increased when there is a lot of activity in a given DCT block. However, a quantitative analysis of these masking effects, at least as they apply to DCT quantization, has not yet been done. In addition, very little is known about suprathreshold effects, and at the low rates typical of MPEG-1, the encoding is normally done under conditions where quantization artifacts are quite visible.

Applications of variable quantization are still based primarily on heuristics and experiments, and there is some question as to just how much benefit can be derived from variable quantization. For still picture coding, advocates of variable quantization claim as much as 30% improvement in compression for a given picture quality [NP77, Gon91]. The JPEG committee recently adopted Part 3 of the JPEG still image compression standard, and this includes variable quantization. The main motivation for this extension to JPEG was, however, to make conversion between JPEG and MPEG easier. Studies reported at the JPEG meetings showed no significant improvement in rate-distortion.

The MPEG syntax provides an efficient code for changing the quantization scale factor. The quantization scale factor is a multiplicative constant applied to all of the AC coefficients, but not to the DC term. This permits the quantization values to be different in different parts of a picture. Although variable quantization is primarily used to control the bit rate, it also provides a mechanism for exploiting activity masking, thereby possibly improving the overall picture quality. Further discussion of variable quantization is deferred to Chapter 14.

---

[10]The mask and stimulus may or may not overlap. When they do not, the process is called *metacontrast*.

4.3   Perception of motion ○

When a viewer watches a movie or television set, he sees a sequence of images in which objects appear to move. The motion is, however, apparent rather than real. Each individual picture in the sequence is a snapshot of the scene, and if the sequence is played at a very slow rate, the objects appear to shift discontinuously from one picture to the next. When the sequence is played at the intended rate, however, the objects appear to move continuously.

Although temporal effects in the human visual system are quite complex, much insight can be gotten from the measurements of spatial-temporal interactions in the HVS.[11] For example, display flicker has been of long-standing interest to display system engineers, and some interesting clues to temporal effects can be gotten from measurements of critical flicker frequency. Measurements of spatial resolution at various temporal frequencies are also relevant, as is pre- and post-masking in the temporal domain. The visual mechanisms involved in the perception of motion, and the ability of the eye to track simple objects moving with uniform velocity (as opposed to complex random motion) are also of interest.[12]

4.3.1   Critical flicker frequency ○

Critical flicker frequency (CFF) is the frequency above which a periodic modulation of the luminance can no longer be seen. Because of its extreme importance in display technology, many detailed measurements have been made of flicker. For very large objects (60 deg field of view), Figure 4.11 [DeL58, Kel61] shows how perception of flicker varies with respect to the intensity of light on the retina (see [Cor70]). In (a), the data are plotted as absolute sensitivity; in (b), as relative sensitivity. The coalescing of the plots in (b) at frequencies below about 4 Hz is expected from Weber's Law. At high temporal frequencies, however, the sensitivity becomes independent of luminance, as can be seen from plot (a).

There is also a dependence of CFF on the size of the object, as can be seen in Figure 4.12 (see [Rog83b]). The experimental observation that small objects have lower critical flicker frequency indicates that temporal resolution is reduced at higher spatial frequencies. One might suspect that

---

[11] A review by Rogowitz [Rog83b] provided us with many ideas for this section.

[12] One variable that is very important in experiments on the HVS is saccadic eye movements. These very small involuntary eye movements are characteristic of normal vision, and play a profound role in HVS properties. For example, when an image is stabilized on the retina by means of an eye tracking device, the subject soon perceives only a uniform gray [Cor70] Presumably, since viewing of MPEG sequences does not involve stabilized images, data for unstabilized stimuli are more relevant. All of the data cited in this section are for unstabilized stimuli.

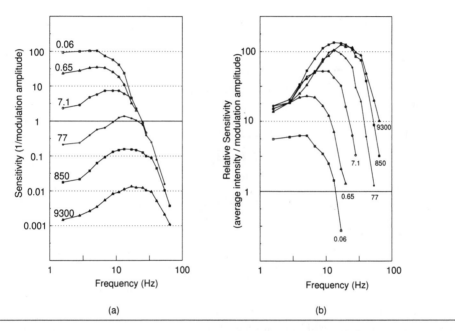

Figure 4.11: Flicker perception: sensitivity as a function of frequency for various luminances: (a) Absolute sensitivity. (b) relative sensitivity. The retinal illuminations (in Trolands) label the curves (after [DeL58, Kel61]).

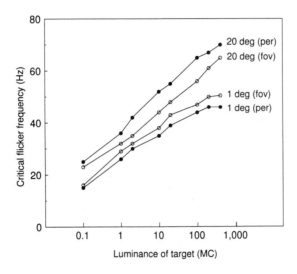

Figure 4.12: Flicker perception: Dependence on target size (after [Rog83b]).

the converse of this is also true; namely, that spatial resolution of vision is reduced at higher temporal frequencies.

4.3.2   Spatial resolution of time-varying scenes ○

Figure 4.13 shows Robson's data [Rob66, Cor70] for the contrast sensitivity as a function of temporal frequency. In this figure data are shown for selected spatial frequencies that bracket the spatial frequency peak in contrast sensitivity. As suspected, there is indeed a reduction in contrast sensitivity at higher temporal frequencies, especially for the highest spatial frequencies.

HVS scientists have modeled this behavior by postulating two types of sensory channels, one for transient signals (fast temporal response) and another for sustained signals (slow temporal response) (see [Rog83b]). Experiments suggest that the contrast sensitivity function is actually the composite response of a number of separate sensory channels tuned fairly narrowly to different spatial frequencies. These are further divided into transient and sustained channels. Sustained channels are broadly tuned, occurring across the full range of spatial frequencies; transient channels respond only at lower spatial frequencies. A qualitative plot of luminance contrast sensitivity as a sum of sustained and transient channels is shown in Figure 4.14. This model is consistent with Robson's data and with the observation that small

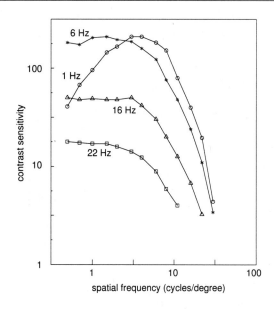

Figure 4.13: Contrast sensitivity as a function of temporal frequency (after [Rob66]).

objects have a lower CFF. It suggests that when viewing objects in motion, high spatial frequency response is attenuated.

4.3.3   Motion sensing ○

There is a considerable body of literature on how the HVS perceives motion (see [AB85, Wan95]), and there has been some debate as to the level of cognition required to perceive object motion. However, models involving only fast low-level visual processing appear to explain many aspects of the phychophysical data on motion perception. These models involve various types of spatio-temporal filters, and postulate filters selective for location, spatial frequency, direction, and speed. Watson and Ahumada [WA85] point out that speed discrimination is virtually absent for moving stimuli at threshold, but is extremely sensitive to speed variations for stimuli above threshold. They use this to postulate a two-stage model for motion sensing. Adelson and Bergen postulate a similar model structure in which the first stage consists of linear filters oriented in space-time and tuned to various spatial frequencies. A second stage then pairs these in quadrature to give motion energy and direction. While the details of these models differ, the general conclusion that motion perception can be explained by low-level visual

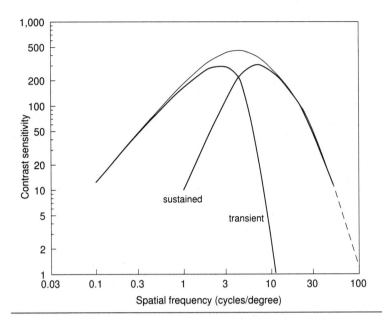

Figure 4.14: Transient and sustained channel contributions to contrast sensitivity (after [Rog83b]).

processing appears valid.

Adelson and Bergen [AB85] provide an interesting illustration for why the eye perceives apparent motion to be the same as real motion. Figure 4.15 illustrates their point of view, showing that low-level spatial and temporal filtering in the HVS provide blurring that makes a sequence of still pictures appear quite similar to the same scene in continuous motion. Both appear to be in continuous motion, provided the picture rate is sufficiently high.

4.3.4   Temporal masking ○

When there is a sudden scene change, a significant time elapses before the eye can adapt to the new detail; during that time, major artifacts in the displayed image may not be visible. This phenomenon is called forward masking; a similar effect, backward masking, occurs just before a scene change.

According to Bloch's law [Owe72], for time intervals less than a certain critical duration, luminance perception is constant if the product of stimulus duration and intensity is kept constant. For displays, this critical duration determines the critical flicker frequency. However, for more complex tasks involving perception of image content, higher levels of cognition are involved,

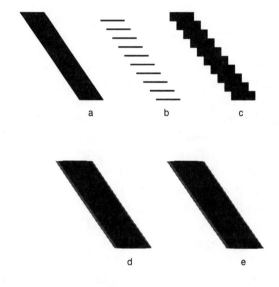

Figure 4.15: Illustration of the effect of spatial and temporal filtering. (a) Motion of a horizontal bar (the time axis is down). (b) Sequence of still snapshots. (c) Display of snapshots. (d) Effect of spatial and temporal low-pass filtering, original scene. (e) Effect of spatial and temporal low-pass filtering on displayed scene.

and the critical duration can be several hundred milliseconds.

4.3.5   Tracking of moving objects ○

The ability of the eye to track moving objects is very important in determining what the eye can resolve. As with scene changes, when an object suddenly starts moving, there is a critical duration before the eye is able to track the object. Once the eye is tracking the object, the ability to resolve spatial patterns improves. [Ger88]. The suppression of contrast sensitivity at low spatial frequencies [Wan95] during saccadic eye movements may be related to this effect.

4.3.6   Coding time-varying scenes ○

These experimental observations have a direct bearing on how an MPEG system should code time-varying scenes. The key to good efficiency in coding changing areas is whether the eye can see the changes. As with spatial variations in intra coding, the basic idea remains the same — if the temporal changes can't be seen, they don't need to be sent. Qualitatively, it seems reasonable to suppose that the eye cannot see temporally changing areas with the same spatial resolution as in stationary areas. Indeed, the concept of trading off spatial and temporal resolution in video coding was first suggested by Seyler [Sey62], and was based on the observation that areas in motion cannot be resolved as well by the eye. It should come as no surprise to the reader that this concept is also very relevant in MPEG video coding.

We have seen above that if an object is in rapid motion, the eye cannot discern high spatial frequencies. There is no reason to code this information until the object slows down; and it is curious, therefore, that the default quantization MPEG uses for the DCT coefficients of the differential data is flat, independent of spatial frequency. This would suggest that MPEG is not taking full advantage of temporal HVS properties. However, most video cameras integrate the light over most of the video field time, and this naturally blurs the images of moving objects. Blurring, of course, attenuates high spatial frequencies, automatically removing them from the transformed data. Slowly moving objects do not blur, but they should retain most of their spatial frequency content. A more conservative quantization is also needed to avoid artifacts in moving objects that the eye can track.

The default flat quantization table for MPEG inter coding may not be appropriate for coding of movies. In movies the image is captured with less integration. The scenes are sharper, but also show much more aliasing. The coding trade-offs here are very complex, in that if the high frequencies are attenuated in inter coding, more frequent updates with intra coding may be needed to restore high frequencies when motion stops.

Scene changes are another place where video coding bit rates can be significantly reduced [SB65, LD94]. If there is a sudden scene change in a video sequence, the new scene normally starts with an intra picture. Since intra pictures typically take about three times as many bits to code as inter pictures, this might lead us to conclude that scene changes are expensive in terms of compressed bit cost. However, since the eye is relatively insensitive to image content for quite a few video field times (15 ms each), the first few pictures before and after the scene change do not need as high a visual fidelity as the scenes following, and this can be used to good advantage to reduce the bit rate near the point of scene change. Just before the scene change the pictures can be repeated with minimal change, and just after, can be coded with less fidelity. Within a few video picture times, however, the fidelity must be restored to normal levels.[13] In addition, these effects should also apply to moving objects in a sequence, indicating that coarser quantization of moving areas can be used, provided that correction of artifacts left behind in the wake of objects are corrected within a few hundred milliseconds.

---

[13] A sequence can also start with B-pictures, and this provides a very simple mechanism for reducing the quality of the first few pictures after a scene change.

5

MPEG Coding Principles

In this chapter we review some of the basic principles used in MPEG coding. We use this chapter to introduce some of the terminology used throughout the book, but our treatment here is necessarily brief. Readers who have no familiarity with data compression will probably want to supplement the treatment here with texts such as [RJ91], [BK95], [Say96], [Hof97] and [PM93].

We first present a very high-level view of a coding system, showing how the system can be divided into modeling and entropy coding sections. Then, we discuss the very important topics of entropy coding and coding models, in that order. Finally, we discuss the specifics of the MPEG-1 coding techniques and present block diagrams for MPEG-1 encoders and decoders. Note that our use of the term *system* in this chapter is unrelated to the MPEG system layer.

5.1 Coding system structure ○

The high-level coding system diagram sketched in Figure 5.1 illustrates the structure of a typical encoder and decoder system. The analog to digital conversion (A/D) determines the basic resolution and precision of the input data, and thus is a very important step in reducing the almost unlimited data that potentially is available from the original scene to a manageable level. However, data reduction does not necessarily stop once the digitization is completed.

Compression systems that do no further data reduction once the picture is digitized are lossless systems; these lossless compression systems rarely compress natural image data by more than a factor of 2 to 3. Compression systems such as MPEG need to achieve considerably more than an order of magnitude higher compression than this, and they do this by means of

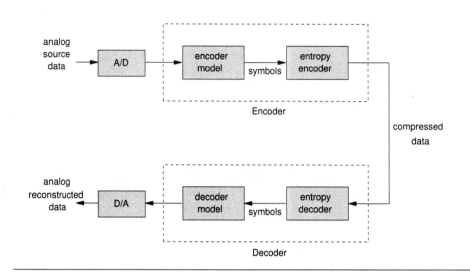

Figure 5.1: Overview of a compression system.

further lossy data reduction after the digitization.

As shown in Figure 5.1, it is convenient to separate a coding system into two parts. The first part is the encoder model that performs lossy data reduction in the process of changing the digital source data into a more abstract representation which is conventionally labeled symbols. The second part then codes these symbols in a process that minimizes the bitstream length in a statistical sense. This second step is called *entropy coding*.[1]

5.1.1   Isolating the model○

The decoder in Figure 5.1 reverses the encoding process, at least to the extent possible. It first losslessly converts the compressed data back to symbols, and then rebuilds a digital picture that is (hopefully) a visually close approximation to the original digital source data. This digital data is fed through a D/A (digital to analog) converter to recreate an analog output signal for the display.

For the moment consider the entropy encoder and decoder to be "black boxes" with the following properties: The entropy encoder accepts a stream of symbols from the encoder model and converts it to compressed data; the entropy decoder decodes that compressed data and returns an identical

---

[1]In [PM93] a slightly different decomposition is used in which an intermediate stage of descriptors is defined. This was necessary because of the two entropy-coding techniques used in JPEG. MPEG uses only one type of entropy coding and this intermediate stage is not needed. We therefore use the more conventional decomposition in this book.

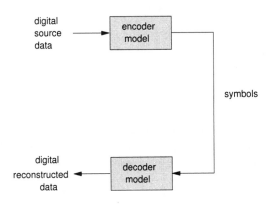

Figure 5.2: Compression system with lossless coding sections removed.

stream of symbols to the decoder model. Since by definition the entropy encoder and decoder are lossless, the compression system can be truncated as shown in Figure 5.2. Although the system may not provide much compression, it is still completely functional in other respects.

5.1.2   Symbols and statistical modeling ○

Symbols are an abstract representation of the data that usually does not permit an exact reversible reconstruction of the original source data (unless specifically designed to be reversible). In general, the coding model involves both data reduction and a recasting of the data into symbols. The symbols chosen to represent the data are usually a less correlated representation that allows an efficient and relatively simple entropy-coding step. The process of choosing a set of symbols to represent the data involves a statistical model of the data — so named, because it is based on a careful analysis of the statistical properties of the data.

Statistical modeling pervades the design of the coding model. For example, one coding model used extensively for lossless coding is DPCM (differential pulse code modulation). In this coding technique a difference is calculated between each pel and a prediction calculated from neighboring pel values already transmitted. Pel differences turn out to be considerably less correlated than the original pel values, and can be coded independently with reasonable efficiency. Better coding efficiency is achieved, however, when correlations between differences are taken into account in the statistical modeling.

An even more effective way of reducing correlation is with the DCT, as has been mentioned in earlier chapters. However, before we get into detailed discussion of DCT based models, we need to develop the topic of entropy coding.

5.2   Entropy coding ○

An entropy encoder losslessly converts a sequence of symbols into compressed data, and the entropy decoder in that figure reverses the process to produce the identical sequence at the decoder. The task of the entropy coder is to encode the sequence of symbols to the shortest possible bitstream.

5.2.1   Principles of entropy coding ○

In general, the symbols fed to the entropy encoder are members of an *alphabet*, a term that requires some explanation. Suppose the task is to compress a string of text consisting of a particular sequence of lowercase letters. If all letters can occur in the sequence, the compression alphabet would then consist of the 26 letters of the normal alphabet, a through z. However, if the text might include both uppercase and lowercase letters, the compression alphabet must have 52 entries (symbols). If spaces are allowed in the text, the space symbol must be added to the alphabet. If any punctuation marks can occur, the compression alphabet must include them as well. If all of the characters that can be expressed by an 8-bit code are allowed, the alphabet must consist of 256 symbols; if two-byte codes are allowed, the alphabet should have 65,536 symbols. In other words, the alphabet is the set of possible symbols, whereas the message — the actual stream of symbols fed to the entropy encoder — is one of many possible streams. It is the task of the entropy encoder to transmit this particular stream of symbols in as few bits as possible, and this requires statistically optimum code lengths for each symbol. The technique for making optimal integer code assignments is called *Huffman coding*.

Entropy encoders are somewhat like intelligent gamblers. The currency is denominated in bits and the objective is to spend as few bits as possible in getting the symbol stream (the message) to the decoder. As with any gambling scheme, symbol probabilities play a major role in how the bets are placed. To win, the more probable symbols must be coded with fewer bits than average. However, the only way this is possible is if the less probable symbols are coded with more bits than average. Therein lies the gamble. Usually, this intelligent gambler wins, but if a message containing only symbols with very low probabilities occurs, the message is expanded rather than compressed. The bet is then lost.

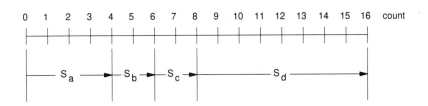

Figure 5.3: Symbol counts on a number line.

5.2.2 Mathematics of entropy coding ◗

According to a fundamental theorem of information theory, the optimum
code length for a symbol, L_s (and also the number of bits of information
transmitted when that symbol is sent) is given by [Sha49]:

$$L_s = \log_2(1/P_s) \qquad (5.1)$$

The entropy, which is simply the average number of bits per symbol, is
given by:

$$\text{entropy} = \sum_s P_s \log_2(1/P_s) \qquad (5.2)$$

where the sum is over all symbols in the alphabet. The entropy is a lower
bound on the average code length in a compressed message.

The following is not a proof of this theorem but may help in understanding it: Suppose the alphabet consists of four possible symbols, a, b, c, and
d. Suppose further that symbol counts, S_a, S_b, S_c, and S_d, are kept of each
symbol for a long message and each symbol has been encountered at least
once. Now, starting with symbol a, stack the symbol counts in succession,
as shown in Figure 5.3.

Each symbol is associated with a particular count interval and thus can
be specified by selecting any number within that interval. Symbol a, for
example, is associated with the interval from 0 to $S_a - 1$ and can be selected
by any number between 0 and 3. Symbol b is associated with the interval
from S_a to $S_a + S_b - 1$, and therefore can be selected by the numbers 4 and
5. Symbol d is associated with the interval from $S_a + S_b + S_c$ to the interval
$S_a + S_b + S_c + S_d - 1$, and can be selected by any number between 8 and
15. Note that there is no overlap in these intervals. The first two columns
of Table 5.1 summarize this.

symbol	count interval	binary count interval	code	probability	code length
a	0..3	0000..0011	00	1/4	2
b	4..5	0100..0101	010	1/8	3
c	6..7	0110..0111	011	1/8	3
d	8..15	1000..1111	1	1/2	1

Table 5.1: Symbol counts, probabilities, and codes.

The third column of Table 5.1 contains the binary representation of the counts, and a very interesting thing can be seen there. For the largest count interval, S_d, only the most significant bit is unique and the lower-order bits can be either zero or one. For the next smaller interval, S_a, the two most significant bits must be zero, but the rest can have either sense. For the two smallest intervals three bits are required to uniquely specify the interval. When the bits that are not unique are discarded, the code words in the 4th column are created. Note that these codes have the required unique prefix property — that is, no short code is a prefix (the leading bits) of a longer code.

If symbol probabilities are calculated from the symbol counts, ideal code lengths can then be calculated from Equation 5.2, as shown in columns 5 and 6 of Table 5.1. These ideal code lengths exactly match the lengths of the code words obtained from the unique leading bits.[2]

5.2.3   Huffman coding ○

Code lengths are readily determined when all probabilities are simple powers of 2, but what does one do when a probability is, for example, 1/37? The code assignment procedure worked out by Huffman [Huf52] solves this problem neatly, producing a compact code — one that is optimum, with unique prefixes, and that uses all of the codes of each length in the table. Huffman coding provides integer length codes of optimum length, and only noninteger length coding techniques such as arithmetic coding [Ris76, CWN87, PM93] can improve upon the coding efficiency of Huffman coding.

5.2.4   Assigning Huffman codes ◗

The procedure for assigning Huffman codes is based on pairing of symbols and groups of symbols. Given the set of symbols, a, b, c, and d, illustrated in

---

[2]This way of partitioning the number line is, in fact, somewhat related to a proof found in [BCW90] for the theorem in Equation 5.2.

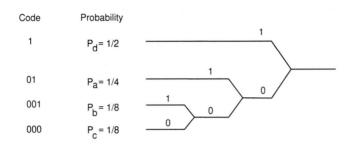

Code	Probability
1	$P_d = 1/2$
01	$P_a = 1/4$
001	$P_b = 1/8$
000	$P_c = 1/8$

Figure 5.4: Huffman coding tree.

Figure 5.3, the code tree in Figure 5.4 can be constructed. In general, a tree such as this is developed by pairing the two symbols or groups of symbols of lowest probability and combining them to form a branch of the tree. The probability of being on that branch is simply the sum of the probabilities of the two symbols (or groups) paired. The tree is further developed until only a single branch — the root of the tree — remains. The two branches of each pair are then assigned a 1 and 0 code bit (the assignment is arbitrary). Code words for a particular symbol are obtained by concatenating the code bits, starting at the root and working back to the leaf corresponding to that symbol. The entire set of code words generated in this manner is called a Huffman code table. Although the example given here is for probabilities that are exact powers of 2, this table generation technique applies generally.

5.2.5   Adaptive entropy coding ◑

MPEG is a nonadaptive coding system, and thus uses fixed code tables. To set up the MPEG code tables, many image sequences were coded and statistics were gathered for each code word in the table. Although ideally the code word lengths should be exactly matched to the statistics, nonadaptive coding is relatively robust when there are minor deviations from the ideal. If a code word happens to be too long for one symbol, the code word assigned to another symbol will be too short. Therefore, when the probabilities for symbols are not quite matched to the assumed values, the increased number of bits for some symbols is at least partly offset by a decreased number of bits for others. If deviations are not minor, however, coding efficiency degrades, and adaptive coding can then provide significantly better compression.

In adaptive entropy coding the code tables are modified to better match the symbol probabilities. In the JPEG still image compression standard, for example, Huffman tables matched to the statistics for a particular picture can be transmitted as part of the compressed picture. JPEG also provides an alternative form of entropy coding, adaptive arithmetic coding, that can even adapt to changing statistics within an image.

There is, of course, a price to be paid for the coding efficiency gains with adaptive coding, in that the system is usually more complex. Only when substantial mismatch between assumed and actual probabilities can occur is there a good justification for adaptive entropy coding.

5.2.6   Symbol probabilities ◑

In Table 5.1 symbol probabilities were calculated from the symbol counts. In doing this the assumption is made that the counts are averages over a large number of "typical" messages. If the number of messages is large enough, all of the symbols will occur sufficiently often that reliable values for the probabilities can be calculated. It sometimes happens, however, that a symbol is theoretically possible, but so rare that it never occurs in any of the messages used to train the code table. The code table must be able to represent any possible message, and this creates a problem known as the zero frequency problem — if the frequency of occurrence is zero, the probability calculated from the ratio of the symbol count to the total count is zero and the ideal code length is therefore infinite. One commonly used solution to the zero frequency problem is to start all symbol counts at one rather than zero.[3]

5.3   Statistical models ○

The statistical model is really distributed throughout the encoder model. The act of decorrelating the data by taking pel differences or by computing a DCT is in some sense part of the statistical model. However, there is another very important task for the statistical model once this decorrelated representation is obtained, and this is the creation of an *efficient* symbol set.

It is certainly possible to code DCT coefficients or pel differences directly, calling these the symbols. However, whenever symbol probabilities get too large, Huffman coding usually becomes inefficient because of the restriction to integer code words. The shortest code length possible is 1 bit,

---

[3]This initialization of counts is one example of a technique for estimating probabilities called Bayesian estimation. When each count is set to the same initial value, the assumption is, in the absence of any information about symbol counts, that all symbols are equiprobable. If some counts are set to larger numbers than others, some prior knowledge is assumed. The initial values used for the counts reflect the degree of confidence in the prior knowledge about probabilities, and fractional values less than 1 imply a great deal of uncertainty about the initial probabilities. Further information on Bayesian estimation may be found in [PM93] and in [Wil91].

corresponding to a symbol probability of 0.5. As the probability rises above 0.5, the code length remains 1, even though ideally it should be shorter.

The solution to this problem is to combine symbols based on DCT coefficients or pel differences, creating a new alphabet in which two or more of the old symbols are represented by a single new symbol. This is a key part of the statistical model.

When combining symbols, a separate symbol is needed for each possible combination of the old symbols. The probability of each symbol is given by the product of the probabilities of the old symbols combined in it. For example, the DCT used in MPEG typically has many zero coefficients, especially at higher spatial frequencies. The probability of a zero value is therefore well above 0.5, and coding the coefficients individually is quite inefficient. Instead, the zero coefficient values are combined in various ways to create new symbols with lower probabilities. Since each possible combination must be given a symbol, combining symbols expands the size of the alphabet. However, the coding efficiency is much improved.

Statistical models can also exploit conditional probabilities. Conditional probabilities are probabilities that depend on a particular history of prior events. For example, the probability of the letter u occurring in text is about 0.02 for typical English text documents [WMB94]. However, if the letter q occurs, u becomes much more probable — about 0.95, in fact. When a probability is a function of past symbols coded we call it a *conditional probability*. When code tables are based on conditional probabilities, the coding is then *conditioned* on prior symbols.

Not surprisingly, many compression systems use conditional probabilities. MPEG, for example, changes the code tables, depending on the type of picture being coded. More details of the statistical models used in MPEG will be found in the following sections on coding models.

5.4   Coding models ○

As anyone who has ever examined a movie film strip knows, the individual pictures in a movie are very similar from one picture to the next. In addition, the two-dimensional array of samples that results from the digitizing of each picture is also typically very highly correlated — that is, adjacent samples within a picture are very likely to have a similar intensity. It is this correlation, both within each picture and from picture to picture that makes it possible to compress these sequences very effectively.

5.4.1   I-, P-, and B-pictures ○

MPEG divides the pictures in a sequence into three basic categories as illustrated in Figure 2.4 in Chapter 2. Intra-coded pictures or I-pictures are coded without reference to preceding or upcoming pictures in the sequence. Predicted pictures or P-pictures are coded with respect to the temporally closest preceding I-picture or P-picture in the sequence. Bidirectionally coded pictures or B-pictures are interspersed between the I-pictures and P-pictures in the sequence, and are coded with respect to the immediately adjacent I- and P-pictures either preceding, upcoming, or both. Even though several B-pictures may occur in immediate succession, B-pictures may never be used to predict another picture.

5.4.2   MPEG coding models ○

Since I-pictures are coded without reference to neighboring pictures in the sequence, the coding model must exploit only the correlations within the picture. For I-pictures, the coding models used by MPEG are similar to those defined by JPEG, and the reader may therefore find the extensive discussion of coding models in [PM93] helpful.

P-pictures and B-pictures are coded as differences, the difference being between the picture being coded and a reference picture. Where the image has not changed from one picture to the next, the difference will be zero. Only the areas that have changed need to be updated, a process known as conditional replenishment. If there is motion in the sequence, a better prediction can be obtained from pels in the reference picture that are shifted relative to the current picture pels. Motion compensation is a very important part of the coding models for P- and B-pictures and will be discussed in detail in Chapter 11.

Given an I-picture or the motion-compensated difference array for a P- or B-picture, two basic coding models are used to complete the coding process: a discrete cosine transform model (DCT-based model) and a predictive model. The DCT-based model is common to the coding of all of the picture types and as discussed earlier, plays a central role in MPEG. The quantization of the DCT coefficients permits the MPEG coding system to take good advantage of the spatial frequency dependency of the human eye's response to luminance and chrominance (see Chapter 4). Furthermore, DCT coefficients are almost perfectly decorrelated, as shown in the pioneering work by Ahmed, Natarajan, and Rao [ANR74]. This means that the coding models do not need to consider conditional statistical properties of coefficients.

The coding of quantized DCT coefficients is lossless — that is, the decoder is able to reproduce the exact same quantized values. The coding models are dependent, however, on the type of picture being coded.

8	16	19	22	26	27	29	34
16	16	22	24	27	29	34	37
19	22	26	27	29	34	34	38
22	22	26	27	29	34	37	40
22	26	27	29	32	35	40	48
26	27	29	32	35	40	48	58
26	27	29	34	38	46	56	69
27	29	35	38	46	56	69	83

Table 5.2: The default luminance quantization table for intra coding.

5.4.3   Quantization in MPEG-1 ◑

As discussed in Chapters 3 and 4, the DCT decomposes the picture data
into underlying spatial frequencies. Since the response of the human visual
system is a function of spatial frequency, the precision with which each
coefficient is represented should also be a function of spatial frequency. This
is the role of quantization.

The quantization table is an array of 64 quantization values, one for
each DCT coefficient. These are used to divide the unquantized DCT coefficients to reduce the precision, following rules described in Section 3.6.1 in
Chapter 3.

There are two quantization tables in MPEG-1, one for intra coding and
one for nonintra coding. The default intra coding table, shown in Table 5.2,
has a distribution of quantizing values that is roughly in accord with the
frequency response of the human eye, given a viewing distance of approximately six times the screen width and a 360x240 pel picture.

The nonintra default MPEG quantization table is flat with a fixed value
of 16 for all coefficients, including the DC term. The constant quantization
value for all coefficients is of interest, for it appears to be at odds with the
notion that higher spatial frequencies should be less visible. However, this
table is used for coding of differential changes from one picture to the next,
and these changes, by definition, involve temporal masking effects.

5.4.4   Coding of I-pictures ◑

In I-pictures the DCT coefficients within a given block are almost completely
decorrelated. However there is still some correlation between the coefficients
in a given block and the coefficients of neighboring blocks. This is especially
true for the block averages represented by the DC coefficients. For this
reason, the DC coefficient is coded separately from the AC by a predictive

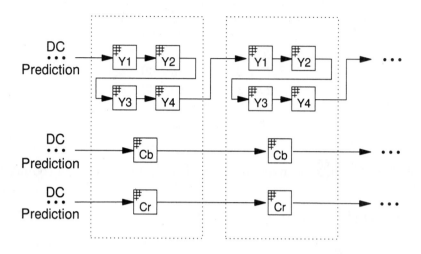

Figure 5.5: Coding order of macroblocks and the DC prediction sequence.

DPCM technique.

5.4.4.1 Coding DC coefficients in I-pictures ◗

As shown in equation 5.3, the DC value of the neighboring block just coded (from the same component), P, is the prediction for the DC value in the current block. The difference, ΔDC, is usually close to zero.

$$\Delta DC = DC - P \qquad (5.3)$$

Note that the block providing the prediction is determined by the coding order of the blocks in the macroblock. Figure 5.5 provides a sketch of the coding and prediction sequence.

The coding of ΔDC is done by coding a size category and additional bits that specify the precise magnitude and sign. Size categories and corresponding code words for luma and chroma are given in Table 5.3.

The size category determines the number of additional bits required to fully specify the DC difference. If size in Table 5.3 is zero, the DC difference is zero; otherwise, *size* bits are appended to bitstream following the code for *size*. These bits supply the additional information to fully specify the sign and magnitude.

The coding of the difference values is done exactly as in JPEG. To code a given difference, the size category is first determined and coded. If the differ-

Y Code	C code	size	magnitude range
100	00	0	0
00	01	1	-1,1
01	10	2	-3...-2,2...3
101	110	3	-7...-4,4...7
110	1110	4	-15...-8,8...15
1110	11110	5	-31...-16,16...31
11110	111110	6	-63...-32,32...63
111110	1111110	7	-127...-64,64...127
1111110	11111110	8	-255...-128,128...255

Table 5.3: DC code tables for luminance and chrominance.

difference		size	additional
decimal	binary		bits
+5	...00101	3	101
+4	...00100	3	100
+3	...00011	2	11
+2	...00010	2	10
+1	...00001	1	1
0	...00000	0	-
-1	...11111	1	0
-2	...11110	2	01
-3	...11101	2	00
-4	...11100	3	011
-5	...11011	3	010

Table 5.4: Examples of DC difference coding.

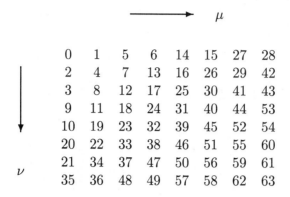

Figure 5.6: Zigzag ordering of DCT coefficients.

ence is negative, 1 is subtracted. Then, *size* low-order bits of the difference are appended to the bitstream. Table 5.4 provides a few examples.

5.4.4.2   Coding AC coefficients in I-pictures ◖

The decorrelation provided by the DCT permits the AC coefficients to be coded independently of one another, and this greatly simplifies the coding process. However, although the coding of a given coefficient is independent of other coefficients, the position of the coefficient in the array is important. Coefficients representing high spatial frequencies are almost always zero, whereas low-frequency coefficients are often nonzero. To exploit this behavior, the coefficients are arranged qualitatively from low to high spatial frequency following the zigzag scan order shown in Figure 5.6. This zigzag scan approximately orders the coefficients according to their probability of being zero.

Even with zigzag ordering, many coefficients are zero in a typical 8×8 block; therefore, it is impossible to code the block efficiently without combining zero coefficients into composite symbols.

A very important example of a composite symbol is the end-of-block (EOB). The EOB symbol codes all of the trailing zero coefficients in the zigzag-ordered DCT with a single code word. Typically, the EOB occurs well before the midpoint of the array, and in fact, is so probable that it is assigned a two-bit code. Note that if each of the coefficients was coded independently, about 60 bits would be needed to code the same information.

EOB coding is illustrated in Figure 5.7. Coding starts with the lowest frequency coefficient, and continues until no more nonzero coefficients remain in the zigzag sequence. The coding of the block is then terminated

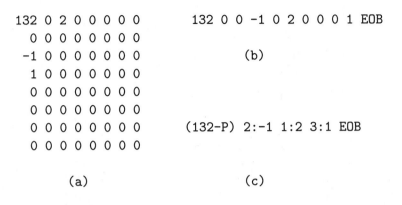

```
132 0 2 0 0 0 0 0        132 0 0 -1 0 2 0 0 0 1 EOB
  0 0 0 0 0 0 0 0
 -1 0 0 0 0 0 0 0                    (b)
  1 0 0 0 0 0 0 0
  0 0 0 0 0 0 0 0
  0 0 0 0 0 0 0 0
  0 0 0 0 0 0 0 0        (132-P) 2:-1 1:2 3:1 EOB
  0 0 0 0 0 0 0 0

         (a)                         (c)
```

Figure 5.7: End-of-block coding in MPEG and JPEG: (a) DCT as a 2-D array. After quantization, all of the high frequencies are zero. (b) zigzag ordered coefficients, but with the trailing zeros replaced by an EOB symbol. (c) zigzag ordered coefficients represented as composite $\Delta DC = 132 - P$, run/level, and EOB symbols.

with the end-of-block code. The end-of-block code is short, only two bits, and thus is very probable.

Runs of zero coefficients also occur quite frequently before the end-of-block. In this case, better coding efficiency is obtained when code words (symbols) are defined combining the length of the zero coefficient run with the amplitude of the nonzero coefficient terminating the run.

Each nonzero AC coefficient is coded using this *run-level* symbol structure. *Run* refers to the number of zero coefficients before the next nonzero coefficient; *level* refers to the amplitude of the nonzero coefficient. Figure 5.7(c) illustrates this. The trailing bit of each run-level code is the bit, s, that codes the sign of the nonzero coefficient. If s is 0, the coefficient is positive; otherwise it is negative.

The MPEG AC code table reproduced in Table5.5 has a large number of run-level codes, one for each of the relatively probable run-level combinations. The EOB symbol, with its two-bit code, is also shown. Combinations of run lengths and levels not found in this table are coded by the escape code, followed by a six-bit code for the run length and an eight- or 16-bit code for the level.

Two codes are listed for run 0/level 1 in Table 5.5, and this may seem a bit puzzling. First, why are there two codes for this? Second, why does the binary code '1 s' (where 's' indicates the sign of the nonzero coefficient) not conflict with the binary '10' end-of-block (EOB) code? The reason for

run/level	VLC	bits
0/1	1s (first)	2
0/1	11s (next)	3
0/2	0100 s	5
0/3	0010 1s	6
0/4	0000 110s	8
0/5	0010 0110 s	9
0/6	0010 0001 s	9
0/7	0000 0010 10s	11
0/8	0000 0001 1101 s	13
0/9	0000 0001 1000 s	13
0/10	0000 0001 0011 s	13
0/11	0000 0001 0000 s	13
0/12	0000 0000 1101 0s	14
0/13	0000 0000 1100 1s	14
0/14	0000 0000 1100 0s	14
0/15	0000 0000 1011 1s	14
0/16	0000 0000 0111 11s	15
0/17	0000 0000 0111 10s	15
0/18	0000 0000 0111 01s	15
0/19	0000 0000 0111 00s	15
0/20	0000 0000 0110 11s	15
0/21	0000 0000 0110 10s	15
0/22	0000 0000 0110 01s	15
0/23	0000 0000 0110 00s	15
0/24	0000 0000 0101 11s	15
0/25	0000 0000 0101 10s	15
0/26	0000 0000 0101 01s	15
0/27	0000 0000 0101 00s	15
0/28	0000 0000 0100 11s	15
0/29	0000 0000 0100 10s	15
0/30	0000 0000 0100 01s	15
0/31	0000 0000 0100 00s	15
0/32	0000 0000 0011 000s	16
0/33	0000 0000 0010 111s	16
0/34	0000 0000 0010 110s	16
0/35	0000 0000 0010 101s	16
0/36	0000 0000 0010 100s	16
0/37	0000 0000 0010 011s	16
0/38	0000 0000 0010 010s	16
0/39	0000 0000 0010 001s	16
0/40	0000 0000 0010 000s	16

Table 5.5: (a) Variable length codes for run-level combinations. The sign bit 's' is '0' for postive and '1' for negative.

run/level	VLC	bits
1/1	011s	4
1/2	0001 10s	7
1/3	0010 0101 s	9
1/4	0000 0011 00s	11
1/5	0000 0001 1011 s	13
1/6	0000 0000 1011 0s	14
1/7	0000 0000 1010 1s	14
1/8	0000 0000 0011 111s	16
1/9	0000 0000 0011 110s	16
1/10	0000 0000 0011 101s	16
1/11	0000 0000 0011 100s	16
1/12	0000 0000 0011 011s	16
1/13	0000 0000 0011 010s	16
1/14	0000 0000 0011 001s	16
1/15	0000 0000 0001 0011 s	17
1/16	0000 0000 0001 0010 s	17
1/17	0000 0000 0001 0001 s	17
1/18	0000 0000 0001 0000 s	17
2/1	0101 s	5
2/2	0000 100s	8
2/3	0000 0010 11s	11
2/4	0000 0001 0100 s	13
2/5	0000 0000 1010 0s	14
3/1	0011 1s	6
3/2	0010 0100 s	9
3/3	0000 0001 1100 s	13
3/4	0000 0000 1001 1s	14
4/1	0011 0s	6
4/2	0000 0011 11s	11
4/3	0000 0001 0010 s	13
5/1	0001 11s	7
5/2	0000 0010 01s	11
5/3	0000 0000 1001 0s	14
6/1	0001 01s	7
6/2	0000 0001 1110 s	13
6/3	0000 0000 0001 0100 s	17
7/1	0001 00s	7
7/2	0000 0001 0101 s	13
8/1	0000 111s	8
8/2	0000 0001 0001 s	13

Table 5.5: (b) Continuation of variable length codes for run-level combinations.

run/level	VLC	bits
9/1	0000 101s	8
9/2	0000 0000 1000 1s	14
10/1	0010 0111 s	9
10/2	0000 0000 1000 0s	14
11/1	0010 0011 s	9
11/2	0000 0000 0001 1010 s	17
12/1	0010 0010 s	9
12/2	0000 0000 0001 1001 s	17
13/1	0010 0000 s	9
13/2	0000 0000 0001 1000 s	17
14/1	0000 0011 10s	11
14/2	0000 0000 0001 0111 s	17
15/1	0000 0011 01s	11
15/2	0000 0000 0001 0110 s	17
16/1	0000 0010 00s	11
16/2	0000 0000 0001 0101 s	17
17/1	0000 0001 1111 s	13
18/1	0000 0001 1010 s	13
19/1	0000 0001 1001 s	13
20/1	0000 0001 0111 s	13
21/1	0000 0001 0110 s	13
22/1	0000 0000 1111 1s	14
23/1	0000 0000 1111 0s	14
24/1	0000 0000 1110 1s	14
25/1	0000 0000 1110 0s	14
26/1	0000 0000 1101 1s	14
27/1	0000 0000 0001 1111 s	17
28/1	0000 0000 0001 1110 s	17
29/1	0000 0000 0001 1101 s	17
30/1	0000 0000 0001 1100 s	17
31/1	0000 0000 0001 1011 s	17
End_of_block	10	2
Escape	0000 01	6

Table 5.5: (c) Continuation of variable length codes for run-level combinations.

this is the dual purpose of the code table — it is really two code tables folded into one. In most situations the end-of-block can occur and the entry labeled "next" is used. However, in nonintra coding a completely zero DCT block is coded in a higher-level procedure and the end-of-block code cannot occur before coding the first run-level. In this particular situation the entry labeled "first" is used.

Codes labeled "first" are used only in nonintra coding. For intra coding, the DC is coded separately and an end-of-block can occur immediately without any nonzero AC coefficients. Consequently, because the first entry, '1 s', conflicts with the EOB code, it is not used.

5.4.5 Coding of P- and B-pictures ◑

The decorrelation property of the DCT is really applicable only to intra-coded pictures. Nonintra pictures are coded relative to a prediction from another picture, and the process of predicting strongly decorrelates the data. Fundamental studies by Rao and coworkers have shown that nonintra pictures will not be optimally decorrelated by the DCT.[4] However, since the correlation is already quite small, any coding efficiency loss in ignoring coefficient correlation has to be small.

The real reason for using the DCT is quantization. Relatively coarse quantization can be used for P- and B- pictures, and the bitrates are therefore very low. Even with a flat default quantization table that may not fully exploit properties of the human visual system, DCT quantization is an effective tool for reducing bit rate.

In P- and B-picture coding the DC coefficient is a differential value and therefore is mathematically similar to the AC coefficients. Consequently, AC and DC coding are integrated into one operation.

The coding of the DCT coefficients in P- and B-pictures is preceded by a hierarchical coding sequence in which completely zero macroblocks (i.e., no nonzero DCT coefficients) are coded by a macroblock address increment that efficiently codes runs of zero macroblocks.

Zero blocks in a nonzero macroblock are coded using the *coded block pattern* (cbp), a six-bit variable in which each bit describes whether a corresponding DCT is active or completely zero. The variable-length codes for the cbp efficiently code the zero blocks within the macroblock. Note that the condition cbp=0 is handled by a macroblock address increment and the macroblock is *skipped*.

---

[4]The work in reference[ANR74] suggests that a sine transform might work better. However, this conclusion is based on artificially created data with zero mean in which DC level shifts do not occur. As we have seen in Chapter 3, the sine transform does not handle DC levels gracefully.

Finally, any nonzero blocks must be coded, and for this the codes in Table 5.5 are used. This table, with the first entry for run 0/level 1 ignored, has already been applied to intra coding of AC coefficients. For P- and B-pictures, the full table is used.

When coding the first run-level, the first entry entry for run 0/level 1 is used. In this case an end-of-block cannot occur before at least one nonzero coefficient is coded, because the zero block condition would have been coded by a cbp code. After that, the end-of-block is possible, and the second entry for run 0/level 1 must be used.

Although the code table is not a true Huffman code when this substitution is made, the approximation is pretty good. It is a very clever way of eliminating the need for two separate code tables.

5.5   Encoder and decoder block diagrams ◗

At this point the various concepts discussed in this chapter and preceding chapters can be merged to create high-level encoders and decoder block diagrams.

5.5.1   Reconstruction module ◗

A reconstruction module, common to both encoders and decoders, is shown in Figure 5.8. This module is used to reconstruct the pictures needed for prediction. Note that the signal definitions in this figure apply to the encoder and decoder block diagrams that follow.

The reconstruction module contains a dequantizer unit (Inv Q), a DC pred unit (for reconstructing the DC coefficient in intra-coded macroblocks) and an IDCT unit for calculating the inverse DCT. The IDCT output is merged with the prediction (zero, in the case of intra-coded macroblocks) to form the reconstruction.

The prediction is calculated from data in the picture store, suitably compensated in the case of nonintra coding for the forward and backward motion displacements.

5.5.2   Encoder block diagram ◗

In addition to the reconstruction module of Figure 5.8, the encoder shown in Figure 5.9 has several other key functional blocks: a controller, a forward DCT, a quantizer unit, a VLC encoder, and a motion estimator.

The modules are fairly self-explanatory: The controller provides synchronization and control, the quantized forward DCT is computed in the FDCT and Q modules, forward and backward motion estimation is carried out in the motion estimator block, and the coding of the motion vectors

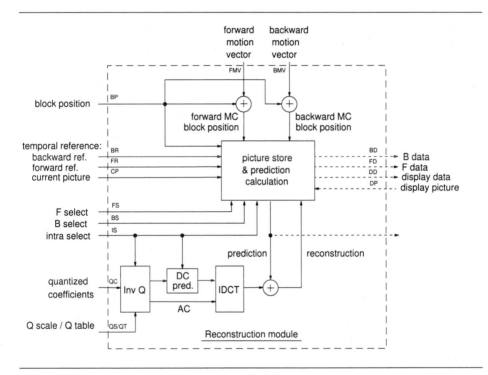

Figure 5.8: Reconstruction module common to both encoder and decoder.

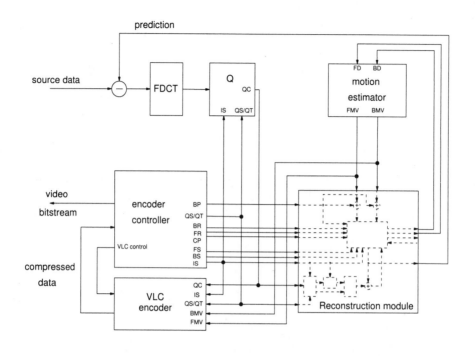

Figure 5.9: MPEG encoder structure.

and DCT data is done in the VLC encoder. The motion estimation block appears to be a simple element, but is perhaps the most complex part of the system.

Note that the reconstruction module is general enough for both encoder and decoder functions. In fact, a simpler version providing only I- and P-picture reconstructions is sufficient for encoding.

5.5.3 Decoder block diagram ◑

As in the encoder, the reconstruction module of Figure 5.8 is the central block of the decoder. In this case motion displacements and DCT data are decoded in the VLC decoder.

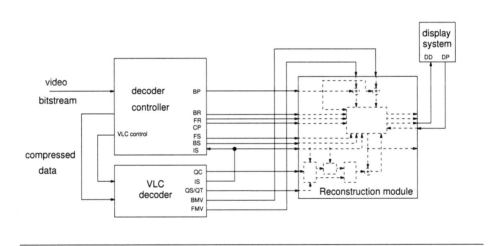

Figure 5.10: MPEG decoder structure.

6

Pseudocode and Flowcharts

Pseudocode is software that follows a set of regular rules, just like any real software environment, but is not a recognized language. It may be presented as full procedures or as code fragments. The MPEG standard makes very heavy use of pseudocode and cannot be understood without a knowledge of the pseudocode syntax. The purpose of this chapter is to review the pseudocode syntax as well as conventions for flowcharts that we use to provide an alternative presentation of the MPEG standard. The pseudocode rules and definitions presented in this chapter are derived from the MPEG standard, but readers familiar with the C language will find that the MPEG pseudocode is basically a subset of C.

The pseudocode in the MPEG-1 standard has no comments — comments are informative and informative material should be explicitly labeled as such if it is placed in a normative section of a standard. While the lack of comments is appropriate in a standard, any software without comments can be rather cryptic at times. The pseudocode in the MPEG-2 standard has the same format, but occasionally does include comments; the document states that these comments can be removed without any effect on the syntax.

Since we don't have the same restrictions in this book, we have attempted to improve on the pseudocode presentation in the standard in several ways: First, as done by MPEG-2, we have extended the pseudocode syntax by adding the normal C language comment structure. This extension allows us to add comments to almost every pseudocode statement. Second, we have collapsed some of the repetitive pseudocode structures in the standard into "generic" procedures. Third, we have modified some of the procedures slightly to make their operation clearer.

We also have defined a set of flowchart rules that we present in this chapter. As will be seen, in addition to the pseudocode, flowcharts are given for all of the MPEG-1 systems and video procedures, in order to provide an alternative for those readers who are unfamiliar with the C language. We think most readers will find the flowcharts to be a useful supplement.

6.1   Pseudocode presentation ◑

Pseudocode statements are always presented in a monospaced boldface type.
Thus, a typical statement would be:

```
typical statement;        /* comment for typical statement   */
```

Pseudocode comments are enclosed by /* and */, just as in the C language. As noted above, MPEG-2 occasionally uses the same comment style
for explanatory comments.

6.2   Flowchart primitives ◑

Figure 6.1 contains the basic structures that will be used in the flowcharts
in this book. Comments are placed to the right of the flowcharts, delineated
by the same /* and */ character pairs used in the pseudocode. The dotted
lines running across the figure illustrate how the flowchart structures and
comments line up.

The oval at the top of a flowchart contains the name of the function or
procedure. The oval at the bottom with the word **Done** finishes the function
or procedure. Executable statements are found in simple boxes that are
sized so as to contain all of the statements.

The term *function* is applied to self-contained code segments with well
defined call and return points where parameters can be passed. Functions
are also *procedures*, but procedure is a more general term that also can be
applied to code fragments that do not involve parameter passing.

Boxes with single lines cn each side represent parameters and data elements present in the syntax. An encoder writes the N-bit data element
specified by a box of this type into the bitstream; the decoder reads the
N-bit data element from the bitstream. In either case the bitstream pointer
is advanced by N bits. The name of the data element is listed to the right of
the box in the center of the flowchart. The width of the box is proportional
to N, thereby giving a graphical presentation of the number of bits. Dotted
lines for the top and bottom edges of a box and a range of values $(M - N)$
inside it indicates an item coded with a variable number of bits. In this case
the box width is determined by the largest number of bits possible.

Boxes with double lines on each side contain the names of functions or
procedures called at that point. If procedures (such as the function **abcd**
shown) are passed with a parameter, v, that parameter is enclosed in parentheses. Function names are usually taken from the MPEG-1 specification;

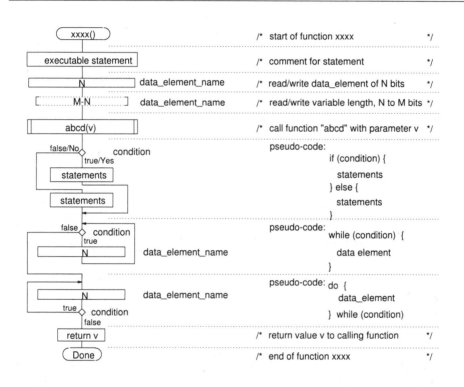

Figure 6.1: Examples of flowchart structures.

test condition	pseudocode	flowchart
Not A?	!A	not A
A equal B?	A == B	A = B
A not equal B?	A != B	A ≠ B
A greater than B?	A > B	A > B
A less than B?	A < B	A < B
A less than or equal B?	A <= B	A ≤ B
A greater than or equal B?	A >= B	A ≥ B
logical OR of A and B	A || B	A OR B
logical AND of A and B	A && B	A AND B

Table 6.1: Notation for tests in pseudocode and flowcharts.

occasionally, however, new procedures are introduced where we felt they would make the presentation clearer.

6.2.1 Decision notation ◑

For decisions we follow the compact notation in [PM93]. A diamond is used to denote a decision, but the diamond is too small for the decision being tested to be place inside as is the normal convention for flowcharts. Instead, the decision is placed to the right of the diamond, thereby giving a significantly more compact flowchart structure. The left, right, or down branches out of the diamonds are labeled with the test result required to take that path. Where the decision is based on a test of a condition, the branches may be labeled **true** and **false** or **not** 0 and 0. As in the pseudocode, a test result of zero is considered false whereas all other answers are considered true. Where the decision is based on the answer to a question, the branches are labeled **Yes** and **No**. Occasionally the branches will be labeled with other conditions such as 4 and **not** 4.

6.2.2 Test notation ◑

At decision points in a flowchart, tests are indicated by comparisons such as A=B?. In the pseudocode the comparisons follow C programming conventions, and are of the form A==B. A list of pseudocode and flowchart tests is given in Table 6.1. The occasional second choice comes from MPEG-2 conventions.

operator	pseudocode	flowchart
absolute value	| | or abs()	| |
addition	+	+
assignment	=	=
cosine	cos	cos
decrement	--	- 1
division (real numbers)	÷	
exponential	exp	exp()
increment	++	+ 1
integer division with rounding	//	round
integer division with truncation	/	/
integer division with truncation towards $-\infty$	DIV	DIV
logical AND (bitwise)	&	AND
logical OR (bitwise)	|	OR
logical NOT	!	not
maximum value of a list	max[...]	max[...]
minimum value of a list	min[...]	min[...]
modulus	%	%
multiplication	* or ×	*
nearest integer	NINT()	NINT()
negation (unary)	−	−
power	^	^
shift left logical	<<	SLL
shift right arithmetic	>>	SRA
sine	sin	sin
subtraction (binary)	−	−

Table 6.2: Notation for operators in pseudocode and flowcharts.

pseudocode and flowchart	result
21 % 8	5
8 % 8	0
9 % 8	1
10 % -2	undefined
-1 % 8	undefined

Table 6.3: Examples of modulus notation and operation.

6.3   Operator notation ◗

A list of operators and their notation in the pseudocode and flowcharts is given in Table 6.2. A brief explanation is given here for those unfamiliar with the pseudocode or flowchart notations. We have included only the operators used in the system and video parts of the standard or in our book.

The absolute value of a number is always positive. Zero and positive numbers are unchanged by the absolute value operation. Negative numbers are negated (i.e., multiplied by negative one) to find their positive absolute value. Addition, multiplication, cosine, sine, and square root follow the standard mathematical conventions. The minus sign is used both for subtraction of one variable from another (binary) and for the negation of a single variable (unary). The context makes it clear which is desired. The logical not is the symbol ! in the pseudocode and spelled out as **not** in our flowcharts.[1] The bitwise operators (logical AND, logical OR, shift right arithmetic, and shift left logical) work on numbers assumed to be in twos complement notation. The shift right arithmetic extents the sign into the most significant bits. The shift left logical operation fills in zeros in the least significant bits.

The modulus operator (%) gives the nonnegative integer remainder when the largest possible integer multiple of the number on the right is subtracted from the number on the left. The modulus operator is only defined for positive numbers. Some examples of tests using the modulus operator are given in Table 6.3.

The C language increment (++) and decrement (--) operators change variables and array indices by one unit. While C programs often use complex expressions with embedded increments and decrements, MPEG pseudocode uses only the simplest form of these operators. Array indexing follows C conventions such that a one unit change in index changes the actual physical storage address by an amount determined by the size of the variable (one

---

[1]We (and others) have missed the ! symbol in a poor reproduction of a standard.

byte for characters, two bytes for short integers, and four bytes for long integers).

The max[...] and min[...] operators return the maximum and minimum values of a list indicated by ....

Three different kinds of integer division are used in the standard. The distinctions are a likely area for confusion and mistakes in implementing the standard. Division with rounding to the nearest integer is denoted by // in the pseudocode and a comment to clarify in the flowcharts. Values at precisely half are expected to round away from zero. If division with truncation is desired, the / is used. All fractional values are truncated towards zero. Occasionally, integer division with truncation towards $-\infty$ is used and denoted as DIV. Table 6.4 gives examples for the three kinds of division.

6.3.1   Branches and loops ◑

The pseudocode fragments in Figure 6.1 illustrate some key branching and looping constructs used in documenting the MPEG Standard. These constructs, as well as other aspects of the pseudo code, follow C language conventions in most respects.

Pseudocode procedures, data elements, and parameter names start with an alphabetical character, but otherwise may use any alphanumeric characters and the underbar character. A procedure call is always of the form procedure_name(). In the C language the parentheses contain the parameters passed to the procedure, and are empty if no parameters are needed. The parentheses are always empty in the pseudocode of the actual standard, but as will be seen, we have found it convenient in some places to use this empty field as a means of passing a bit length. If the function returns a value v, this is indicated by a return v statement in the pseudocode; the equivalent flowchart will have a box with return v near the bottom.

Within a procedure each statement is terminated with a semicolon. (The standard is not consistent on this point, but the pseudocode in this book will be.) Curly braces, { and }, are used to group statements into logical blocks when more than one statement is executed in the block. If curly braces are absent, only one statement or data element follows a branch test.

The uppermost diamond in Figure 6.1 is the decision for an if ...else ... set of branches. If the test is true, the branch is to one set of statements; if false, it is to an alternate set of statements. The corresponding pseudocode is shown in the comments field. Note the use of curly braces to group the statements for each branch path.

The next fragment of pseudocode in Figure 6.1 is for the while loop. In this case the test is made first; if the condition is true, the statements

operation	result
rounding	
4//4	1
3//4	1
2//4	1
1//4	0
(-1)//4	0
(-2)//4	-1
(-3)//4	-1
(-4)//4	-1
truncation toward 0	
4/4	1
3/4	0
2/4	0
1/4	0
(-1)/4	0
(-2)/4	0
(-3)/4	0
(-4)/4	-1
truncation toward $-\infty$	
4 DIV 4	1
3 DIV 4	0
2 DIV 4	0
1 DIV 4	0
(-1) DIV 4	-1
(-2) DIV 4	-1
(-3) DIV 4	-1
(-4) DIV 4	-1

Table 6.4: Three types of integer division.

within the curly braces are executed. In this particular illustration, a data element is read from or written to the bitstream.

The do-while loop occurs next in Figure 6.1. In the do-while loop the while test occurs after the statements within the curly braces of the loop are executed; the statements within the loop are therefore executed at least once.

Another important pseudocode looping structure is the for loop, consisting of for(expr1, expr2, expr3) { ...loop statements ...} . In this construct expr1 initializes a counter variable, expr2 is a condition that must be true for the loop statements in curly braces to be executed, and expr3 modifies the counter after the loop statements have been executed. The for loop can be constructed from a while loop by preceding the loop with counter initialization (expr1), making expr2 the while loop test, and making expr3 one of the statements executed inside the while loop. Note that the condition in expr2 is tested before any for loop statements are executed. Loop counters generally start with zero.

6.4   Data element procedures ◑

The reading or writing of data elements in the bitstream is done as a series of procedural calls, one for each data element. In the actual standard, the call in the pseudocode takes the form:

```
data_element_name()
```

In the pseudocode in this book we have revised and extended this notation slightly. If the number of bits of the data element, m, is fixed, it is included in the call as a parameter:[2]

```
data_element_name(m);      /* r/w m bit data element          */
```

If the data element length is variable, from m to n bits, the call is:

```
data_element_name(m-n);    /* r/w variable m to n bits        */
```

If the data element is an array, we follow the MPEG conventions of brackets indicating its array nature. The comment field is then used to communicate the size of each element.

```
data_element_name[];       /* array of 8-bit data             */
```

---

[2]The pseudocode in the standard lists the number of bits and the data type in columns to the right of the data_element_name. We created room for comments by collecting the data types into syntax summary tables.

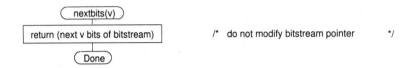

Figure 6.2: Flowchart for nextbits() procedure.

Counters normally start at 0, and the **n+1** element in the array is therefore designated by **n** inside brackets.

```
data_element_name[n];       /* r/w n+1 element of 8-bit array  */
```

Two dimensional arrays have two sets of brackets.

```
data_element_name[m][n];    /* r/w m+1,n+1 8-bit array element */
```

Three dimensional arrays have three sets of brackets and are not needed in this book until the MPEG-2 discussion in Chapter 9.

```
data_element_name[l][m][n];/* r/w l+1,m+1,n+1 8-bit array element */
```

The pseudocode usually applies equally to both encoder and decoder. However, in the decoder the procedural call implies a read of the data element, whereas in the encoder it implies a write. The r/w in the comments reflects this dual role. The bitstream pointer that keeps track of the current position in the bitstream is always incremented by the number of bits of the data element being read or written.

6.5   Nextbits() procedure ◑

The pseudocode must, at certain points, look ahead in the bitstream for a number of bits. This is done by the **nextbits()** procedure shown in Figure 6.2. In the standard the parameter passed with **nextbits()** is empty, whereas in this book the parameter contains the number of look-ahead bits needed. Thus, in the pseudocode in this book the call is to **nextbits(v)** rather than **nextbits()**.

Note that the bitstream pointer is not changed when the v bits are fetched. This allows the decoder to compare the upcoming bits against expected values of different lengths. The encoder must decide which symbol or parameter to code and, thus, which bit sequence to write to the bitstream.

6.6   Encoder versus decoder operation ◑

In the decoder the `nextbits(v)` procedure described in the preceding section simply fetches the value of the next v bits and the decoder does whatever it is told to do (assuming a compliant bitstream); in the encoder additional external information is needed to determine what the next v bits should be. A substantial amount of processing may be required to determine these v bits, and, therefore, MPEG encoders are usually considerably more intelligent than decoders.

Encoder decisions determine the structure of the bitstream. They are related to constraints needed for a particular application and by factors such as rate control, a topic which will be addressed in Chapter 15. Decoder decisions are driven from the bitstream, but the decoder must present the output such that the compressed data buffer will not overflow or underflow. As will be seen, timing information and other parameters are included in the compressed bitstream that tell the decoder how full the buffer should be at particular points in the bitstream, given an idealized decoder model. The decoder must use that information to maintain the compressed data buffer while keeping the system synchronized.

6.7   Other low-level procedures ◑

The `bytealigned()` procedure flowcharted in Figure 6.3 returns a 0 if the bitstream pointer is not byte aligned, and a 1 if it is aligned. The procedure is used to synchronize the input pointer to byte boundaries, as part of the process of looking for the next start code. The start codes are byte aligned, but may be proceeded by any number of zero stuffing bits. Insertion and erasure errors can destroy this alignment, and decoders must be prepared to deal with such cases.

The `Sign()` function is used to determine the sign of a variable such as a DCT coefficient (see Equation 3.8). As illustrated in the flowchart in Figure 6.4, `Sign()` returns a $+1$ if the variable is greater than zero, 0 if the variable is zero, and -1 if the variable is less than zero. There is no pseudocode for this function because it is defined in the set of MPEG arithmetic operators. Since it looks like a function, however, we have provided a flowchart.

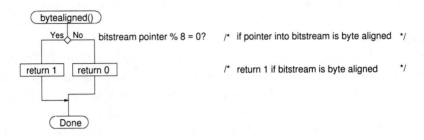

Figure 6.3: Flowchart for **bytealigned()** procedure.

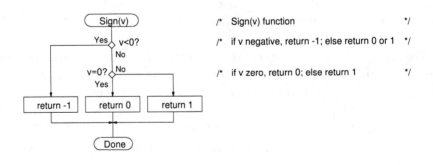

Figure 6.4: Flowchart for **Sign()** function.

7

MPEG System Syntax

An MPEG compressed video bitstream cannot stand alone. Aside from the fact that video alone is not considered a valid bitstream, the presentation time information coded in the system layer is essential to meet real time compliance requirements. Even though both audio and video are needed for many applications, the video and system layers together do make up a complete MPEG stream. This relationship between video and systems is even stronger in MPEG-2.

This chapter reviews the MPEG systems standard. An overview is presented first; this is followed by detailed pseudocode and flowcharts for the system syntax.

7.1 Start codes ○

The system and video layers contain unique byte-aligned 32-bit patterns called *start codes*. These codes, as the name would suggest, start certain high-level segments of the bitstream. With some care to avoid accidental emulation in some types of data, these start codes can be located without parsing the bitstream.

The start codes all have a three-byte prefix of 23 zero bits followed by a '1' bit; the final byte (called the *start code value* in MPEG-2) then identifies the particular start code. Start codes may be preceded with as many zero bytes (or zero bits) as desired. The byte alignment is easily maintained in the system layer because the headers are always multiples of bytes, regardless of the combinations of optional fields.

7.1.1 Next_start_code() function ◑

The byte alignment of start codes is achieved by stuffing as many '0' bits as needed to get to byte alignment. Zero bytes may then be stuffed if desired.

Start code name	hexa-decimal	binary
video start codes:		
picture_start_code	00000100	00000000 00000000 00000001 00000000
slice_start_code 1	00000101	00000000 00000000 00000001 00000001
...	...	...
slice_start_code 175	000001AF	00000000 00000000 00000001 10101111
reserved	000001B0	00000000 00000000 00000001 10110000
reserved	000001B1	00000000 00000000 00000001 10110001
user_data_start_code	000001B2	00000000 00000000 00000001 10110010
sequence_header_code	000001B3	00000000 00000000 00000001 10110011
sequence_error_code	000001B4	00000000 00000000 00000001 10110100
extension_start_code	000001B5	00000000 00000000 00000001 10110101
reserved	000001B6	00000000 00000000 00000001 10110110
sequence_end_code	000001B7	00000000 00000000 00000001 10110111
group_start_code	000001B8	00000000 00000000 00000001 10111000
system start codes:		
iso_11172_end_code	000001B9	00000000 00000000 00000001 10111001
pack_start_code	000001BA	00000000 00000000 00000001 10111010
system_header_start_code	000001BB	00000000 00000000 00000001 10111011
packet start codes:		
reserved stream	000001BC	00000000 00000000 00000001 10111100
private_stream_1	000001BD	00000000 00000000 00000001 10111101
padding stream	000001BE	00000000 00000000 00000001 10111110
private_stream_2	000001BF	00000000 00000000 00000001 10111111
audio stream 0	000001C0	00000000 00000000 00000001 11000000
...	...	...
audio stream 31	000001DF	00000000 00000000 00000001 11011111
video stream 0	000001E0	00000000 00000000 00000001 11100000
...	...	...
video stream 15	000001EF	00000000 00000000 00000001 11101111
reserved stream 0	000001F0	00000000 00000000 00000001 11110000
...	...	...
reserved stream 15	000001FF	00000000 00000000 00000001 11111111

Table 7.1: MPEG start codes in numeric order.

```
next_start_code(){                 /* from ISO 11172 Parts 1 & 2  2.3  */
  while (!bytealigned())           /* if not byte aligned               */
    zero_bit(1);                   /* r/w '0'                           */
  while (nextbits(24)!=0x000001)   /* while not start code prefix       */
    zero_byte(8);                  /* r/w '0000 0000'                   */
}                                  /* end next_start_code() function    */
```

Figure 7.1: The next_start_code() function.

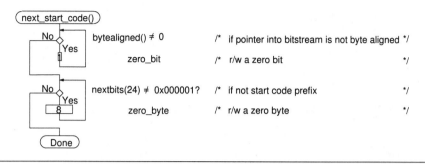

Figure 7.2: Flowchart for **next_start_code** procedure.

The pseudocode for the **next_start_code()** procedure is found in Figure 7.1. The equivalent flowchart is given in Figure 7.2. **Zero_byte** stuffing before the start codes is one of the mechanisms used to avoid decoder buffer overflow.

7.1.2   System and video start codes ◑

Table 7.1 lists the complete set of start codes in numeric order in both hexadecimal and binary formats. The video start codes (0x00000100 through 0x000001B8) are found only in the video syntax layers. The system start codes (0x000001B9 through 0x000001FF) are found only in the system syntax. Most of the system codes are the **packet_start_codes** (0x000001BC through 0x000001FF).

The **packet_start_codes** are made up from a 24-bit **packet_start_code_prefix** (0x000001) and an eight-bit "**stream_id**". Each elementary stream in a ISO 11172 multiplexed stream has a unique **stream_id**. The 32 audio streams are useful for multiple language support. They also might be used in pairs for stereo audio.

The video start codes have a **sequence_error_code** that may replace any video start code to flag that the upcoming video bitstream contains errors. Since this code is not mentioned in Part 1 (systems), it may not be used in the system layer even though there is not an equivalent systems error start code.

7.2   System overview ○

Figure 7.3 shows an overview of an ISO 11172 stream. The highest level consists of one or more packs followed by a four-byte **iso_11172_end_code**. The optional nature of all but the first pack is shown by the dashed lines in the figure. This first expanded view is not labeled on the right since

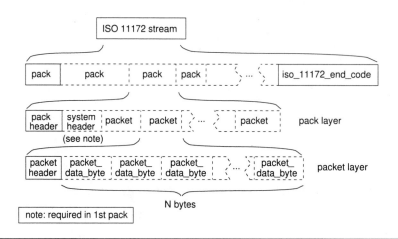

Figure 7.3: System layer overview.

it is technically not a layer, even though it is sometimes called that in the standard.[1] Bitstreams such as shown in Figure 7.3 stand alone and MPEG-1 system syntax provided no mechanism for multiplexing system layer bitstreams. Of course, such data can always be multiplexed by higher-level protocols or packet switching networks that are outside the scope of the standard.

A more detailed view of a pack is shown in the next layer in Figure 7.3. Each pack starts with a 12-byte pack header. Essential system clock information and the rate at which the bytes are delivered to the *system target decoder* (STD) (see Section 2.1.1) are communicated in the pack header. Following the pack header is a variable-length system header. The system header is a place to put global information about total, audio, and video rate bounds for all of the packets, whether all of the packet parameters are constrained, whether the video and/or audio are locked to their respective clocks, and bounds on the buffer needed in a STD. Only the first pack is required to contain a system header. An arbitrary number of packets follow. It is legal to create a pack without any packets.

A packet always starts with a packet header, as shown in the expanded packet at the bottom of Figure 7.3. Part of the packet header is a packet_length containing the length of the rest of the packet. The packet header may also contain the STD buffer size and time stamp information.

---

[1]The word *layer* in the MPEG standard has multiple meanings and can be confusing. Even though there are only two system syntax layers (pack and packet), the pseudocode and a subclause in the standard are labeled ISO/IEC 11172 layer. In addition, in Part 2 (video), the entire system syntax is lumped together as the *system layer*.

```
iso11172_stream(){        /* from ISO 11172-1 2.4.3.1      */
  do {                    /* do pack                       */
    pack();               /* encode/decode packs           */
  } while (nextbits(32)==0x000001BA);  /* while pack_start_code */
  iso_11172_end_code(32); /* r/w 0x000001B9                */
}                         /* end iso11172_stream() function */
```

Figure 7.4: The iso11172_stream() function.

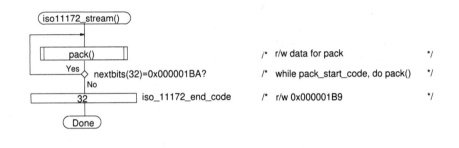

Figure 7.5: Flowchart for iso11172_stream() function.

The video, audio, padding or private streams follow the packet header as packet_data_bytes. All of the streams in a given packet are the same type, as specified by the stream_id (the final byte of the packet_start_code).

7.3   ISO/IEC 11172 stream ◑

The pseudocode in Figure 7.4 defines the syntax for the ISO 11172 stream. This is equivalent to the code in ISO 11172-1, subclause 2.4.3.1, except for the minor changes in conventions already described in Chapter 6 on flowchart and pseudocode conventions. Figure 7.5 contains the flowchart for this function. Note that the pack procedure is located between curly braces in the do-while loop, thereby requiring that at least one pack must be present in an ISO/IEC 11172 stream. All valid bitstreams end with an iso_11172_end_code (0x000001B9).

7.4   Pack layer ◑

The pseudocode for the pack layer syntax is given in Figure 7.6. The corresponding flowchart is illustrated in Figure 7.7. All packs start with a pack_start_code (0x000001BA). A fixed four-bit '0010' pattern helps byte-align

```
pack(){                         /* from ISO 11172-1  2.4.3.2         */
  pack_start_code(32);          /* r/w 0x000001BA                    */
  '0010';                       /* r/w 4-bit fixed pattern           */
  system_clock_reference(3);    /* r/w bits 32 to 30 of SCR          */
  marker_bit(1);                /* r/w '1'                           */
  system_clock_reference(15);   /* r/w bits 29 to 15 of SCR          */
  marker_bit(1);                /* r/w '1'                           */
  system_clock_reference(15);   /* r/w bits 14 to 0 of SCR           */
  marker_bit(1);                /* r/w '1'                           */
  marker_bit(1);                /* r/w '1'                           */
  mux_rate(22);                 /* r/w mux rate                      */
  marker_bit(1);                /* r/w '1'                           */
  if (nextbits(32)==0x000001BB) /* if system_header_start_code       */
    system_header();            /* r/w system header                 */
  while(nextbits(32)>=0x000001BC)/* while packet_start_code_prefix*/
    packet();                   /* r/w packets                       */
}                               /* end pack() function               */
```

Figure 7.6: The pack() function.

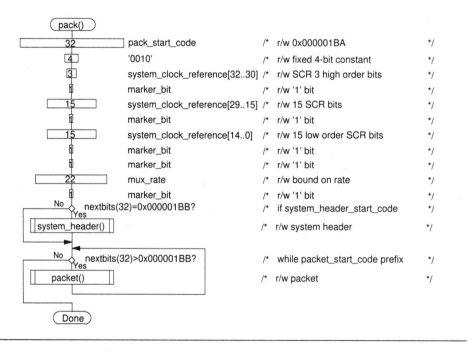

Figure 7.7: Flowchart for pack() function.

the next data element, the `system_clock_reference` (SCR). The 33-bit SCR is broken up into three pieces with marker bits ('1') following each of the three-bit, 15-bit, and 15-bit sections. The marker bits guarantee that start codes are not accidentally emulated.

The `system_clock_reference` fields in the pack header can be used to determine when stream bytes should enter the system target decoder. The value encoded in the 33-bit SCR field indicates the time when the last byte of the SCR (bits 7 to 0) should enter the STD. The *system_clock_frequency* is measured in Hz and must meet a constraint that it is 90,000 Hz $\pm$ 4.5.

$$SCR(i) = NINT(\text{system\_clock\_frequency} \times time(i))\%2^{33} \qquad (7.1)$$

where i is the byte count from the beginning of the pack to the last byte of the SCR and $time(i)$ is the time when this byte entered the STD. $NINT$ rounds to the nearest integer.

The next three bytes are the `mux_rate`, bracketed by marker bits. The `mux_rate` is measured in units of 50 bytes/s and indicates the rate at which bytes arrive at the STD. A pack could be composed of only these first 12 bytes unless it is the first pack. In the first pack the `system_header` is a requirement rather than an option. Since all system headers in a given ISO 11172 stream are required to be identical, the extra copies are redundant. They can be useful to assist special features such as fast forward playback.

An arbitrary number of packets finish up the pack. These packs can be identified by means of the leading four-bytes, since the values are always greater or equal to 0x000001BC. (The `packet_start_codes` cover the range from 0x000001BC to 0x000001FF.)

7.4.1   System header ◑

The pseudocode for processing the system header is given in Figure 7.8, and the flowchart corresponding to this code is given in Figure 7.9. The system header is required in the first pack and although allowed in any pack, its values cannot change. It is a requirement that all elements stay the same. The `system_header()` data elements are mostly informative, but they do place legal constraints on the bitstream. If new elementary streams are remultiplexed later, or the STD buffer size bound is changed, then the system stream should be terminated with a `iso_11172_end_code`, and a whole new stream should follow.[2]

---

[2]The pack layer is analogous to the sequence layer in MPEG video: the sequence header parameters must be the same for all pictures within a sequence (with the exception of downloadable quantization matrices). If vbv_buffer_size or the bit_rate changes, then the old sequence should be terminated and a new sequence should be started. It's a good idea to make packs and video sequences coincident for this reason, since STD buffer fullness is directly related to VBV.

```
system_header(){              /* from ISO 11172-1  2.4.3.2        */
  system_header_start_code(32); /* r/w 0x000001BB               */
  header_length(16);          /* r/w number of bytes in header  */
  marker_bit(1);              /* r/w '1'                         */
  rate_bound(22);             /* r/w upper bound on mux_rate     */
  marker_bit(1);              /* r/w '1'                         */
  audio_bound(6);             /* r/w upper bound on no. of audio */
  fixed_flag(1);              /* r/w '1' if fixed bit rate       */
                              /* r/w '0' if variable bit rate    */
  CSPS_flag(1);               /* r/w '1' if constrained bitstream */
                              /* r/w '0' if unconstrained        */
  system_audio_lock_flag(1);  /* r/w '1' if SCR locked to audio  */
                              /* r/w '0' if not                  */
  system_video_lock_flag(1);  /* r/w '1' if SCR locked to video  */
                              /* r/w '0' if not                  */
  marker_bit(1);              /* r/w '1'                         */
  video_bound(5);             /* r/w upper bound on no. of video */
  reserved_byte(8);           /* r/w '1111 1111'                 */
  while (nextbits(1)=='1'){  /* while not a start code          */
    stream_id(8);             /* r/w stream ID                   */
    '11';                     /* r/w 2-bit fixed pattern         */
    STD_buffer_bound_scale(1); /* r/w '0' if audio; '1' if video */
                              /* r/w choice if other streams     */
    STD_buffer_size_bound(13);/* r/w bound on buffer size        */
  }                           /* end while not a new start code  */
}                             /* end system_header() function    */
```

Figure 7.8: The system_header() function.

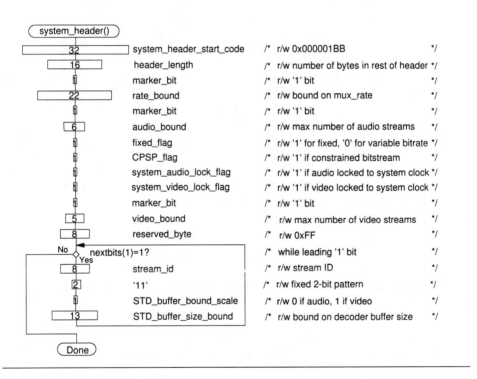

Figure 7.9: Flowchart for **system_header()** function.

The `system_header_start_code` (0x000001BB) always starts a system header. It is followed by a two-byte `header_length`. This length counts the rest of the bytes in the system header. Its minimum value is six (6+0*3) and its maximum value is 165 (6+53×3) bytes. (Only 53 `stream_ids` are valid in the pack layer. The rest are reserved and still undefined for ISO 11172 bits streams.)

The 22-bit `rate_bound` is bracketed by marker bits to prevent accidental start code creation. Its value is equal to or greater that any `mux_rate` defined in the packet headers for this stream. Zero is not an allowed value for it.

The six-bit `audio_bound` tells a decoder how many active audio streams are arriving to the STD or will be presented within the same time period of the pack. Note that a recently terminated audio stream from a previous pack() could be partially presented within the current pack()'s arrival time. [3] The `audio_bound` is a number from 0 to 32 inclusive and may be greater than the actual number of active audio streams. The video CD (White Book)[4] has only one stream (e.g. English), hence `audio_bound`=1, but the digital video disk (DVD) would have at least three (e.g. English, French, Spanish) and maybe five [VCD94].

The `fixed_flag` is the system layer mechanism for signaling variable and fixed-rate operation. In the video layer variable rate is signaled by setting `vbv_delay` to 0xFFFF, as will be described more fully in Chapter 8. If the `fixed_flag` is set (indicating fixed bit rate operation), the values encoded in all SCR fields in MPEG-1 streams are given by:

$$SCR(i) = NINT(c1 \times i + c2)\%(2^{33}) \qquad (7.2)$$

where i is the index of the final byte of any SCR field. $c1$ and $c2$ are real-valued constants. In fact, $c1$ is proportional to `mux_rate` and $c2$ acts like a phase to account for small initial delays. Some examples of SCR calculations are given in Section 7.7 later in this chapter.

The `CSPS_flag` (constrained system parameter stream flag) is the system layer analog to the video layer `constrained_parameters_flag`. Packet rate and STD size restrictions are given in 2.4.6 of 11172-1.

The `system_audio_lock_flag` indicates that the encoder used a simple relationship between the `system_clock_frequency` (nominally 90 kHz) and the coded audio sample rate to compute the SCR. In many implementations audio is the master clock since it is far more reliable than video clocks. Locking `system_clock_frequency` to the audio rate ensures that at least

---

[3] A decoder could derive for itself how many streams there are since each packet header comes with a stream_id field. However, the decoder wouldn't know this until it has parsed a decent way into the stream.

[4] The White Book is a proprietary compact disk digital video specification developed by JVC, Matsushita, Philips, and Sony. For further information, contact these companies.

the audio maintains a linear relationship with time. Without the lock, there is an extra indirection since audio and video are parametrically linked to a third signal, rather than to each other. The `system_video_lock_flag` indicates that the encoder used the video sample rate to compute the SCR.

The `marker_bit` prevents accidental start code emulation.

The five-bit `video_bound` is greater than or equal to the number of video streams that will be presented to the STD during the time of the pack. It is analogous to the `audio_bound`.

The `reserved_byte` is reserved for future extensions of the system syntax. It has a value of 1111 1111 until future ISO standards redefine it.

The `stream_id` is usually the same as the final byte of the `packet_start_code` and can be found indented in Table 7.2 under the heading of `stream_id` or packet_start_code byte. [5] The `stream_id` contains information about the type of stream (audio, video, padding, or private) and the stream number to which the next two data elements (`STD_buffer_bound_scale` and `STD_buffer_size_bound`) apply. Two special values for the `stream_id` allow these two following data elements to apply to all audio (0xB8) or video (0xB9) streams (see Table 7.2 under the heading stream_id only).

The `STD_buffer_bound_scale` defines the dynamic range of the next data element, `STD_buffer_size_bound`. If `STD_buffer_bound_scale` is 0, (required for audio streams) the `STD_buffer_size_bound` is in units of 128. If 1, (required for video streams) the `STD_buffer_size_bound` is in units of 1024. Video bit rates are usually three to 20 times greater than audio bit rates, so the 1024-byte granularity instead of the 128-byte granularity is reasonable.

The `STD_buffer_size_bound` is meant to be informative to a decoder. It places an upper limit on the size of any `STD_buffer_size` that may be encountered within the MPEG-1 stream, not just the immediate pack.

7.5 Packet layer ◗

The pseudocode for processing the packets is given in Figure 7.10, and the flowcharts corresponding to this code are in Figures 7.11 and 7.12. A packet always starts with a packet header as shown in the expanded packet at the bottom of Figure 7.3. Each packet header starts with a `packet_start_code` (0x000001 prefix) plus a one-byte `stream` ID. The two-byte packet length is next in the packet header. It contains the length of the packet starting with the first byte after the packet length. Except for private stream number 2, the next bytes may be up to 16 bytes of stuffing data (0xFF). These bytes

---

[5]Some popular elementary stream types, such as Dolby AC-3 streams (recently officially called *Dolby Digital*), are private_streams. The first byte of the `packet_data` of these private packets identifies which audio stream number it actually belongs to.

```
packet(){                        /* from ISO 11172-1  2.4.3.3         */
  packet_start_code_prefix(24);  /* r/w 0x000001                      */
  stream_id(8);                  /* r/w stream ID                     */
  packet_length(16);             /* r/w number of bytes in packet     */
  if(stream_id(8)!=0xBF){        /* if not private_stream_2           */
    while(nextbits(8)==0xFF)     /* while byte padding desired         */
      stuffing_byte(8);          /* r/w '11111111'                    */
    if (nextbits(2)=='01'){      /* if buffer scale/size next         */
      '01';                      /* r/w 2-bit fixed pattern           */
      STD_buffer_scale(1);       /* r/w buffer scale                  */
      STD_buffer_size(13);       /* r/w buffer size                   */
    }                            /* end if buffer scale/size next     */
    if (nextbits(4)=='0010'){    /* if only PTS                       */
      '0010';                    /* r/w 4-bit fixed pattern           */
      presentation_time_stamp(3); /* r/w bits 32 to 30 of PTS         */
      marker_bit(1);             /* r/w '1'                           */
      presentation_time_stamp(15); /* r/w bits 29 to 15 of PTS        */
      marker_bit(1);             /* r/w '1'                           */
      presentation_time_stamp(15); /* r/w bits 14 to 0 of PTS         */
      marker_bit(1);             /* r/w '1'                           */
    }                            /* end if only PTS                   */
    elseif (nextbits(4)=='0011'){ /* elseif both PTS and DTS          */
      '0011';                    /* r/w 4-bit fixed pattern           */
      presentation_time_stamp(3); /* r/w bits 32 to 30 of PTS         */
      marker_bit(1);             /* r/w '1'                           */
      presentation_time_stamp(15); /* r/w bits 29 to 15 of PTS        */
      marker_bit(1);             /* r/w '1'                           */
      presentation_time_stamp(15); /* r/w bits 14 to 0 of PTS         */
      marker_bit(1);             /* r/w '1'                           */
      '0001';                    /* r/w 4-bit fixed pattern           */
      decoding_time_stamp(3);    /* r/w bits 32 to 30 of DTS          */
      marker_bit(1);             /* r/w '1'                           */
      decoding_time_stamp(15);   /* r/w bits 29 to 15 of DTS          */
      marker_bit(1);             /* r/w '1'                           */
      decoding_time_stamp(15);   /* r/w bits 14 to 0 of DTS           */
      marker_bit(1);             /* r/w '1'                           */
    }                            /* end if both PTS and DTS           */
    else                         /* else private stream 2             */
      '0000 1111';               /*   r/w 8-bit fixed pattern         */
  }                              /* end if not private stream 2       */
  for (i=0; i<N;i++){            /* do i=0 to N-1 by 1                */
    packet_data_byte(8);         /*   r/w 8-bit packet data           */
  }                              /* end do loop                       */
}                                /* end packet() function             */
```

Figure 7.10: The packet() function.

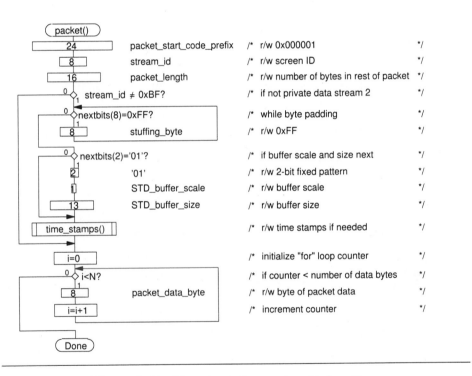

Figure 7.11: Flowchart for **packet()** function.

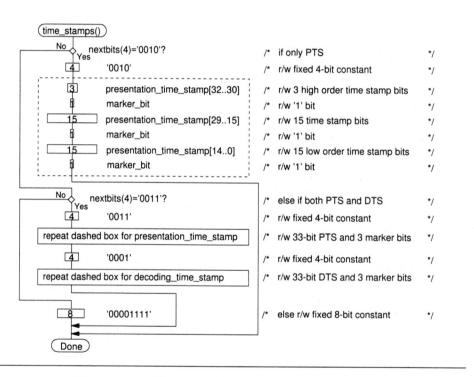

Figure 7.12: Flowchart for time_stamp() function.

can be used to arrange for packets to fall on specific byte boundaries, as might be required by digital storage media. If the STD buffer size for the packet stream ID is known, it is coded next, along with a scale to indicate whether the units are to be multiplied by 128 (0) or 1024 (1).[6]

The presentation time stamp (PTS), either alone or in conjunction with the decoding time stamp (DTS), may be sent next. In addition to helping to preserve byte-alignment, the fixed patterns '0010', '0011', and '0001' are also used as a flag or small start code to uniquely signal the particular time stamp field that follows. If no time information is coded, then a 0x0F byte is coded. The format of the PTS and DTS is the same as the SCR, namely the 33 bits are split into three segments with marker bits.

In the flowchart version of the packet function (Figure 7.11) reading and writing time stamps has been moved to a procedure `time_stamps()` shown in Figure 7.12. A dashed box surrounds the first opportunity to code the `presentation_time_stamp`. The second opportunity is shown with a box containing the words "repeat dashed box for `presentation_time_stamp`". Another box is used for the `decoding_time_stamp`. In each case the 33-bit time stamp is broken into three segments with marker bits separating the segments. Comparison with the pseudocode (Figure 7.10) can be used to resolve any confusion or questions.

N, the number of data bytes following the packet header, is the difference between the `packet_length` and the number of packet header bytes following the packet length. If $N=0$, the packet contains no `packet_data_bytes`; packets without `packet_data_bytes` are used for stuffing.

7.6   Summary of system data elements ◑

Table 7.2 summarizes all of the data elements and parameters that appear in the system syntax pseudocode. The first collection of columns shows whether a data element is set in the ISO 11172 stream (I), pack layer including stream header (P), or packet layer (T).[7] The second collection of columns tells when the data element is used or referenced. A solid black circle indicates that a data element is named. An open circle shows that a data element is implied such as when its value is used in a condition test or a generic name stated rather than a specific name.

The data type (dt) has been collected into this summary table rather

---

[6]Note that each elementary stream has its own STD buffer size. Implementors may choose to create a buffer which is the sum of the video and audio STD buffer sizes, but the standard is not specified that way.

[7]We had to adopt the convention of using the last letter when two items started with the same first letter. Note that we are using uppercase for the system layers. Later we will use lowercase for the video layers.

system data element name	set I P T	used I P T	dt	# of bits	value range
audio_bound	. ● .	. . .	U	6	0 . . . 32
CSPS_flag	. ● .	. . .	U	1	0,1
decoding_time_stamp (DTS)	. . ●	. . .	U	33	0 . . . 0x1FFFFFFFF
fixed_flag	. ● .	. . .	U	1	0,1
header_length	. ● .	. . .	U	16	6 . . . 165
iso_11172_end_code	● . .	. . .	B	32	0x000001B9
marker_bit	. ● ●	. . .	B	1	'1'
mux_rate	. ● .	. . .	U	22	1 . . . 0x3FFFFF
pack_start_code	. ● .	○ . .	B	32	0x000001BA
packet_data_byte	. . ●	. . .	U	8	0 . . . 255
packet_length	. . ●	. . .	U	16	1 . . . 0xFFFF
N	. . .	. . ●			0 . . . 0xFFFE
packet_start_code_prefix	. . ●	. ○ .	B	24	0x000001
presentation_time_stamp (PTS) ...	. . ●	. . .	U	33	0 . . . 0x1FFFFFFFF
rate_bound	. ● .	. . .	U	22	1 . . . 0x3FFFFF
reserved_byte	. ● .	. . .	B	8	'1111 1111'
STD_buffer_bound_scale	. ● .	. . .	U	1	0,1
STD_buffer_scale	. . ●	. . .	U	1	0,1
STD_buffer_size	. . ●	. . .	U	13	0 . . . 0x1FFF
STD_buffer_size_bound	. ● .	. . .	U	13	0 . . . 0x1FFF
stream_id only	. ● .	. . .	U	8	0xB8, 0xB9
STD bounds for all audio	. ○ .	. . .	B	8	'1011 1000'
STD bounds for all video	. ○ .	. . .	B	8	'1011 1001'
stream_id or packet_start_code byte	. ● ●	. . ●	U	8	0xBC . . . 0xFF
ISO 11172-2 video stream - # 0 .	. ○ ○	. . ○	B	8	'1110 0000'
. . .	. ○ ○	. . ○	B	8	'1110 xxxx'
ISO 11172-2 video stream - # 15	. ○ ○	. . ○	B	8	'1110 1111'
ISO 11172-3 audio stream - # 0 .	. ○ ○	. . ○	B	8	'1100 0000'
. . .	. ○ ○	. . ○	B	8	'110x xxxx'
ISO 11172-3 audio stream - # 31	. ○ ○	. . ○	B	8	'1101 1111'
padding stream	. ○ ○	. . ○	B	8	'1011 1110'
private_stream_1	. ○ ○	. . ○	B	8	'1011 1101'
private_stream_2	. ○ ○	. . ●	B	8	'1011 1111'
reserved stream	. . .	. . .	B	8	'1011 1100'
reserved data stream - # 0	. . .	. . .	B	8	'1111 0000'
. . .	. . .	. . .	B	8	'1111 xxxx'
reserved data stream - # 15	. . .	. . .	B	8	'1111 1111'
stuffing_byte	. . ●	. . ○	B	8	'1111 1111'
system_audio_lock_flag	. ● .	. . .	U	1	0,1
system_clock_reference (SCR)	. ● .	. . .	U	33	0 . . . 0x1FFFFFFFF
system_header_start_code	. ● .	. ● .	B	32	0x000001BB
system_video_lock_flag	. ● .	. . .	U	1	0,1
video_bound	. ● .	. . .	U	5	0 . . . 16

Table 7.2: System syntax data element summary.

than listed next to the pseudocode is as done in the standard. The MPEG system data elements are one of two basic types: unsigned integer (U) or bit string (B). Unsigned integers are embedded in the bitstream with the most significant bit first. Bit strings are bit sequences with fixed size and are transmitted left bit first.[8] Start codes, for example, are bit strings. The data types at the system level are close enough to the definitions in Table 8.11 that it can be used as a reference, although the application of these types in MPEG systems and MPEG video is not completely consistent.[9]

The number of bits used to represent each data element in the system layers is listed in the column to the right of the data type. This duplicates the information given in our pseudocode in the parameter passed in parentheses. In the flowcharts this information is shown by the size of the box and a number inside each box representing a data element. The value range is listed in the rightmost column.

In the ISO 11272 stream, the `iso_11172_end_code` is the only data element set. The `pack_start_code` is the only data element used, e.g., referenced in a condition test.

The pack layer has many more data elements set than are used. Most of its data elements are used in calculations that don't appear in the syntax, such as variable bit rate control and buffer management. The `stream_id` in the pack layer can set bounds for all of the audio and/or all of the video streams, as well as for specific streams.

Even thought the `stream_id` is explicitly set and used in the packet layer, only the `private_stream_2` is referenced by name. Note that while the `packet_length` is never used directly, it is needed in the calculation of N. The fact that N is calculated from `packet_length` is indicated by listing N underneath and indented.

7.7 SCR calculation examples ●

To illustrate how practical restraints are put on MPEG systems syntax by real-world applications, calculations for the `system_clock_reference` (SCR) are given below for video CD and DVD (digital video disk).

In the video CD (White Book) packs are equivalent to sectors, and the SCR for a sector is:

$$SCR(sn) = C + 1200 \times sn \tag{7.3}$$

---

[8]The standard uses longer mnemonics for U and B, "uimsbf" and "bslbf." respectively.

[9]In the video syntax three more data types are defined: variable-length codes (V), fixed one-bit ('1') and fixed zero-bit ('0'). There, the bit string is usually used for fixed patterns of more than one bit and one-bit flags are either listed as "U" or left blank. In the video syntax the `marker_bit` is a fixed 1-bit data type and not U or B. The time stamps and system clock are treated as bit strings instead of unsigned integers, perhaps because they are broken into three bit-specific segments.

where sn is the sector number. C helps to account for the actual position of the SCR time stamp with respect to the sector or pack().[10] The constant 1200 can be derived from the number of system clocks per CD-ROM sector or pack().

$$1200 = (90,000 \text{ system clocks/s})/(75 \text{ sectors/s}) \qquad (7.4)$$

White Book sectors are always 2352 bytes long. However there are 28 bytes worth of hidden header data above the MPEG layer, so the MPEG pack() size is actually 2324 bytes long. Since there is no track buffer (as is the case of DVD), the delivery rate into STD is based on 75 sectors/s × 2352 bytes/s, not 75 × 2324.[11] The generic "$c1$" constant in MPEG's formula reflects this scaling.

$$c1 = \texttt{mux\_rate} \times 50 \times (2352/2324) \times (1/90000) \qquad (7.5)$$

The last byte of the SCR is located at the eight byte from the beginning of the pack which starts 28 bytes into the sector.

$$C = (28 + 8) \times (90000/(75 \times 2352)) \approx 19 \text{ system cycles} \qquad (7.6)$$

Thus, C is just a small delta from zero.

DVD's track buffer, on the other hand, allows a seamless translation of the disc's native transfer rate into the $\texttt{mux\_rate}$ of the MPEG system decoder. The SCR value for the sixth pack() in a disc volume is given below.[12] Packs are equivalent to DVD sectors. The SCR will be 36 bytes into the sixth sector.

$$\begin{aligned} i \quad &= 36 + 5 \text{ sectors} \times 2352 \text{ bytes/sector} \\ &= 11796 \text{ bytes} \end{aligned} \qquad (7.7)$$

$$SCR(i) \quad = \left[\frac{(11796 \text{ bytes} \times 90000)}{(2352 \text{ bytes/sector} \times 75 \text{ sectors/s})} \right] \%(2^{33})$$

$$= 6018 \qquad (7.8)$$

---

[10]The White Book suggests that C should be zero in most cases. However, C should not be zero since that would make SCR a multiple of the sector number. This does not matter as long as things are consistent from pack to pack, since that is how bitstream conformance is usually measured.

[11]Figure VI.8 in White Book 2.0 states so.

[12]Note that disc volume was used instead of stream (or more precisely ISO11172_stream()) because multiple streams (organized as files) can coexist in a disc volume.

8

MPEG-1 Video Syntax

In this chapter we describe the six syntactical layers of the MPEG-1 video bitstream – the video sequence, group of pictures, picture, slice, macroblock, and block layers. This chapter deals primarily with the mechanics of moving or translating parameter values (data elements, in MPEG parlance) to and from the coded bitstream. How the parameters are actually used in quantization, coding, and reconstruction of the pels is described elsewhere.

As in preceding chapters, we present the syntax for each layer in two different ways, pseudocode and flowcharts. The pseudocode uses essentially the same formatting conventions as in the standard, but with comments added to explain each step. The flowcharts present the same material graphically. Chapter 6 provides a tutorial covering the pseudocode and flowchart conventions used in this chapter.

Although we have made every effort to provide a correct description of MPEG, any serious implementer should work from the actual standards documents. These are the *only* official sources for information about MPEG. Since these documents evolve over time, an official copy of the standard should be ordered from your national standards body or the ISO Information Technology Task Force (ITTF) in order to obtain the latest version and errata sheets.

8.1   Video sequence overview ○

Figure 8.1 gives an overview of the six layers in MPEG-1 video sequences. The system layer, not shown in Figure 8.1,[1] provides a wrapper around the video sequence layer. Among other things, the system layer packs and packets synchronize and multiplex the audio and video bitstreams into an integrated data stream.

---

[1]The system layer is discussed briefly in Chapter 2 and in detail in Chapter 7.

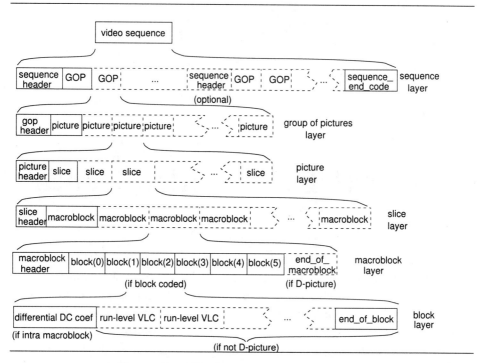

Figure 8.1: The layers of a video stream.

A video sequence always starts with a sequence header. The sequence header is followed at by least one or (optionally) more groups of pictures (GOP) and ends with a `sequence_end_code`. Additional sequence headers may appear between any groups of pictures within the video sequence. The optional nature of the second (extra) sequence header and the extra GOPs is illustrated by broken lines in Figure 8.1. Most of the parameters in these optional additional sequence headers must remain unchanged from their values recorded in the first sequence header. A possible reason for including extra sequence headers is to assist in random access playback or video editing.

Below the sequence layer in Figure 8.1 is an expanded view of a group of pictures. A group of pictures always starts with a group of pictures header and is followed by at least one picture.

Each picture in the group of pictures has a picture header followed by one or more *slices*. In turn, each slice is made up of a slice header and one or more groups of DCT blocks called *macroblocks*. The first slice starts in the upper left corner of the picture and the last slice (possibly the same slice) ends in the lower right corner. For MPEG-1 the slices must completely fill the picture.

The macroblock is a group of six 8×8 DCT blocks — four blocks contain luminance samples and two contain chrominance samples. Each macroblock starts with a macroblock header containing information about which DCT blocks are actually coded. All six blocks are shown in Figure 8.1 even though in practice some of the blocks may not be coded.

DCT blocks are coded as intra or nonintra, referring to whether the block is coded with respect to a block from another picture or not. If an intra block is coded, the difference between the DC coefficient and the prediction is coded first. The AC coefficients are then coded using the variable-length codes (VLC) for the run-level pairs until an **end_of_block** terminates the block. If a nonintra block is coded, DC and AC coefficients are coded together using the same run-level VLC codes. This will be explained in more detail in following sections.

The four MPEG-1 picture types are: I-pictures (intra-coded), P-pictures (nonintra or predictive-coded), B-pictures (bidirectionally predictive-coded) and D-pictures (DC-coded pictures). I-pictures are coded independently of other pictures, i.e., exclusively with intra DCT blocks. P-pictures are coded with respect to a preceding picture, and B-pictures are coded with respect to both preceding and following pictures. D-pictures contain only DC coefficient information and because of this, there are no run-level pairs and no end-of-block. Instead, an **end_of_macroblock** data element is used to complete each macroblock. The function of the **end_of_macroblock** is to avoid long strings of zeros that might emulate start codes; therefore, this data element is used only in D-pictures.[2]

To make comparison with the pseudocode in the standard easier, the material in the rest of this chapter is organized in the same order as the MPEG video standard. Section headings from the standard are noted for cross-reference.

8.2   Start codes ◑

The sequence header, group of pictures header, picture header, and slice header all start with unique byte-aligned 32-bit patterns called start codes. Other start codes are defined for system use, user data, and error tagging. The start codes all have a three-byte prefix of 23 zero bits followed by a 1 bit; the final byte (called the *start code value* in MPEG-2) then identifies the particular start code.

The byte alignment of start codes is achieved by stuffing as many zero bits as needed to get to byte alignment. Zero bytes may then be stuffed if desired. The description of the **next_start_code()** procedure for doing this is described in Chapter 7 (see Figures 7.1 and 7.2). **Zero_byte** stuffing

---

[2]Since video sequences are not allowed to mix D-pictures with other types, D-pictures are rarely used.

Start code name	hexa-decimal	binary
extension_start_code	000001B5	00000000 00000000 00000001 10110101
group_start_code	000001B8	00000000 00000000 00000001 10111000
picture_start_code	00000100	00000000 00000000 00000001 00000000
reserved	000001B0	00000000 00000000 00000001 10110000
reserved	000001B1	00000000 00000000 00000001 10110001
reserved	000001B6	00000000 00000000 00000001 10110110
sequence_end_code	000001B7	00000000 00000000 00000001 10110111
sequence_error_code	000001B4	00000000 00000000 00000001 10110100
sequence_header_code	000001B3	00000000 00000000 00000001 10110011
slice_start_code 1	00000101	00000000 00000000 00000001 00000001
...	...	...
slice_start_code 175	000001AF	00000000 00000000 00000001 10101111
user_data_start_code	000001B2	00000000 00000000 00000001 10110010

Table 8.1: MPEG video start codes.

before the start codes is one of the mechanisms used to avoid decoder buffer overflow.

Table 8.1 lists the set of video start codes in alphabetical order in both binary and hexadecimal format.

A numeric ordering of the video and system start codes can be found in Table 7.1. The video start codes have a range of values from 0x00000100 to 0x000001B8. System start codes have a range of values from 0x000001B9 to 0x000001FF. They are reserved for use in the system layer and must not occur inside a video sequence. The sequence_error_code is available for tagging a sequence that has unrecoverable errors and therefore should not occur in a decodable video bitstream.

8.3   Video sequence layer ◑

The pseudocode in Figure 8.2 defines the syntax for a video sequence. This is equivalent to the code in ISO 11172-2, subclause 2.4.2.2, except for the minor changes in conventions already described in Chapter 6 on flowchart and pseudocode conventions. Figure 8.3 contains the flowchart for this function

.

The first procedure in Figure 8.3, next_start_code(), locates the start code for the video sequence. The bitstream is defined to be a video sequence and the start code of the video sequence is the sequence header code, 0x000001B3. By definition, this is the first code encountered by next_start_code() after it achieves byte alignment in the bitstream and skips over any padded zero bytes. A sequence header is always followed by at least one group of pictures.

```
video_sequence(){          /* from ISO 11172-2 2.4.2.2          */
  next_start_code();       /* find next byte aligned start code */
  do {                     /* do sequence(s)                    */
    sequence_header();     /* r/w sequence header               */
    do{                    /* do group(s) of pictures (GOP)     */
      group_of_pictures(); /* r/w group(s) of pictures          */
    } while (nextbits(32)==group_start_code); /* while 0x000001B8 */
  }while (nextbits(32)==sequence_header_code  /* while 0x000001B3 */
  sequence_end_code(32);   /* r/w 0x000001B7                    */
}                          /* end video_sequence() function     */
```

Figure 8.2: Video_sequence() function.

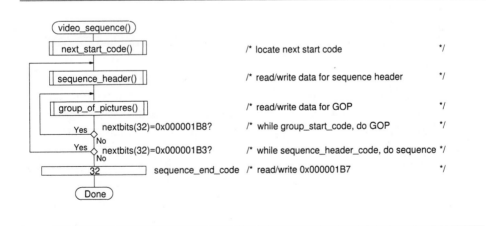

Figure 8.3: Flowchart for the video_sequence() function.

More than one sequence header can occur in a video sequence. A group of pictures must follow a sequence header and (as will be seen shortly) the final step of the `sequence_header()` process positions the bitstream pointer at the group of pictures start code. After each group of pictures, however, a test must be made for the type of start code coming next in the bitstream. If the next start code is for a group of pictures, processing of groups of pictures continues. Otherwise, either a sequence header occurs (followed by at least one more group of pictures) or the sequence terminates with the sequence end code.

The pseudocode for processing the sequence header is given in Figure 8.4, and flowcharts corresponding to this code are in Figure 8.5 and 8.6.

The first 95 bits of the sequence header define 10 data elements that have fixed-length sizes. (The numbers in the data element boxes in the flowcharts are the number of bits needed to represent that parameter. The MPEG names for these data elements are given in the center of the flowchart.) The first data element is the 32-bit `sequence_header_code`. This is followed immediately by two 12-bit parameters, `horizontal_size` and `vertical_size`, that define the width and height of the digital picture. Certain restrictions apply to their valid values for MPEG-1 bitstreams. For example, neither is allowed to be zero, and `vertical_size` must be an even value. Note that it would take two illegal values, a `horizontal_size` of zero and a `vertical_size` of one, to generate an accidental start code prefix. The sequence header also contains a marker bit, a single bit with a value of 1, that prevents the accidental creation of a start code. Marker bits are used quite frequently in the MPEG syntax.

The `pel_aspect_ratio` is a four-bit parameter that provides information about the pel shape. Table 8.2 lists the pel aspect ratios defined for MPEG-1, along with the associated video sources.

`Picture_rate` is a 4-bit parameter that selects one of the common picture rates listed in Table 8.3. Note that the NTSC format referred to in this table is a composite signal that, among other things, uses a different color space (YIQ) than MPEG (YCbCr). Note also that MPEG-2 calls noninterlaced video *progressive scan*.

The `bit_rate` data element has a length of 18 bits, and informs the decoder about the compressed data rate. The rate is in units of 400 bits/s, and the value of `bit_rate` is related to the true bit rate, R, by:

$$\text{bit\_rate} = \lceil R/400 \rceil \qquad (8.1)$$

where $\lceil \ldots \rceil$ is a ceiling function that rounds up to an integer value. A value of zero is not allowed for `bit_rate`, and the maximum possible value, 0x3FFFF, signals that the bit rate is variable.

Following the `marker_bit` is the 10-bit `vbv_buffer_size` data element,

```
sequence_header(){              /* from ISO 11172-2  2.4.2.3      */
  sequence_header_code(32);   /* r/w 0x000001B3                  */
  horizontal_size(12);        /* r/w picture width              */
  vertical_size(12);          /* r/w picture height             */
  pel_aspect_ratio(4);        /* r/w sample aspect ratio        */
  picture_rate(4);            /* r/w frame rate                 */
  bit_rate(18);               /* r/w bit rate                   */
  marker_bit(1);              /* r/w '1'                        */
  vbv_buffer_size(10);        /* r/w video buffer verifier buf.size*/
  constrained_parameters_flag(1); /* r/w '1' if constrained     */
                                  /* r/w '0' if unconstrained   */
  load_intra_quantizer_matrix(1); /* r/w flag for intra quantizer */
  if (load_intra_quantizer_matrix)    /* if flag set            */
    intra_quantizer_matrix[0..63];    /*   r/w 64 8-bit values  */
  load_non_intra_quantizer_matrix(1); /* r/w flag for nonintra Q */
  if (load_non_intra_quantizer_matrix)/* if flag set            */
    non_intra_quantizer_matrix[0..63];/*   r/w 64 8-bit values  */
  next_start_code();                  /* find next start code   */
  if (nextbits(32)==extension_start_code){ /* if 0x000001B5     */
    extension_start_code(32);         /* r/w extension start code */
    while (nextbits(24)!=0x000001){ /* while not start code prefix*/
      sequence_extension_data(8);   /*   r/w byte of data       */
    }
    next_start_code();                /* find next start code   */
  }                                   /* sequence extension data end*/
  if (nextbits(32)==user_data_start_code){/* if 0x000001B2      */
    user_data_start_code(32);         /* r/w user data start code */
    while (nextbits(24)!=0x000001){ /* while not start code prefix*/
      user_data(8);                   /*   r/w byte of user data */
    }                                 /* start code prefix occurs */
    next_start_code();                /* find next start code   */
  }                                   /* user data done         */
}                                     /* end sequence_header() function*/
```

Figure 8.4: The sequence_header() function.

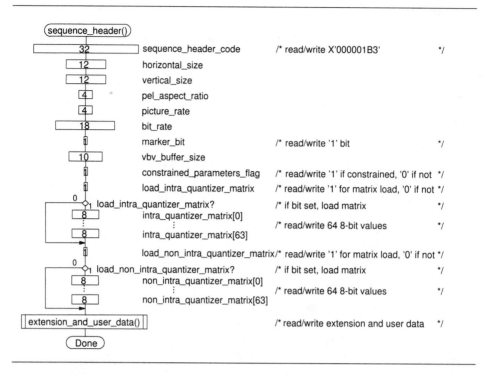

Figure 8.5: Flowchart for sequence_header() function.

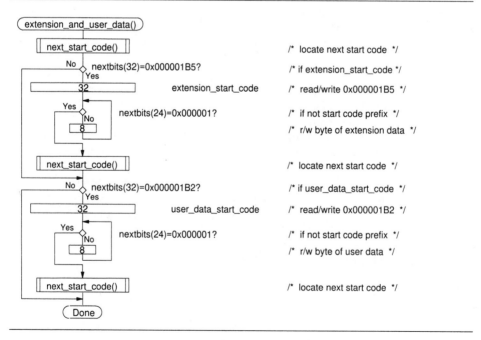

Figure 8.6: Flowchart for extension and user data processing.

pel_aspect_ratio	height/width	video source
0000	forbidden	
0001	1.0000	computers (VGA)
0010	0.6735	
0011	0.7031	16:9, 625-line
0100	0.7615	
0101	0.8055	
0110	0.8437	16:9, 525-line
0111	0.8935	
1000	0.9157	CCIR Rec. 601, 625-line
1001	0.9815	
1010	1.0255	
1011	1.0695	
1100	1.0950	CCIR Rec. 601, 525-line
1101	1.1575	
1110	1.2015	
1111	reserved	

Table 8.2: Ratio of height to width for the 16 pel_aspect_ratio codes.

picture_rate	nominal picture rate	typical applications
0000		Forbidden
0001	23.976	Movies on NTSC broadcast monitors
0010	24	Movies, commercial clips, animation
0011	25	PAL, SECAM, generic 625/50Hz component video
0100	29.97	Broadcast rate NTSC
0101	30	NTSC profession studio, 525/60Hz component video
0110	50	Noninterlaced PAL/SECAM/625 video
0111	59.94	Noninterlaced broadcast NTSC
1000	60	Noninterlaced studio 525 NTSC rate
1001		
...		Reserved
1111		

Table 8.3: Picture rate in pictures per second and typical applications.

horizontal_size ≤ 768 pels.
vertical_size ≤ 576 lines.
number of macroblocks ≤ 396.
(number of macroblocks)×picture_rate≤ 396 × 25.
picture_rate ≤ 30 pictures per second.
vbv_buffer_size ≤ 160.
bit_rate ≤ 4640.
forward_f_code ≤ 4.
backward_f_code ≤ 4.

Table 8.4: Constrained parameters bounds.

a parameter that tells the decoder what the lower bound is for the size of the compressed data buffer. The buffer size is in units of 2048 bytes, and the actual buffer size, in bits, is:

$$B = 8 \times 2048 \times \text{vbv\_buffer\_size} \qquad (8.2)$$

The decoder's buffer must be at least this large in order to successfully decode the video sequence. Further discussion of this data element is deferred to Chapter 15.

The constrained_parameters_flag is a one-bit data element that may be set to one only when the horizontal_size, vertical_size, picture_rate, vbv_buffer_size, bit_rate, and motion parameters all satisfy the restrictions on their values listed in Table 8.4. The last two items in this table refer to bounds on the size of motion displacements, and will be discussed in Section 8.5 and in Chapter 11.

The final two bits in the sequence header control the loading of the intra and nonintra quantization matrices. The load_intra_quantizer_matrix flag, when set to one, signals that 64 eight-bit values for the intra_quantizer_matrix follow. If this flag is zero, the table is not present in the bitstream. In an identical way, the load_non_intra_quantizer_matrix bit flags whether or not the non_intra_quantizer_matrix is to be loaded.

Figure 8.6 contains the flowchart for the extension_and_user_data() procedure. After the next start code is located with the function next_start_code(), extension and then user data are processed.[3] The first test in extension_and_user_data() checks the next 32 bits for the extension_start_code. For MPEG-1 data streams this test must always fail, since the

---

[3]Unless zero bytes are padded to the bitstream, the call to next_start_code() is actually an unnecessary step. The upcoming bitstream is already byte-aligned, since the sequence header is exactly 12 bytes (no quantization matrices loaded), 76 bytes (one quantization matrix loaded), or 140 bytes (both quantization matrices loaded).

```
group_of_pictures(){          /* from ISO 11172-2  2.4.2.4       */
  group_start_code(32);       /* r/w 0x000001B8                  */
  time_code(25);              /* r/w SMPTE time code             */
  closed_gop(1);              /* r/w '1' if closed, '0' if open  */
  broken_link(1);             /* r/w normally '0', '1' if broken */
  next_start_code();          /* find next start code            */
  if (nextbits(32)==extension_start_code){/* if 0x000001B5       */
    extension_start_code(32);       /* r/w extension start code  */
    while (nextbits(24)!=0x000001){ /* while not start code prefix*/
      group_extension_data(8);      /*   r/w byte of data        */
    }                               /* group extension data done */
    next_start_code();              /* find next start code      */
  }
  if (nextbits(32)==user_data_start_code){/* if 0x000001B2       */
    user_data_start_code(32);       /* r/w user data start code  */
    while (nextbits(24)!=0x000001){ /* while not start code prefix*/
      user_data(8);                 /*   r/w byte of data        */
    }                               /* group user data done      */
    next_start_code();              /* find next start code      */
  }
  do {                              /* do picture(s)             */
    picture();                      /*   encode/decode picture   */
  } while (nextbits(32)==picture_start_code)/* while 0x00000100  */
}                                   /* end group_of_pictures function */
```

Figure 8.7: The group_of_pictures() function.

presence of an **extension_start_code** signals an MPEG-2 bitstream. The **user_data_start_code** is always allowed. Note that user and extension data are always delineated by byte-aligned start codes.

The only constraint on user and extension data is that the 23 zero bits of the start code prefix may not occur in this part of the bitstream, even on nonbyte boundaries. MPEG-2 uses extension data in the sequence header to provide additional precision for an expanded range of parameter values.

8.4   Group of pictures layer ◑

The group of pictures layer starts with nine required data elements. These may be followed by optional extension and user data before the picture layer is coded. The pseudocode in Figure 8.7 specifies the group of pictures layer syntax; the corresponding flowchart is in Figure 8.8.

In the **group_of_pictures()** function a 25-bit **time_code** follows the 32-bit **group_start_code**. This time code contains six data elements as

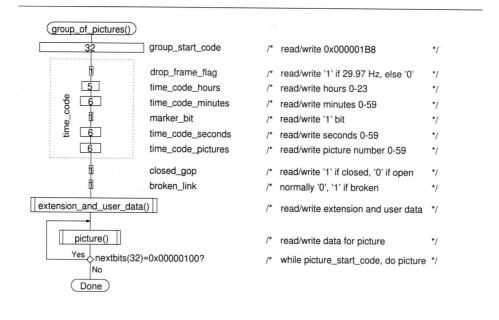

Figure 8.8: Flowchart for the group_of_pictures() function.

illustrated inside the dotted box in the flowchart. These data elements are the same as defined in the IEC standard for video tape recorders [IEC86].

The time code data elements, time_code_hours , time_code_minutes, and time_code_seconds give the hours-minutes-seconds time interval from the start of the sequence. time_code_pictures counts the pictures within a second. This time code information applies to the first picture in the group of pictures.

drop_frame_flag is a one-bit flag that is 0 unless the picture rate is 29.97 Hz. For that particular picture rate, it is used to correct a 30 Hz picture rate to a true 29.97 Hz rate by dropping one count out of every 1000. (29.97/30 = .999). The count dropping is done as follows: At the start of each minute, pictures 0 and 1 are omitted from the count. This means that a count of 1798 instead of 1800 (60x30) is registered every minute. Since the true period is 1798.2 for a 29.97 Hz clock, the count is now 0.2 pictures below the the required value, or 2 pictures every 10 minutes. Therefore, at 10-minute intervals, pictures 0 and 1 are not dropped. The arithmetic is not quite precise, in that 29.97 is an approximation to 30/1001 or 29.97002997. This gives a 75 ms residual error in a 24 hour period [Cal85].

Two one-bit flags follow the time_code. The closed_gop flag is set if the group of pictures can be decoded without reference to I- or P-pictures outside of the group of pictures. The broken_link flag is used to signal that

picture_coding_type	picture type
000	forbidden
001	I-picture
010	P-picture
011	B-picture
100	D-picture
101	reserved
. . .	. . .
111	reserved

Table 8.5: Picture type codes.

editing has disrupted the original sequence of groups of pictures.

Since the data elements in the group of pictures header occupy only 27 bits, the next_start_code() call must read or write at least five zero bits to reestablish byte alignment. At that point, any extension or user data is read before processing the pictures in the group. In MPEG-1 bitstreams, no extension data is allowed, and the group of pictures header must be present. In MPEG-2 the group of pictures header is optional.

8.5 Picture layer ◑

The picture layer is where the picture_coding_type is signaled and (when needed) the forward and backward motion vector data elements are established that define the scale and precision of the motion vectors. The pseudocode in Figure 8.9 defines the picture layer, and the corresponding flowchart is in Figure 8.10.

The temporal_reference is the picture count in the sequence, modulo 1024 (i.e., the ten least significant bits). The temporal_reference starts at zero, and is used to determine display order.

The picture_coding_type data element is the next three bits, and the meaning of these bits is defined by Table 8.5. The vbv_delay that follows gives the number of bits that must be in the compressed data buffer before the idealized decoder model defined by MPEG decodes the picture. This parameter plays an important role in preventing buffer overflow and underflow and will be discussed further in Chapter 15.

If the picture type is 2 (P-picture) or 3 (B-picture), then the forward motion vector scale information for the full picture is sent. For B-pictures, the backward motion vector scale information is also needed. The dashed line in the flowchart indicates that the first test could have skipped the second test since a path for not 2 or 3 is guaranteed to fail a test for 3.

```
picture(){                       /* from ISO 11172-2 2.4.2.5        */
  picture_start_code(32);        /* r/w 0x00000100                  */
  temporal_reference(10);        /* r/w picture count modulo 1024   */
  picture_coding_type(3);        /* r/w picture type                */
  vbv_delay(16);                 /* r/w VBV buffer delay            */
  if (picture_coding_type==2)||(picture_coding_type==3){
                                 /* if P or B type, need forward mv */
    full_pel_forward_vector(1); /* r/w 1=full pel, 0=half pel      */
    forward_f_code(3);           /* r/w fwd motion vector range     */
  }
  if (picture_coding_type==3){/* if B-picture, need backward mv   */
    full_pel_backward_vector(1); /* r/w 1=full pel, 0=half pel      */
    backward_f_code(3);          /* r/w bkwd mot. vector range      */
  }
  while (nextbits(1)=='1'){      /* while '1', extra information    */
    extra_bit_picture(1);        /* r/w '1'                         */
    extra_information_picture(8);/* r/w byte of extra information */
  }
  extra_bit_picture(1);          /* r/w '0' to end extra information */
  next_start_code();             /* find next start code            */
  if (nextbits(32)==extension_start_code){/* if 0x000001B5          */
    extension_start_code(32);       /* r/w extension start code   */
    while (nextbits(24)!=0x000001){ /* while not start code prefix*/
      picture_extension_data(8);    /*   r/w byte of data         */
    }
    next_start_code();           /* find next start code            */
  }
  if (nextbits(32)==user_data_start_code){ /* if 0x000001B2       */
    user_data_start_code(32);       /* r/w user data start code   */
    while (nextbits(24)!=0x000001){ /* while not start code prefix*/
      user_data(8);                 /*   r/w byte of user data    */
    }
    next_start_code();           /* find next start code            */
  }
  do {                           /* do slice(s)                     */
    slice();                     /*   process a slice               */
  } while (nextbits(32)==slice_start_code)/* while 0x00000101-1AF */
}                                /* end picture() function          */
```

Figure 8.9: The picture() function.

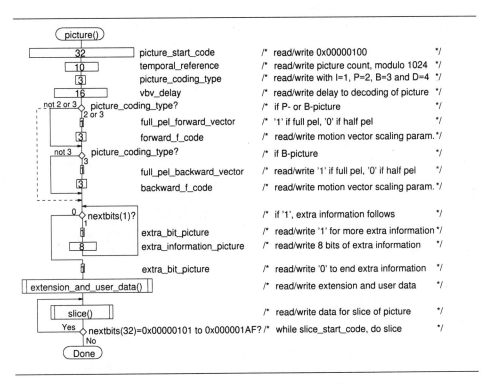

Figure 8.10: Flowchart for `picture()` function.

```
slice(){                        /* from ISO 11172-2 2.4.2.6      */
  slice_start_code(32);         /* r/w 0x00000101-0x000001AF     */
  quantizer_scale(5);           /* r/w quantizer scale           */
  while (nextbits(1)=='1'{      /* while '1', extra slice info.  */
    extra_bit_slice(1);         /*   r/w '1'                     */
    extra_information_slice(8); /*   r/w byte of extra information*/
  }                             /* end - extra slice info.       */
  extra_bit_slice(1);           /* r/w '0' to end extra slice info. */
  do {                          /* do macroblock(s)              */
    macroblock();               /*    process a macroblock       */
  } while (nextbits(23)!=0)     /*    do while not 23 zeros      */
  next_start_code();            /* find next start code          */
}                               /* end - slice() function        */
```

Figure 8.11: The `slice()` function.

The `extra_information_picture` data element is reserved and its function is currently undefined. However, if present in the bitstream, the decoder must be able to parse `extra_information_picture` and discard it.

The `next_start_code()` function locates the next byte boundary and reads or writes zero padding bytes. The `extension_and_user_data()` function processes the extra bytes of extension data (MPEG-2 only) and user data, preceded by the appropriate start code. As many slices as needed to make up the picture are then decoded/encoded.

8.6   Slice layer ◗

The pseudocode in Figure 8.11 defines the slice layer syntax. The corresponding flowchart is in Figure 8.12.

The final byte of the `slice_start_code` has a value in the range from 1 to X'AF' inclusive. Defining the top of the picture to be at macroblock row 1, this value is the macroblock row in which the slice starts. The horizontal position at which the slice starts within the macroblock row is determined by data elements in the macroblock layer header, as will be discussed in the next section of this chapter.

The `quantizer_scale` initializes the quantizer scale factor. The `extra_bit_slice` data element that follows is always zero in an MPEG-1 bitstream; the value of '1' is reserved for future ISO standards. However, in future standards, as long as `extra_bit_slice` is set, eight bits of `extra_information_slice` are appended. The actual macroblock coding follows an `extra_bit_slice` of zero. Finally, `next_start_code()` is called to byte-align the next slice and possibly add stuffed zero bytes.

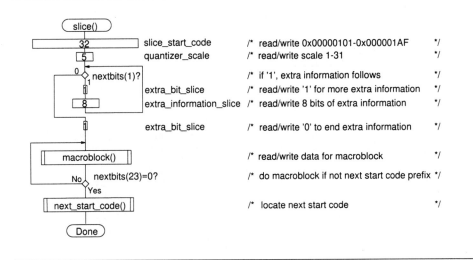

Figure 8.12: Flowchart for the slice() function.

Since each slice is coded independently of all other slices, the slice layer provides for a limited degree of error recovery.

8.7   Macroblock layer ◑

The macroblock header provides information about the position of the macroblock relative to the position of the macroblock just coded. It also codes the motion vectors for the macroblock and identifies the blocks in the macroblock that are coded. The pseudocode for the macroblock layer is found in Figure 8.13, and the corresponding flowchart is in Figure 8.14.

The macroblock() procedure starts with an optional 11-bit macroblock_stuffing code that may be repeated as many times as needed.

The macroblock address is the index of the macroblock in the picture, where index values are in raster scan order starting with macroblock 0 in the upper left corner. At the start of coding of a picture, the macroblock address is set to −1 and the macroblock_address_increment establishes the change in macroblock address to reach the macroblock being coded. When the address increment is greater than 1, intervening macroblocks are skipped. The rules for when and how this is done will be discussed in Section 8.7.1.

If the macroblock_address_increment is greater than 33, it coded as a sequence of one or more macroblock_escape codes that increment the address by 33, followed by the appropriate code from Table 8.6 for the remaining increment. The codes in Table 8.6 are ordered for a decoder.

```
macroblock(){                          /* from ISO 11172-2 2.4.2.7  */
  while (nextbits(11)=='00000001111')/* while macroblock stuffing */
    macroblock_stuffing(11);          /* r/w '00000001111'         */
  while (nextbits(11)=='00000001000')/* while macroblock escape   */
    macroblock_escape(11);            /* r/w '00000001000'         */
  macroblock_address_increment(1-11);/* r/w VLC for mb address     */
  macroblock_type(1-6);               /* r/w VLC for mb type        */
  if (macroblock_quant)               /* if quant. scale change     */
    quantizer_scale(5);               /*    r/w new quantizer scale */
  if (macroblock_motion_forward){     /* if forward motion vector   */
    motion_horizontal_forward_code(1-11); /* r/w VLC for fwd h. mv*/
    if (forward_f!=1)&&
        (motion_horizontal_forward_code!=0)) /* if fwd. h. mv     */
      motion_horizontal_forward_r(1-6);/* r/w residual of h. mv   */
    motion_vertical_forward_code(1-11);/* r/w VLC for fwd v. mv    */
    if (forward_f!=1)&&
        (motion_vertical_forward_code!=0)) /* if fwd. v. mv        */
      motion_vertical_forward_r(1-6);/* r/w residual of v. mv     */
  }                                   /* end if forward motion vect*/
  if (macroblock_motion_backward){    /* if backward motion vector */
    motion_horizontal_backward_code(1-11);/* r/w VLC for bkwd h.mv*/
    if (backward_f!=1)&&
        (motion_horizontal_backward_code!=0)) /* if bkwd h. mv    */
      motion_horizontal_backward_r(1-6); /* r/w residual of h. mv */
    motion_vertical_backward_code(1-11); /* r/w VLC for bkwd v.mv */
    if (backward_f!=1)&&
        (motion_vertical_backward_code!=0)) /* if bkwd v. mv      */
      motion_vertical_backward_r(1-6); /* r/w residual of v. mv   */
  }                                   /* end if backward motion vec*/
  if (macro_block_pattern)            /* if any blocks coded        */
    coded_block_pattern(3-9);         /* r/w coded block pattern    */
  for (i=0; i<6; i++)                 /* for the 6 blocks           */
    block(i);                         /*   r/w block data           */
  if (picture_coding_type==4)         /* if D-picture               */
    end_of_macroblock(1);             /*   r/w '1' - end of mb      */
}                                     /* end macroblock() function  */
```

Figure 8.13: The macroblock() function.

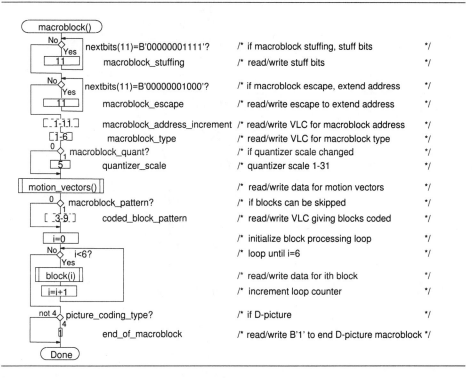

Figure 8.14: Flowchart for `macroblock()` function.

increment value	macroblock_ address_increment
macroblock_escape	0000 0001 000
macroblock_stuffing	0000 0001 111
33	0000 0011 000
32	0000 0011 001
31	0000 0011 010
30	0000 0011 011
29	0000 0011 100
28	0000 0011 101
27	0000 0011 110
26	0000 0011 111
25	0000 0100 000
24	0000 0100 001
23	0000 0100 010
22	0000 0100 011
21	0000 0100 10
20	0000 0100 11
19	0000 0101 00
18	0000 0101 01
17	0000 0101 10
16	0000 0101 11
15	0000 0110
14	0000 0111
13	0000 1000
12	0000 1001
11	0000 1010
10	0000 1011
9	0000 110
8	0000 111
7	0001 0
6	0001 1
5	0010
4	0011
3	010
2	011
1	1

Table 8.6: Variable length codes for macroblock_address_increment.

macro-block _intra	macro-block _pattern	macro-block _motion _backward	macro-block _motion _forward	macro-block _quant	macro-block _type VLC code
I-pictures					
1	0	0	0	0	1
1	0	0	0	1	01
P-pictures					
0	0	0	1	0	001
0	1	0	0	0	01
0	1	0	0	1	0000 1
0	1	0	1	0	1
0	1	0	1	1	0001 0
1	0	0	0	0	0001 1
1	0	0	0	1	0000 01
B-pictures					
0	0	0	1	0	0010
0	0	1	0	0	010
0	0	1	1	0	10
0	1	0	1	0	0011
0	1	0	1	1	0000 11
0	1	1	0	0	011
0	1	1	0	1	0000 10
0	1	1	1	0	11
0	1	1	1	1	0001 0
1	0	0	0	0	0001 1
1	0	0	0	1	0000 01
D-pictures					
1	0	0	0	0	1

Table 8.7: VLC for macroblock_type in I-, P-, B-, and D-pictures.

cbp		block #	coded_block
decimal	binary	0123 4 5	_pattern
		YYYY CbCr	VLC code
0	000000	.... . .	forbidden
1	000001	 c	0101 1
2	000010	.... c .	0100 1
3	000011	.... c c	0011 01
4	000100	...c . .	1101
5	000101	...c . c	0010 111
6	000110	...c c .	0010 011
7	000111	...c c c	0001 1111
8	001000	..c. . .	1100
9	001001	..c. . c	0010 110
10	001010	..c. c .	0010 010
11	001011	..c. c c	0001 1110
12	001100	..cc . .	1001 1
13	001101	..cc . c	0001 1011
14	001110	..cc c .	0001 0111
15	001111	..cc c c	0001 0011
16	010000	.c.. . .	1011
17	010001	.c.. . c	0010 101
18	010010	.c.. c .	0010 001
19	010011	.c.. c c	0001 1101
20	010100	.c.c . .	1000 1
21	010101	.c.c . c	0001 1001
22	010110	.c.c c .	0001 0101
23	010111	.c.c c c	0001 0001
24	011000	.cc. . .	0011 11
25	011001	.cc. . c	0000 1111
26	011010	.cc. c .	0000 1101
27	011011	.cc. c c	0000 0001 1
28	011100	.ccc . .	0111 1
29	011101	.ccc . c	0000 1011
30	011110	.ccc c .	0000 0111
31	011111	.ccc c c	0000 0011 1

Table 8.8: (a) MPEG-1 coded_block_pattern (cbp) VLC codes. Blocks labeled "." (bit=0) are skipped, whereas blocks labeled "c" (bit=1) are coded.

cbp		block #	coded_block
decimal	binary	0123 4 5	_pattern
		YYYY CbCr	VLC code
32	100000	c... . .	1010
33	100001	c... . c	0010 100
34	100010	c... c .	0010 000
35	100011	c... c c	0001 1100
36	100100	c..c . .	0011 10
37	100101	c..c . c	0000 1110
38	100110	c..c c .	0000 1100
39	100111	c..c c c	0000 0001 0
40	101000	c.c. . .	1000 0
41	101001	c.c. . c	0001 1000
42	101010	c.c. c .	0001 0100
43	101011	c.c. c c	0001 0000
44	101100	c.cc . .	0111 0
45	101101	c.cc . c	0000 1010
46	101110	c.cc c .	0000 0110
47	101111	c.cc c c	0000 0011 0
48	110000	cc.. . .	1001 0
49	110001	cc.. . c	0001 1010
50	110010	cc.. c .	0001 0110
51	110011	cc.. c c	0001 0010
52	110100	cc.c . .	0110 1
53	110101	cc.c . c	0000 1001
54	110110	cc.c c .	0000 0101
55	110111	cc.c c c	0000 0010 1
56	111000	ccc. . .	0110 0
57	111001	ccc. . c	0000 1000
58	111010	ccc. c .	0000 0100
59	111011	ccc. c c	0000 0010 0
60	111100	cccc . .	111
61	111101	cccc . c	0101 0
62	111110	cccc c .	0100 0
63	111111	cccc c c	0011 00

Table 8.8: (b) MPEG-1 coded_block_pattern (cbp) VLC codes continued. Blocks labeled "." (bit=0) are skipped, whereas blocks labeled "c" (bit=1) are coded.

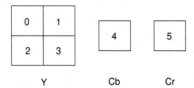

Figure 8.15: Block indices in a macroblock. Block 0 corresponds to the msb of the `pattern_code`.

The `macroblock_type` is specified by a variable-length code from one to six bits in length, and the decoded value sets the following five control bits:

`macroblock_quant:` if set, a new five-bit quantization scale is sent.

`macroblock_motion_forward:` if set, a forward motion vector is sent. Details of this part of the procedure are given below.

`macroblock_motion_backward:` if set, a backward motion vector is sent. Details of this part of the procedure are given below.

`macroblock_pattern:` if set, the `coded_block_pattern` variable length code from Table 8.8 follows to code the six `pattern_code` bits in the variable cbp. These bits identify the blocks in the macroblock that are coded, as shown in Figure 8.15.

`macroblock_intra:` if set, the blocks in the macroblock are coded as intra blocks without reference to blocks from other pictures.

These five control bits determine the rest of the macroblock processing steps. Table 8.7 lists the variable-length codes and control-bit values of `macroblock_type` for each of the four picture types. Note that some of the control bits are forced to default values for I-, P-, and D-pictures.

Once the `pattern_code` bits are known, any blocks where the bit is set are coded. The test for coding of a block is made in the `block()` procedure that will be described in the next section.

For D-pictures (`picture_coding_type` = 4), only the DC coefficients are coded. For these D-pictures, the 1-bit `end_of_macroblock` data element is needed to prevent accidental start code emulation.

motion value magnitude	motion_code VLC
0	1
1	01s
2	001s
3	0001 s
4	0000 11s
5	0000 101s
6	0000 100s
7	0000 011s
8	0000 0101 1s
9	0000 0101 0s
10	0000 0100 1s
11	0000 0100 01s
12	0000 0100 00s
13	0000 0011 11s
14	0000 0011 10s
15	0000 0011 01s
16	0000 0011 00s

Table 8.9: Variable length codes for the motion codes. "s" is 0 for positive motion values and '1' for negative motion values.

Figure 8.16 is a flowchart for coding of motion vectors for a macroblock with motion compensation. The control bits, macroblock_motion_forward and macroblock_motion_backward have already been established by the decoding of macroblock_type. If a given motion compensation control bit is set, additional parameters for that type of motion vector are sent.

The variable-length codes for motion_code are given in Table 8.9. The forward, backward, horizontal, and vertical combinations of motion codes all use this same table.   For example, if macroblock_motion_forward is set, then the one- to 11-bit variable length code for the motion_horizontal_forward_code is coded. Note that the s bit in this table provides the sign of the motion value. If the forward_f code is more than 1 and motion_horizontal_forward_code is not zero, then one to six additional bits are needed for motion_horizontal_forward_r. A similar structure then applies for vertical forward motion. The interpretation of these parameters will be discussed in detail in Chapter 11.

The same coding structure is used to code backward motion vectors. Note that forward motion vectors are used in both P-pictures and B-pictures, whereas backward motion vectors are used only in B-pictures.

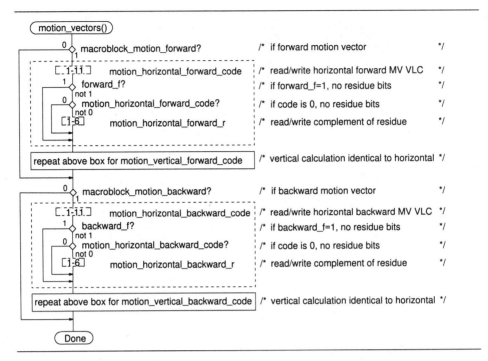

Figure 8.16: Flowchart for motion vector coding.

8.7.1 Skipping macroblocks ◑

First and last macroblocks in a slice must always be coded, but in P- and B-pictures macroblocks in between can sometimes be skipped by simply incrementing the macroblock address. I-pictures may never skip macroblocks.

Nonintra coded macroblocks are skipped only when the predicted values are good enough that the DCT coefficients of all six blocks are zero. In P-pictures the motion vectors of skipped macroblocks are set to zero before calculating the prediction. In B-pictures the motion vectors and the type of prediction (forward, backward, or bidirectional prediction) are propagated to the skipped macroblock when calculating the prediction. This will be discussed in more depth in Chapter 11.

8.8 Block layer ◑

The block layer is the lowest layer of the video sequence, and it is in this layer that the actual 8×8 DCT blocks are coded. Note that the coding depends on whether the block is a luminance or chrominance block, and on whether the macroblock is intra or nonintra. Note also that when the coding is nonintra, blocks where all DCT coefficients are zero are simply skipped.

```
block(i){                       /* from ISO 11172-2  2.4.2.8     */
  if (pattern_code[i]){         /* if ith block coded            */
    if (macroblock_intra){      /* if intra-coded macroblock     */
      if (i<4){                 /* if luminance block            */
        dct_dc_size_luminance(2-7);/* r/w VLC for Y size         */
        if (dc_size_luminance!=0)  /* if Y size not zero         */
          dct_dc_differential(1-8);/* r/w size bits of diff. DC  */
      }                         /* end if luminance block        */
      else{                     /* else chrominance block        */
        dct_dc_size_chrominance(2-7);/* r/w VLC for Cr or Cb size */
        if (dc_size_chrominance!=0)/* if Cr or Cb size not zero  */
          dct_dc_differential(1-8);/* r/w size bits of diff. DC  */
      }                         /* end else chrominance block    */
    }                           /* end if intra-coded macroblock */
    else {                      /* else not intra-coded macroblock */
      dct_coeff_first(2-28);    /* r/w VLC 1st run-level         */
    }                           /* end else not intra-coded mb   */
    if (picture_coding_type!=4){/* if not D-picture              */
      while (nextbits(2)!='10') /* while not end-of-block        */
        dct_coeff_next(3-28);   /* r/w VLC next run-level        */
      end_of_block(2);          /* r/w '01'                      */
    }                           /* end if not D-picture          */
  }                             /* end if ith block coded        */
}                               /* end block(i) function         */
```

Figure 8.17: block() function.

The pseudocode for coding the ith block of a macroblock is in Figure 8.17, and the corresponding flowchart is in Figure 8.18.

The block() procedure starts with a test of the ith element of pattern_code. When this ith element is zero, the quantized coefficients are zero for the entire ith block. Conversely, if the ith element of pattern_code is nonzero, some coefficients in the ith block are not zero and the block is coded. Note that for intra-coded macroblocks pattern_code[i] is always one.

If macroblock_intra (set by macroblock_type) is set, the DC coefficient is coded separately from the AC coefficients. Note that the coding is of the difference between the DC coefficient of the block and the prediction made from the DC coefficient of the block just coded from the same component.

Code tables for coding the DC coefficients are given in Section 5.4.4.1 of Chapter 5. Referring to these tables, the parameter dct_dc_size_luminance (*size* in Table 5.3) is coded with a VLC of two to seven bits for each of the four luminance blocks. If dct_size_luminance is nonzero for a block, the bits needed to code dct_dc_differential for that block follow. Similarly,

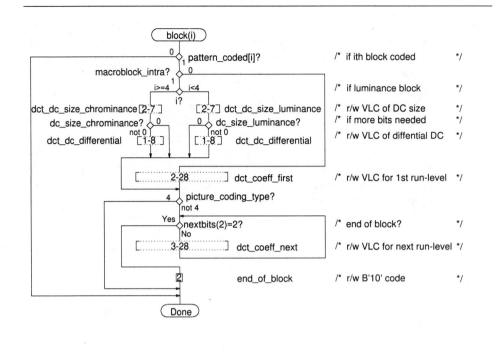

Figure 8.18: Flowchart for `block()` function.

for the two chrominance blocks, `dct_dc_size_chrominance` is used to code the DC coefficients; in each case, if size is nonzero, the rest of the information is coded in `dct_dc_differential`.

The tables for AC coding and nonintra DCT coding are given in Section 5.4.4.2. If the block is nonintra coded, the DC coefficient is already a differential value and is coded using the run-level AC coefficient coding structure. Since the block is skipped if the EOB occurs immediately, the end-of-block symbol cannot occur before the first run-level is coded. Conceptually, therefore, it is removed from the table and the short code pair, '1 s' is assigned to run-level 0-1 (labeled "first" in Table 5.5). However, for the rest of the run-level coding in the block, the EOB can occur and the longer code pair, '11 s', is used for run-level 0-1 (labeled "next" in Table 5.5).

For D-pictures (`picture_coding_type` = 4) only DC coefficients are coded and the AC coefficient and end-of-block are bypassed. For all other picture types, the rest of the run-levels are coded with the `dct_coeff_next` until block coding is terminated with an `end_of_block` code.

video data element name	set v g p s m b	used v g p s m b	dt	# of bits	value range
backward_f_code	. . ● . . .	. . . . . .	U	3	1...7
backward_f	. . ○ . . .	. . . . ● .			1,2,4,8,16,32,64
bit_rate	●	. . . . . .	U	18	1...0x3FFFF
broken_link	. ●	. . . . . .	U	1	0,1
closed_gop	. ●	. . . . . .	U	1	0,1
coded_block_pattern	. . . . ● .	. . . . . .	V	3-9	1...63
pattern_code[0]	. . . . ○ .	. . . . ●			0,1
...	. . . . ○ .	. . . . ●			0,1
pattern_code[5]	. . . . ○ .	. . . . ●			0,1
constrained_parameters_flag ...	●	. . . . . .	U	1	0,1
dct_coeff_first	. . . . ● .	. . . . . .	V	2-28	
dct_coeff_next	. . . . ● .	. . . . . .	V	3-28	
dct_dc_differential	. . . . ● .	. . . . . .	U	1-8	-255...-1,1...255
dct_dc_size_chrominance	. . . . ● .	. . . . . .	V	2-8	
dc_size_chrominance	. . . . ○ .	. . . . ●			0...8
dct_dc_size_luminance	. . . . ● .	. . . . . .	V	2-7	
dc_size_luminance	. . . . ○ .	. . . . ●			0...8
end_of_block	. . . . ● .	. . . . ○	V	2	'10'
end_of_macroblock	. . . . ● .	. . . . . .	B	1	'1'
extension_start_code	● ● ● . . .	● ● ● . . .	B	32	0x000001B5
extra_bit_picture	. . ● . . .	. . ○ . . .	U	1	'0'
extra_bit_slice	. . . ● . .	. . . ○ . .	U	1	'0'
extra_information_picture	. . ● . . .	. . . . . .	-	8	reserved
extra_information_slice	. . . ● . .	. . . . . .	-	8	reserved
forward_f_code	. . ● . . .	. . . . . .	U	3	1...7
forward_f	. . ○ . . .	. . . . ● .			1,2,4,8,16,32,64
full_pel_backward_vector	. . ● . . .	. . . . . .	U	1	0,1
full_pel_forward_vector	. . ● . . .	. . . . . .	U	1	0,1
group_extension_data	. ●	. . . . . .	-	8	reserved
group_start_code	. ●	. ●	B	32	0x000001B8
horizontal_size	●	. . . . . .	U	12	1...4095
intra_quantizer_matrix[0]	●	. . . . . .	U	8	8
...	●	. . . . . .	U	8	1...255
intra_quantizer_matrix[63]	●	. . . . . .	U	8	1...255
load_intra_quantizer_matrix ...	●	●	U	1	0,1
load_non_intra_quantizer_matrix	●	●	U	1	0,1
macroblock_address_increment	. . . . ● .	. . . . . .	V	1-11	1...33
macroblock_escape	. . . . ● .	. . . . ○ .	V	11	'00000001000'

Table 8.10: (a) Video syntax data element summary.

video data element name	set v g p s m b	used v g p s m b	dt	# of bits	value range
macroblock_stuffing	. . . . ● .	. . . . ○ .	V	11	'00000001111'
macroblock_type	. . . . ● .	. . . . . .	V	1-6	0...63
macroblock_intra	. . . . ○ .	. . . . . ●	-	1	0,1
macroblock_motion_backward	. . . . ○ .	. . . . . ●	-	1	0,1
macroblock_motion_forward	. . . . ○ .	. . . . . ●	-	1	0,1
macroblock_pattern	. . . . ○ .	. . . . . ●	-	1	0,1
macroblock_quant	. . . . ○ .	. . . . . ●	-	1	0,1
marker_bit	●	. . . . . .	B	1	'1'
motion_horizontal_backward_code	. . . . ● .	. . . . . ●	V	1-11	-16...16
motion_horizontal_backward_r	. . . . ● .	. . . . . .	U	1-6	0...63
motion_horizontal_forward_code	. . . . ● .	. . . . . ●	V	1-11	-16...16
motion_horizontal_forward_r	. . . . ● .	. . . . . .	U	1-6	0...63
motion_vertical_backward_code	. . . . ● .	. . . . . ●	V	1-11	-16...16
motion_vertical_backward_r	. . . . ● .	. . . . . .	U	1-6	0...63
motion_vertical_forward_code	. . . . ● .	. . . . . ●	V	1-11	-16...16
motion_vertical_forward_r	. . . . ● .	. . . . . .	U	1-6	0...63
non_intra_quantizer_matrix[0]	●	. . . . . .	U	8	1...255
...	●	. . . . . .	U	8	1...255
non_intra_quantizer_matrix[63]	●	. . . . . .	U	8	1...255
pel_aspect_ratio	●	. . . . . .	U	4	1...14
picture_coding_type	. . ● . . .	. . ● . ● ●	U	3	1...4
picture_extension_data	. . ● . . .	. . . . . .	-	8	reserved
picture_rate	●	. . . . . .	U	4	1...8
picture_start_code	. . ● . . .	. . ● . . .	B	32	0x00000100
quantizer_scale	. . ● ● . .	. . . . . .	U	5	1...31
sequence_end_code	●	●	B	32	0x000001B7
sequence_extension_data	●	. . . . . .	-	8	reserved
sequence_header_code	●	●	B	32	0x000001B3
slice_start_code 1	. . ● . . .	. . ● . . .	B	32	0x00000101
...	. . ● . . .	. . ● . . .	B	32	0x000001xx
slice_start_code 175	. . ● . . .	. . ● . . .	B	32	0x000001AF
temporal_reference	. . ● . . .	. . . . . .	U	10	0...1023
time_code	. ●	. . . . . .	-	25	
user_data	● ● ● . . .	. . . . . .	-	8	0...255
user_data_start_code	● ● ● . . .	● ● ● . . .	B	32	0x000001B2
vbv_buffer_size	●	. . . . . .	U	10	0...1023
vbv_delay	. . ● . . .	. . . . ○ .	U	16	0...0xFFFF
vertical_size	●	. . . . . .	U	12	2,4,...,4094

Table 8.10: (b) Video syntax data element summary.

data element type	original pseudo-code mnemonic	summary chart mnemonic
bit string, left bit first	bslbf	B
unsigned integer, most significant bit first	uimsbf	U
variable length code, left bit first	vlclbf	V
fixed 1-bit	"1"	'1'
fixed 0-bit	"0"	'0'

Table 8.11: Data types in the video bitstream.

8.9 Summary of data element syntax ◑

Table 8.10 is a summary chart listing in alphabetical order all of the video data elements set in the six video layers. Parameters derived from these elements are indented under their main parent, thereby helping to explain how the information is used in the syntax pseudocode. The six columns labeled "set" have a column each for the video sequence (v), GOP (g), picture (p), slice (s), macroblock (m), and block (b) layers. The filled circles mark the layer(s) in which given data elements are set and open circles indicate when derived variables are set. Only a few data elements like extension_start_code, user_data_start_code, user_data, and quantizer_scale can be set in more than one layer. The columns labeled "used" have the same six layers, and a filled circle in a particular column indicates that the data element is used in the pseudocode defined for that layer. An open circle means that the data element was implicitly, but not explicitly, used. The number of bits column identifies the range of bits the given data element can have. The final column shows the value range allowed. Hexadecimal numbers start with 0x, the usual C language convention. Binary numbers are enclosed by single quotes, i.e., '01001'.

As summarized in Table 8.11, an MPEG video data element is usually one of three basic data types (dt): unsigned integer (U), variable-length coded (V), or bit string (B). Unsigned integers are the only integer format embedded in the bitstream; if signs are needed, they are coded as separate bits. Variable-length codes are always of the unique prefix class, such that the completeness of the code can be determined from the value at any point during the reading of the bitstream. Bit strings are bit sequences with predetermined patterns and fixed size. Start codes, for example, are bit strings.

Two other data types are defined, fixed 1-bit and fixed 0-bit. These are single bits of known value inserted into the bitstream. For example, the marker_bit is a fixed 1-bit data type.

Compressed data (hexadecimal format):
000001B302001014FFFFE0A0000001B880080040000001
00000FFFF800000101FA96529488AA25294888000001B7

Compressed data (binary format):
00000000 00000000 00000001 10110011 00000010
00000000 00010000 00010100 11111111 11111111
11100000 10100000 00000000 00000000 00000001
10111000 10000000 00001000 00000000 01000000
00000000 00000000 00000001 00000000 00000000
00001111 11111111 11111000 00000000 00000000
00000001 00000001 11111010 10010110 01010010
10010100 10001000 10101010 00100101 00101001
01001000 10001000 00000000 00000000 00000001
10110111

Figure 8.19: Two flat macroblocks compressed as an I-picture.

The data element type also determines the order in which the bits are inserted or removed from the bitstream. The "bslbf" mnemonic stands for "bit string, left bit first". This means that the bit string is inserted into the coded bitstream with the left bit first (i.e. it looks same in a bit string printed from left (earlier in time) to right as it does as a binary number). The "uimsbf" mnemonic stands for "unsigned integer, most significant bit first". This means that the value is converted to an unsigned (binary) integer with the designated number of bits. At that point, the most significant bit is to the left, next to the preceding data element. The "vlclbf" mnemonic (variable-length code, left bit first) means that the variable codes are inserted into the data stream in the same order that they are printed in the tables, leftmost bit first. For those familiar with the "big endian, little endian" notation, these definitions correspond to "big endian" conventions.

Many data elements are set, but appear to never be used. Often this means that the information is intended for system control (how this is done is not always defined in the standard) or is used in the IDCT calculations, motion compensation, quantization changes, bit rate control, and pel reconstruction. These topics are covered in the next few chapters.

8.10   MPEG-1 video data stream example ●

A simple source image consisting of two macroblocks, with YCbCr values all set to a constant, 128 (decimal), was compressed to create a simple MPEG-

Sequence header: 0x000001B302001014FFFFE0A0

```
00000000 00000000 00000001 10110011   sequence_header_code
00000010 0000                         horizontal_size=32 pels
0000 00010000                         vertical_size=16 pels
0001                                  pel_aspect_ratio=1
0100                                  picture_rate=4
11111111 11111111 11                  bit_rate=0x3ffff (variable)
1                                     marker bit=1
00000 10100                           vbv_buffer_size=20
0                                     constrained_parameters_flag=0
0                                     load_intra_quantizer_matrix=0
0                                     load_nonintra_quantizer_matrix=0
```

Figure 8.20: Parsed sequence header.

1 video data stream. The compressed data stream is shown in Figure 8.19 both in hexadecimal format and in binary format. In hexadecimal format it is easy to pick out the start codes, since they always fall on byte boundaries. The quantizer_scale was set to 31 in the slice header, then forced to 5 for the first macroblock and 8 for the second. Rate was variable.

The sequence header for this simple image is shown in hexadecimal format at the top of Figure 8.20. Then the data elements are shown to the right of the sequence header binary bits. The extra spaces are inserted at byte boundaries.

The group_of_pictures header is shown in Figure 8.21. After the group_start_code comes the SMPTE time code. For this illustration the drop_frame_frame is set and the time code hours, minutes, seconds, and pictures are all zero. This is a closed group of pictures and no links have been broken. The last five zero bits are stuffed bits to align the next start code on a byte boundary.

Figure 8.22 shows the picture header both in hexadecimal format and parsed into data elements. The picture number in display order is always reset to zero at the start of a group of pictures. Since this first picture is an intra-coded (I-) picture, its picture_coding_type is 1. The vbv_delay set to 0xFFFF indicates variable rate coding. I-pictures have no motion vectors. Extra_bit_picture is always '0' since the extra_picture bytes are reserved for future standards. The final two zero bits are stuffing bits to align the next start code on byte boundaries.

Figure 8.23 parses the slice header. The final byte of the slice_start_code gives the slice_vertical_position. Macroblock_address is reset to

```
Group_of_pictures header   0x000001B880080040

00000000 00000000 00000001 10111000  group_start_code
                                      time_code:
1                                        drop_frame_flag
00000                                    time_code_hours=0
00 0000                                  time_code_minutes=0
1                                        marker_bit
000 000                                  time_code_seconds=0
00000 0                                  time_code_pictures=0
1                                     closed_gop=1
0                                     broken_link=0
00000                                 stuffed bits to byte boundary
```

Figure 8.21: Parsed GOP header.

```
Picture header: 0x00000100000FFFF8

00000000 00000000 00000001 00000000  picture_start_code
00000000 00                           temporal_reference=0
001                                   picture_coding_type=1 (I-pict.)
111 11111111 11111                    vbv_delay=0xFFFF (variable rate)
0                                     extra_bit_picture=0
00                                    stuffing bits to byte boundary
```

Figure 8.22: Parsed picture header.

```
Slice header: 0x00000101FA

00000000 00000000 00000001 00000001  slice_start_code
                                        slice_vertical_position=1
                                        macroblock_address=-1
11111                                 quantizer_scale=31
0                                     extra_bit_slice=0
10                                    belong to macroblock layer
```

Figure 8.23: Parsed slice header.

−1 by a slice header. The `quantizer_scale` is set to 31 here. This was an arbitrary choice since it is immediately reset in the macroblock. `Extra_bit_slice` is always set to zero because `extra_information_slice` which would follow a 1 bit is still reserved. The final two bits of the byte belong to the macroblock layer.

The macroblock, block, and end of video sequence compressed data is shown in binary format at the top of Figure 8.24. Since this is a I-picture, the `macro_block_address_increment` must always be 1, There are only two choices for `macroblock_type`, with quantizer scale change and without, and all blocks are coded (i.e., `cbp`='111111'). The next five bits change the `quantizer_scale` to five. The first macroblock has four luminance blocks with a DC difference of 0 followed by an immediate end-of-block and the same for the two chrominance blocks. Table 5.3 gives both the luminance and chrominance DC difference codes. Table 5.5 gives the AC coefficient VLC codes including the end_of_block. The second macroblock changes the `quantizer_scale` to 8 but is in all other respects identical to the first macroblock. Two bits are stuffed after this second (and final) macroblock to align the `sequence_end_code` on a byte boundary.

```
macroblock, block, and sequence_end_code compressed data:
      10 10010110 01010010 10010100 10001000 10101010 00100101
00101001 01001000 10001000 00000000 00000000 00000001 10110111
```

```
                  macroblock 1
1                                       macroblock_address_increment=1
0 1                                     macroblock_type=iq
                                          macroblock_intra=1
                                          macroblock_quant=1
                                          cpb='1111 11'
00101                                   quantizer_scale=5
10 0                                    dct_dc_size_luminance=0
10                                      end_of_block Y1
100                                     dct_dc_size_luminance=0
10                                      end_of_block Y2
100                                     dct_dc_size_luminance=0
10                                      end_of_block Y3
100                                     dct_dc_size_luminance=0
10                                      end_of_block Y4
00                                      dct_dc_size_chrominance=0
10                                      end_of_block Cb5
00                                      dct_dc_size_chrominance=0
10                                      end_of_block Cr6
                  macroblock 2
1                                       macroblock_address_increment=1
01                                      macroblock_type=iq
                                          macroblock_intra=1
                                          macroblock_quant=1
                                          cpb='1111 11'
010 00                                  quantizer_scale=8
100                                     dct_dc_size_luminance=0
10                                      end_of_block Y1
1 00                                    dct_dc_size_luminance=0
10                                      end_of_block Y2
100                                     dct_dc_size_luminance=0
1 0                                     end_of_block Y3
100                                     dct_dc_size_luminance=0
10                                      end_of_block Y4
00                                      dct_dc_size_chrominance=0
10                                      end_of_block Cb5
00                                      dct_dc_size_chrominance=0
10                                      end_of_block Cr6
00                                      stuffed bits to byte boundary
00000000 00000000 00000001 10110111    sequence_end_code
```

Figure 8.24: Two macroblocks and end of video sequence, binary format.

9

MPEG-2 Overview

This chapter contains an overview of the MPEG-2 standard. The MPEG-2 system syntax is reviewed in Section 9.4. Scalable MPEG-2 features are discussed briefly in Section 9.8. Chapter 10 provides more details of the MPEG-2 video main profile syntax.[1]

Readers should be aware that MPEG-2 is a large, complex standard. The descriptions we provide here and in the next chapter are only an introduction to this important standard.

9.1 Advantages of MPEG-2 over MPEG-1 ○

MPEG-2 offers little benefit over MPEG-1 for programming material that was initially recorded on film. In fact, nearly all movies and television programs with high production values (budgets) are shot at 24 celluloid frames per second. What MPEG-2 does offer is a more efficient means to code interlaced video signals, such as those that originate from electronic cameras (vacuum tubes or CCDs). MPEG-1 was frozen in 1991, and was intended only for progressive (i.e., noninterlaced) video pictures. MPEG-2 was initiated that same year with the goal of defining a syntax suitable for interlaced video. MPEG-2 video syntax was frozen in April 1993; in 1995, two years later, the three primary documents (systems, video, audio) which comprise the MPEG-2 standard finally reached international standard status.

Although MPEG-2 encoders may be focused on interlaced video coding, they often possess more mature and powerful coding methods that can also be applied to create better MPEG-1 bitstreams. Indeed, MPEG-2 encoders

---

[1]Another book in this digital multimedia standards series [HPN97] is devoted to MPEG-2. MPEG-2 scalability features and applications in advanced TV are discussed in much greater depth there.

need to be at least 50% more powerful than an MPEG-1 encoder capable of processing the same sample rate.

9.2   MPEG-2 applications ○

MPEG-1 was targeted primarily at bit rates of around 1.5 Mbits/s and was, therefore, particularly suitable for storage media applications such as CD-ROM retrieval. As suggested by its title, *Information technology — Generic coding of moving pictures and associated audio information*, MPEG-2 is aimed at more diverse applications such as television broadcasting, digital storage media, digital high-definition TV (HDTV), and communication. Some of the possible applications listed in the MPEG-2 video standard are: broadcast satellite service (BSS) to the home; cable TV (CATV), distribution on optical and copper wire networks; cable digital audio distribution (CDAD); digital sound broadcasting (DSB) on terrestrial and satellite links; digital terrestrial television broadcasting (DTTB); electronic cinema (EC); electronic news gathering (ENG); fixed satellite service (FSS); home television theater (HTT); interactive storage media (ISM) such as optical disks; interpersonal communications (IPC) such as videoconferencing and videophone; multimedia mailing (MMM); networked database services (NDB) via ATM; news and current affairs (NCA); remote video surveillance (RVS); satellite news gathering (SNG); and serial storage media (SSM) such as digital video tape recorders (VTR).

Both MPEG-1 and MPEG-2 can be applied to a large number of bit rates and sample rates. White Book video (on standard 650 MByte compact discs) has created a widespread impression that MPEG-1 has picture dimensions of only 352 pels/line × 240 lines/frame. On DSS, both MPEG-1 and MPEG-2 are applied to the full 480 lines/frame of North American and Japanese televisions (NTSC).

There are no true intermediate levels between MPEG-1 and MPEG-2 such as MPEG++, MPEG 1.5, or MPEG 1.8. There is only MPEG-1 and the defined profiles and level combinations of MPEG-2 such as the very popular *main profile at main level* (MP@ML). If the reader should encounter a name suggesting an intermediate level, it was probably created to distinguish the MPEG under consideration from the low-resolution MPEG associated with compact discs.

Since MPEG-2 encoders had not yet reached maturity when the DSS service began in spring 1994, DSS applied MPEG-1 to interlaced video for its first year of operation. Later, it migrated the service to MPEG-2.

9.3   Joint development with ITU-T ○

MPEG-1 was developed wholly as a ISO/IEC effort. MPEG-2 was developed jointly with the International Telecommunications Union - Terminal Sector (ITU-T), formerly called the Consultative Committee of the International Telephone and Telegraph (CCITT). The MPEG-2 systems and video recommendations/international standards share common text and are identical except for title pages and forewords. The full references to the first three parts are:

ITU-T Rec. H.222.0 | ISO/IEC 13818-1:1996 Information technology — Generic coding of moving pictures and associated audio — Part 1: *Systems*.[2]

ITU-T Rec. H.262 | ISO/IEC 13818-2:1996 Information technology — Generic coding of moving pictures and associated audio — Part 2: *Video*.

ISO/IEC 13818-3:1995 Information technology — Generic coding of moving pictures and associated audio — Part 3: *Audio*.[3]

9.4   MPEG-2 systems ○

MPEG-2 systems [Mac94] evolved from MPEG-1 systems. MPEG-1 systems was designed to work with digital storage media that have minimal errors. Traditionally, software processed the system layers and larger packets were desirable to minimize software processing requirements.

MPEG-2 added requirements. It also had to work in ATM networks and therefore needed improved error resilience. It had to handle more programs simultaneously without requiring a common time base. Since backwards compatibility with MPEG-1 was also a requirement, MPEG-1 requirements were automatically retained.

The MPEG-2 systems solved these diverse requirements by defining two different data stream types: the *program stream* (PS) and the *transport stream* (TS). Both types utilize the same *packetized elementary stream* (PES) structure.

Each PES contains exactly one video or audio elementary stream. The PES have pack headers containing system level clocks, optional encryption (scrambling), packet priority levels, trick mode indications that assist fast forward and slow motion, and packet sequence numbering. Since these functions are supported at the PES level, switching between program and transport streams preserves these features.

The program stream provides MPEG-1 compatibility. In fact, program stream decoders are required to be able to decode MPEG-1 system streams.

---

[2]The word *generic* in the title is sometimes missing, probably because it was not part of the MPEG-1 title.

[3]Note that audio is not a joint effort with ITU-T.

PS packets tend to have a variable length, typically 1k−2k bytes, but can be as large as 64 kbytes. The environment is assumed to be relatively error free. The program stream carries a single program consisting of multiple video and audio elementary streams with a common time base, the system clock reference (SCR) time stamps.

The transport stream can handle single or multiple programs. The programs do not have to share a common time clock, but may have independent program clock reference (PCR) time stamps. The transport stream is independent of the transmission data link. Transmission may be in lossy or noisy environments. Each packet is exactly 188 bytes and the header syntax is new. The short packet size was chosen to facility the use of forward error correction and to help isolate errors.

The synchronization of elementary video and audio streams is accomplished with presentation time stamps (PTS) that indicate when the decoded pictures and audio segments are to be presented. When the video is not in display order (i.e., B-pictures are present), another time stamp, the decoding time stamp (DTS), may differ from the PTS. These time stamps let the system target decoder (STD) operate without concern for buffer underflow or overflow if the encoder followed the MPEG-2 rules when encoding the video and audio information.

9.5   Fields, frames, and pictures ○

Figure 1.2 illustrates a frame of analog video containing two fields. MPEG-1 is often used for noninterlaced video formats of approximately 288×352 at 24 to 30 pictures/s. (Larger horizontal and vertical dimensions are allowed, however.) These video images are small enough to contain only a single field, or perhaps came from film in which 24 frames/s were captured as 24 pictures/s. If the original moving picture contains interlaced fields, it is preprocessed into a single picture.

At the higher target bitrates and picture rates of MPEG-2, fields and interlaced video become important. The image types expand from I-, P-, B-, and D-pictures to I-field picture, I-frame picture, P-field picture, P-frame picture, B-field picture, and B-frame picture. D-pictures are allowed only in the MPEG-1 compatibility mode.

In an interlaced analog frame composed of two fields, the top field occurs earlier in time than the bottom field. (Field 1 in Figure 1.2 is the top field and field 2 is the bottom field.) In MPEG-2 coded frames may be composed of any adjacent pairs of fields and the bottom field is not necessarily later in time. The parity of a field defines whether it is the top field or bottom field, with bottom assigned to 0 and top to 1.

MPEG-2 defines three types of coded frames, I-frames, P-frames, and

B-frames. A coded I-frame consists of a I-frame picture, a pair of I-field pictures or an I-field picture followed by a P-field picture. A coded P-frame consists of a P-frame picture or a pair of P-field pictures. A coded B-frame consists of a B-frame picture or a pair of B-field pictures.

A P-field picture followed by a backward-predicted B-field is considered a B-frame with two B-fields, since both fields in the P- and B-frames must be of the same coding type. If the first field in coded order in an I-frame is an I-field picture, the second field can be a P-field, provided that prediction is formed from the first (intra) field.

Progressive video consists of frames in which the raster lines are sequential in time. By definition, therefore, progressive frames are not made up of two fields. MPEG-1 allows only progressive pictures.

One confusing aspect of the words field and frame particularly needs to be guarded against. If frame is used as a noun, it can mean either a frame that is coded as a single picture (either a progressive frame or a frame created by interleaving two fields) or a frame that is coded as two separate field pictures. If frame is used as an adjective as in "frame picture", it always means that the image is coded as one picture.

9.6   Chrominance sampling ○

MPEG-2 defines three chrominance sampling formats relative to the luminance. These are labeled 4:2:0, 4:2:2, and 4:4:4.

In *4:2:0 format* the chrominance is sampled 2:1 both vertically and horizontally as in MPEG-1, but with the alignment shown in Figure 9.1. As can be seen in this figure, MPEG-2 alignment is not the same as MPEG-1 alignment.

In *4:2:2* format the chrominance is subsampled horizontally but not vertically, with the chrominance aligned with the luminance as shown in Figure 9.2(c). This is the sampling used in CCIR 601. Figure 9.2 illustrates the sampling of the two interlaced fields in a 4:2:0 format frame.

The *4:4:4 format* has the same sampling for all three components, and the decomposition into interlaced fields is the same for all three components. It is particularly useful for RGB images or for higher-quality video production environments.

9.7   Video formats ○

MPEG-2 has a means to specify various encoder source video formats: Component video, PAL, NTSC, SECAM, MAC, and the default "unspecified video format". PAL (phase alternating line), NTSC (National Television System Committee), SECAM (sequentiel couleur a memoire), and MAC

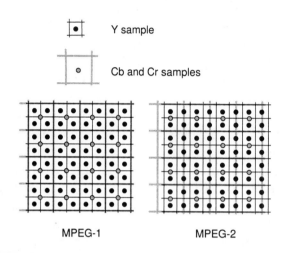

Figure 9.1: Luminance and chrominance samples in a 4:2:0 frame.

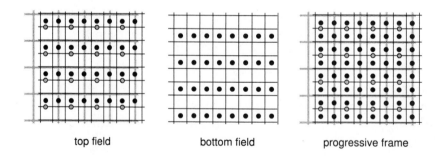

Figure 9.2: Luminance and chrominance samples in an interlaced 4:2:0 frame. Independent of whether the top field or the bottom field is coded first, the chroma is associated with the top field.

(multiplexed analog components) formats must be converted to the YCbCr format input to the encoder.

PAL is used in the United Kingdom, Germany, and much of Europe. NTSC is the U.S. analog television standard and is also used in Canada, Mexico, Japan, and many other countries. SECAM is used in France and the countries of the former Soviet Union. MAC is a European invention which time division multiplexes the three components (Y, Cb, Cr) onto a single carrier. By contrast, NTSC, PAL, and SECAM employ subcarriers to modulate chrominance information, placing them in the frequency division multiplexing category. The MAC multiplex period is usually one scan line. For example, 400 cycles of Y would be followed by 100 cycles of Cb and another 100 cycles of Cr. Thus, a scan line would look like:

```
YYY[..]YYYYY, sync pulse, CbCb[..]CbCb, sync pulse,  CrCr[..]CrCr
```

MAC was part of the pre-MPEG European scheme to migrate broadcast video from PAL/SECAM to HDTV. Variants include D-MAC (digital MAC), but D-MAC was abandoned in 1993 in favor of MPEG-2. The method was very similar to predigital ATV proposals for the United States (circa 1989).

9.8   Scalability ○

Not all applications have a single, well-defined end user. Services such as asynchronous transfer mode networks (ATM) and HDTV with TV backward compatibility need to deliver more than one resolution and quality. MPEG-2 has three types of scalability enhancements (profiles) that allow complete images to be decoded from only part of the bitstream. The three types are SNR (signal-to-noise ratio), spatial, and high. Only the high scalable profile is required to support support the chrominance format of 4:2:2 in addition to 4:2:0.

The compressed images are assembled into several layers.[4] The standalone base layer uses the nonscalable MPEG-1 syntax or even ITU-T H.261 syntax. One or two enhancement layers are then used to get to higher resolution or quality. This generally takes substantially fewer bits than independent compressed images at each resolution and quality would take.

9.8.1   SNR scalability ○

SNR scalability maintains the same luminance resolution in the lower layer and a single enhancement layer. For two-quality service for standard TV

---

[4]These layers are not to be confused with the six syntax layers.

(SDTV) or HDTV, both layers would have the same resolution and format. For video production and distribution, the lower layer might be a distribution image in 4:2:0 format and the enhancement layer a production image in 4:2:2 format. In transmission environments with errors, the base layer can be protected with more error correction and the enhancement layer allowed to be less resilient to errors. The base layer could be coded with MPEG-1 for backward compatibility.

9.8.2   Spatial scalability ○

Spatial scalability generally starts with a base layer at a lower resolution and adds an enhancement layer at higher resolution. The input source video is preprocessed to create the lower-resolution image. This is independently coded, perhaps with MPEG-1 coding. In the enhancement layer the differences between an interpolated version of the base layer and the source image are coded. There is significant flexibility in the choice of video formats for each layer.

9.8.3   Temporal scalability ○

Temporal scalability provides an extension to higher temporal picture rates while maintaining backward compatibility with lower-rate services. The lower temporal rate is coded by itself as the basic temporal rate. Then, additional pictures are coded with temporal prediction relative to the base layer. Some systems may only decode the base layer; others may decode both layers and multiplex the output to achieve the higher temporal rate. The enhancement layer can be transmitted with less error protection.

9.8.4   Data partitioning extension ○

ATM networks, terrestrial broadcast, magnetic media, and other applications sometimes have two channels available for transmission and/or storage. The video bitstream is split so that one channel contains all of the key headers, motion vectors, and low-frequency DCT coefficients. The second channel transmits less critical information such as high frequency DCT coefficients, possibly with less error protection. Partitioned data is not backward compatible with other MPEG-2 bitstreams.

9.9   Profiles ○

MPEG-2 defines five distinct profiles: Simple profile (SP), main profile (MP), SNR scalable profile (SNR), spatially scalable (Spt), and high profile

parameter	CPB	MPEG-1	SP	MP
chroma format	4:2:0	4:2:0	4:2:0	4:2:0
picture type	I, P, B, D	I, P, B, D	I, P	I, P, B
slices	all mb	all mb	all mb	all mb
scalable	no	no	no	no
intra DC precision	8	8	8, 9, 10	8, 9, 10

Table 9.1: Some parameter values for constrained-parameters bitstream (CPB) MPEG-1, generic MPEG-1, simple profile (SP) MPEG-2, and main profile (MP) MPEG-2.

(HP). This book is only concerned with the required functionality of the two nonscalable profiles, SP and MP.

Table 9.1 lists parameter values for constrained-parameters bitstream (CPB) MPEG-1, generic MPEG-1, simple profile (SP) MPEG-2, and main profile (MP) MPEG-2. These four MPEG definitions must be able to use the 4:2:0 format. The restriction to only I- and P-pictures distinguishes the SP MPEG-2 from MP MPEG-2 (MP MPEG-2 allows B-pictures). CPB and generic MPEG-1 are able to decode D-pictures.

None of the four MPEG definitions in Table 9.1 is scalable. In all four, all of the macroblocks (mb) in a picture must be explicitly coded or explicitly skipped; there can be no gaps in the slice pattern. MPEG-2 has a further restriction that a slice only contains macroblocks from one block-row. Note that in MPEG-2 the fixed eight bits of precision for the quantized DC coefficient in MPEG-1 is broadened to three choices in MPEG-2: Eight, nine, or 10 bits.

9.10   Levels ◐

Even with profiles to define specific subsets of the MPEG-2 video syntax and functionality, the parameter ranges are too large to insist on compliance over the full ranges. Therefore, levels are used to put constraints on some of the parameters (or their combinations). Applications are encouraged to insist upon full implementation of the allowed range of values of a particular profile at a particular level.

Four levels are defined in MPEG-2: low (LL), main (ML), high-1440 (H-14), and high (HL). Table 9.2 shows the level bounds for the main profile. Table 9.3 shows the only level defined for the simple profile. Note that the simple and main profiles define the main level exactly the same. All of the bounds except VBV buffer size are upper bounds. VBV buffer size is a lower bound.

Level	Parameter	Bound
High (MP@HL)	samples/line	1920
	lines/frame	1152
	frames/sec	60
	luminance rate	62,668,800 samples/s
	bitrate	80 Mbits/s
	VBV buffer size	9,781,248 bits
High – 1440 (MP@H-14)	samples/line	1440
	lines/frame	1152
	frames/sec	60
	luminance rate	47,001,600 samples/s
	bitrate	60 Mbits/s
	VBV buffer size	7,340,032 bits
Main (MP@ML)	samples/line	720
	lines/frame	576
	frames/sec	30
	luminance rate	10,368,000 samples/s
	bitrate	15 Mbits/s
	VBV buffer size	1,835,008 bits
Low (MP@LL)	samples/line	352
	lines/frame	288
	frames/sec	30
	luminance rate	3,041,280 samples/s
	bitrate	4 Mbits/s
	VBV buffer size	475,136 bits

Table 9.2: Level definitions for main profile.

Level	Parameter	Bound
Main (SP@ML)	samples/line	720
	lines/frame	576
	frames/sec	30
	luminance rate	10,368,000 samples/s
	bitrate	15 Mbits/s
	VBV buffer size	1,835,008 bits

Table 9.3: Level definition for the simple profile.

Level	Parameter	Range
High and High-1440	frame_rate_code	1,...,8
	f_code[0][0] (forward horizontal)	1,...,9
	f_code[1][0] (backward horizontal)	1,...,9
	horizontal vector range	-2048,...,2047.5
	frame f_code[0][1] (forward vertical)	1,...,5
	frame f_code[1][1] (backward vertical)	1,...,5
	frame vertical vector range	-128,...,127.5
	field f_code[0][1] (forward vertical)	1,...,4
	field f_code[1][1] (backward vertical)	1,...,4
	field vertical vector range	-64,...,63.5
Main	frame_rate_code	1,...,5
	f_code[0][0] (forward horizontal)	1,...,8
	f_code[1][0] (backward horizontal)	1,...,8
	horizontal vector range	-1024,...,1023.5
	frame f_code[0][1] (forward vertical)	1,...,5
	frame f_code[1][1] (backward vertical)	1,...,5
	frame vertical vector range	-128,...,127.5
	field f_code[0][1] (forward vertical)	1,...,4
	field f_code[1][1] (backward vertical)	1,...,4
	field vertical vector range	-64,...,63.5
Low	frame_rate_code	1,...,5
	f_code[0][0] (forward horizontal)	1,...,7
	f_code[1][0] (backward horizontal)	1,...,7
	horizontal vector range	-512,...,511.5
	frame f_code[0][1] (forward vertical)	1,...,4
	frame f_code[1][1] (backward vertical)	1,...,4
	frame vertical vector range	-64,...,63.5
	field f_code[0][1] (forward vertical)	1,...,3
	field f_code[1][1] (backward vertical)	1,...,3
	field vertical vector range	-32,...,31.5

Table 9.4: Level constraints on some ranges.

Profile@Level in bitstream	SP@ML	MP@LL	MP@ML	MP@H-14	MP@HL
MP@HL	·	·	·	·	●
MP@H-14	·	·	·	●	●
MP@ML	·	·	●	●	●
MP@LL	●	●	●	●	●
SP@ML	●	·	●	●	●
ISO/IEC 11172-2@HL	·	·	·	·	?
ISO/IEC 11172-2@H-14	·	·	·	?	?
ISO/IEC 11172-2@ML	·	·	?	?	?
ISO/IEC 11172-2@LL	·	?	?	?	?
MPEG-1 constrained-parameters	●	●	●	●	●
D-pictures@HL	·	·	·	·	●
D-pictures@H-14	·	·	·	●	●
D-pictures@ML	●	·	●	●	●
D-pictures@LL	●	●	●	●	●

Table 9.5: Decoder forward compatibility.

Table 9.4 gives the constraints on some data elements for the different levels. These data elements will be further defined in Chapter 10. The important point here is that the ranges narrow as the levels move from HL and HL-14 to ML and LL. These ranges are independent of the profiles except that a simple profile implementation has to set backward f_code[1][0] and f_code[1][1] to 15 (not used) because it has no B-pictures.

Less complex levels and profiles are subsets of the more complex levels and profiles. Provision has been made in the syntax to drop the subsetting requirement for future bitstreams, but these reserved levels and profiles are not yet defined. Table 9.5 summarizes via solid circles the forward compatibility requirement on SP and MP decoders. Note that the simple profile at main level (SP@ML) is also required to decode the main profile at low level (MP@LL). All decoders must handle MPEG-1 constrained-parameters bitstreams and D-pictures up to their claimed level. MPEG decoders belong to one or more conformance points. A conformance point, or class, is always a combination of level and profile—never a level or profile by itself.

Generic MPEG-1 (ISO/IEC 11172-2) compliant bitstreams are listed with a question mark at appropriate levels rather than a solid circle because of some ambiguity as to whether the generic MPEG-1 or just the constrained-parameters MPEG-1 is required. Even if there was no requirement that MP@ML decoders be capable of decoding MPEG-1 bitstreams with main level parameters, this is still is very much a real-world requirement for settop boxes and DVD players. The base layer for MPEG-2 multilayer

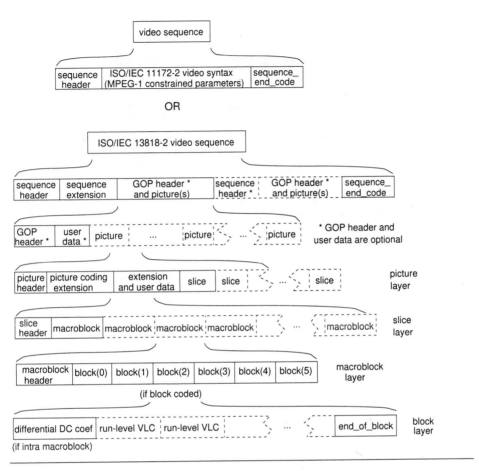

Figure 9.3: MPEG-2 main profile video sequence.

bitstreams is either the main profile, the simple profile, or an MPEG-1 compliant bitstream. The data partitioned bitstreams are incompatible.

9.11 Main profile video sequence overview ○

Figure 9.3 shows an overview of MPEG-2 main profile video bitstreams. Every MPEG video sequence starts with a sequence header and ends with a sequence_end_code. If the sequence header is not followed immediately by an extension_start_code, the syntax is governed by the ISO/IEC 11172-2 rules (MPEG-1 constrained parameters). A decoder can always choose to implement more than the levels and profiles require, but it must decode MPEG-1 constrained parameters video bitstreams to be compliant with MPEG-2 video.

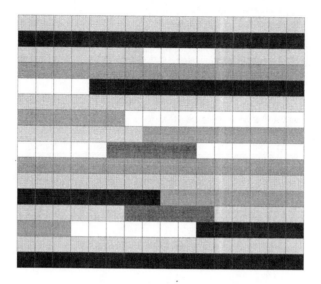

Figure 9.4: MPEG-2 slice structure.

If the sequence header is followed by the sequence extension, then the syntax is governed by the ISO/IEC 13818-2 rules (MPEG-2). In MPEG-1 the extension data is reserved for later versions of the standard. After the required sequence extension, optional (illustrated with dashed lines on the top and bottom edges) extension data and user data fields are allowed. Each extension data field has a unique identifier. The optional fields can be in any order, but each identifier can only occur once between the sequence header and the next GOP header or picture header. Group of picture (GOP) headers and picture(s) repeat until the sequence_end_code occurs. Note that the sequence header and sequence extension may be repeated.

The next blowup shows that the group of pictures header is always optional. It may be followed immediately by user data, but never by extension data. At least one picture follows each GOP header.

A picture header is always followed by the picture coding extension, other extensions and user data (optional), and picture data. Picture data is made up of slices, and each slice consists of a slice header and macroblock data. The MPEG-2 slice structure is shown in Figure 9.4. This should be compared to the MPEG-1 slice structure in Figure 2.7.

Each macroblock has a header, followed by data for the coded DCT blocks. For the required 4:2:0 chroma format up to six blocks follow the header. For the 4:2:2 or 4:4:4 optional formats an additional two or six more chrominance blocks may be coded.

The bottom blowup shows a coded block. If it is an intra-coded block, the differential DC coefficient is coded first. Then the rest of the coefficients are coded as runs and levels until the end_of_block terminates the variable-length codes for that block.

9.12   Differences between MPEG-1 and MPEG-2 video ◑

All currently defined MPEG-2 profiles and levels require decoders to handle MPEG-1 constrained parameters bitstreams within their level capabilities. For most data elements, MPEG-2 is a direct superset of MPEG-1. However, some MPEG-1 data elements do not have a direct equivalent in MPEG-2. This section documents those elements.[5]

The IDCT mismatch is handled differently in MPEG-1 and MPEG-2. MPEG-1 makes each nonzero coefficient odd after inverse quantization. MPEG-2 adjusts only one coefficient (the highest vertical and horizontal frequency) if the sum of the inverse quantized coefficients is even.

MPEG-1 pictures have the chrominance samples horizontally and vertically positioned in the center of a group of four luminance samples. MPEG-2 co-locates the chrominance samples on luminance samples.

MPEG-1 always has a fixed picture rate. MPEG-2 has a low-delay mode in which a "big picture" can take more than the normal single picture time inferred from picture_rate.

MPEG-1 increments the temporal reference by one (modulo 1024) for each coded picture. The group of pictures header resets it to zero. MPEG-2 follows the same rules except for big pictures in low-delay operation.

MPEG-1 codes the four-bit pel_aspect_ratio in the sequence header. MPEG-2 specifies the display aspect ratio in the data element aspect_ratio_information. The frame size and display size are used to derive the pel aspect ratio.

A MPEG-1 slice can cross macroblock row boundaries. Consequently, a single slice can cover the entire picture. MPEG-2 slices begin and end in the same macroblock row. The standard does not require the MPEG-2 slices to cover the entire picture. However, all profiles do require this.

MPEG-2 allows both half-pel and full pel motion vector units. MPEG-2 always uses half-pel motion vector units. MPEG-1 codes the f_code values for the motion vectors in the forward_f_code and backward_f_code data elements in the picture header. MPEG-2 sets these data elements to '111'

---

[5]This section is based on clause D.9 in ISO/IEC 13818-2:1995 which lists the differences between MPEG-2 and MPEG-1 video.

and used the picture extension data elements `f_code[s][t]` to communicate this information to the decoder.

MPEG-1 allows use of the 11-bit '00000001111' `macroblock_stuffing` VLC code as often as desired before each `macroblock_address_increment`. This code is not used in MPEG-2 and stuffing can only be obtained by stuffing extra zero bytes before start codes.

MPEG-1 codes run-level values that do not have a preassigned VLC code with the six-bit escape code '000010' followed by fixed length codes of either 14 bits for levels from $-127 \leq$ `level` ≤ 127 or 22 bits for levels from $-255 \leq$ `level` ≤ 255. MPEG-2 uses the same six-bit escape code followed by 18-bit fixed length codes for levels from $-2047 \leq$ `level` ≤ 2047.

MPEG-1 allows D-pictures (`picture_coding_type=4`). MPEG-2 does not allow D-pictures except within MPEG-1 constrained parameter bitstreams.

MPEG-1 constrained parameters bitstreams can have a `horizontal_size` of up to 768. MPEG-2 main profile limits the horizontal size to 720 pels.

MPEG-1 `bit_rate` and `vbv_delay` are set to all '1's (18 and 16 bits respectively) to indicate variable bitrate. All other values imply constant bitrate. MPEG-2 may still set `vbv_delay` to all '1's, but other values do not necessarily mean constant bitrate. Constant bitrate is considered a special case of variable bitrate. The value in `bit_rate` is only an upper bound on the actual bitrate and may not be the actual bitrate. Computations on all the values of `vbv_delay` are needed to determine if the operation is constant bitrate.

MPEG-1 only defines the video buffering verifier (VBV) for constant bitrate conditions. The system target decoder (STD) is used for variable bitrate conditions. MPEG-2 only defines the VBV for variable bitrate conditions and then regards constant bitrate as a special case.

10

MPEG-2 Main Profile Syntax

MPEG-2 video syntax for the main profile is a superset of the MPEG-1 video syntax. This nonscalable profile is aimed at higher bit rates than MPEG-1 and is able to handle interlaced video formats.[1] In the discussions in this chapter the data element ranges are restricted to main profile at main level (MP@ML). Higher MPEG-2 levels allow a broader range of values, but have no other effect on the syntax.

10.1 MPEG-2 start codes ◑

MPEG-2 uses exactly the same video start codes as MPEG-1 in its video sequences (see Table 8.1). If the `sequence_extension` (signaled by the `extension_start_code`) occurs immediately after the sequence header, the sequence is an MPEG-2 video sequence. This sequence extension is not allowed in MPEG-1 and, therefore, is a means by which a decoder can differentiate between MPEG-1 and MPEG-2 video bitstreams. MPEG-2 also requires a picture extension immediately after the picture header and defines a number of optional sequence and picture extensions.

The parameter ranges allowed in MPEG-2 sometimes exceed the precision defined in the MPEG-1 syntax. Parameter range expansion is usually accomplished in MPEG-2 by defining the MPEG-1 data element to be the low-order bits and using an extension to communicate the high-order bits.

```
video_sequence() {                  /* from ISO 13818-2 6.2.2     */
  next_start_code();                /* find next start code       */
  sequence_header();                /* r/w sequence header        */
  if (nextbits(32)==extension_start_code) {
                                    /* if 0x000001B5, MPEG-2      */
    sequence_extension();           /* r/w sequence extension     */
    do {                            /* do sequences               */
      extension_and_user_data_V();  /* r/w seq. ext. & user data  */
      do {                          /* do pictures                */
        if (nextbits(32)==group_start_code) {
                                    /* if 0x000001B8              */
          group_of_pictures_header(); /*   r/w GOP header         */
          extension_and_user_data_G();/*   r/w GOP user data      */
        }                           /* end if 0x000001B8          */
        picture_header();           /* r/w picture header         */
        picture_coding_extension();/* r/w picture coding extension*/
        extension_and_user_data_P();/* r/w picture ext.&user data */
        picture_data();             /* r/w picture data           */
      } while ((nextbits(32)==picture_start_code) ||
              (nextbits(32)==group_start_code))
                    /* while 0x00000100 or 0x000001B8, do pictures */
      if (nextbits(32)!=sequence_end_code){ */if not 0x000001B7   */
        sequence_header();          /*   r/w sequence header      */
        sequence_extension();       /*   r/w sequence extension   */
      }                             /* end if not 0x000001B7      */
    }while (nextbits(32)!=sequence_end_code)
                            /*while not 0x000001B7, do sequences  */
  } else {                          /* else not MPEG-2            */
    /* ISO/IEC 11172-2  */          /*   MPEG-1 bitstream         */
  }                                 /* end else not MPEG-2        */
  sequence_end_code(32);            /* r/w 0x000001B7             */
}                                   /* end video_sequence()       */
```

Figure 10.1: MPEG-2 `video_sequence()` function.

10.2   MPEG-2 video sequence ●

The pseudocode in Figure 10.1 defines the syntax for an MPEG-2 video sequence. In an MPEG-2 bitstream if the first sequence header is followed immediately by the `extension_start_code`, (0x000001B5), then every sequence header must be followed by the sequence extension. Optional extension and user data follow after the sequence extension.[2]

Unlike MPEG-1, no GOP headers are required. If the `group_start_code` occurs, it signals a GOP header that may be followed by optional user data. (No extension data is defined in the `extension_and_user_data_G()` function.) At least one picture must follow the sequence header and the optional GOP header. After any GOP header, the first picture is required to be a coded I-picture. After a sequence header the first picture must be a coded I- or P-picture.[3] Of course, if the first picture in a video sequence has nonintra macroblocks, it can only be partially decoded. In this case, display refresh may take several coded pictures.

Picture headers, picture coding extensions, and optional `extension_and_user_data_P()` precede picture data. When all sequence headers, GOPs, and pictures have been coded, the sequence ends with the `sequence_end_code` as in MPEG-1, (0x000001B7).

If the first sequence header is not also followed by the `extension_start_code`, then the rest of the video sequence bitstream is governed by MPEG-1 (ISO/IEC 11172-2) rules; this is indicated by the comment,

```
/* ISO/IEC 11172-2 */
```

in Figure 10.1.[4]

10.2.1   MPEG-2 sequence header ●

The pseudocode for processing the MPEG-2 sequence header is given in Figure 10.2. The reader may want to compare it to the MPEG-1 sequence header in Figure 8.4. As in MPEG-1, if the sequence header is repeated in the video sequence, all values except those related to the quantization matrix remain the same.

---

[1]MPEG-2 system syntax has already been briefly described in Chapter 9.

[2]The standard's notation for the extension and user data function is `extension_and_user_data(i)` where i=0, 1, and 2 for video sequence, group of pictures, and picture extensions respectively. We have converted this to `extension_and_user_data_X()` where X is V, G, and P. (Our version of pseudocode usually uses the parameter field for the number of bits of a data element.)

[3]Thus, a decoder cannot assume that the data following a sequence header can be decoded independently of data preceding the sequence header.

[4]This convention is used in the MPEG-2 standard.

```
sequence_header(){                       /* from ISO 13818-2 6.2.2.1 */
  sequence_header_code(32);              /* r/w 0x000001B3            */
  horizontal_size_value(12);             /* r/w width (12 lsb)        */
  vertical_size_value(12);               /* r/w height (12 lsb)       */
  aspect_ratio_information(4);           /* r/w sample aspect ratio   */
  frame_rate_code(4);                    /* r/w frame rate            */
  bit_rate_value(18);                    /* r/w bit rate (18 lsb)     */
  marker_bit(1);                         /* r/w '1'                   */
  vbv_buffer_size_value(10);             /* r/w VBV buf.size (10 lsb) */
  constrained_parameter_flag(1);         /* r/w '1' if MPEG-1 constr. */
                                         /* r/w '0' if MPEG-2         */
  load_intra_quantiser_matrix(1);        /* r/w flag intra Q matrix   */
  if (load_intra_quantiser_matrix)       /* if flag set               */
    intra_quantiser_matrix[1..64](8);    /*   r/w 64 8-bit values     */
  load_non_intra_quantiser_matrix(1);    /* r/w flag nonintra Q mat.  */
  if (load_non_intra_quantiser_matrix)   /* if flag set               */
    non_intra_quantiser_matrix[1..64](8);/* r/w 64 8-bit values       */
  next_start_code();                     /* find next start code      */
}                                        /* end sequence_header()     */
```

Figure 10.2: MPEG-2 sequence_header() function.

The sequence_header_code (0x000001B3) is the same as in MPEG-1. The next pair of data elements, horizontal_size_value and vertical_size_value, are examples of _value being appended to a data element name. It always means that these bits are the least significant bits (lsb) of an MPEG-1 parameter whose range has been extended. In the required extension that comes later is a data element with the same name root, but with _extension appended. This contains the most significant bits (msb) of the parameter.

Equation 10.1 shows how to calculate a data element with 14-bit precision by shifting the _extension left by the number of bits in _value (12) before adding it to the _value term.

$$\text{horizontal\_size} = \ (\text{horizontal\_size\_extension} << 12)$$
$$+ \text{horizontal\_size\_value} \qquad (10.1)$$

In this example horizontal_size_value is found in the sequence header and horizontal_size_extension is found in the mandatory sequence_extension() function.

In MPEG-1 start code emulation is avoided by not allowing horizontal_size or vertical_size to be zero. These data elements also cannot be zero in MPEG-2; therefore, horizontal_size and vertical_size cannot be multiples of 4096.

display aspect ratio (V:H)	sampling aspect ratio (H:V)	MP	aspect_ratio_information
forbidden	forbidden	.	0000
–	1:1 (square pels)	●	0001
3:4	–	●	0010
9:16	–	●	0011
1:2.21	–	.	0100
reserved	–	.	0101
...	...	.	...
reserved	–	.	1111

Table 10.1: MPEG-2 aspect_ratio_information codes.

The next parameter in the sequence header is aspect_ratio_information. This parameter is much like the MPEG-1 pel_aspect_ratio, but has two interpretations, depending on whether the value applies to the entire display or a section of the display. If the sequence_display_extension is not present in the bitstream, the value applies to the entire display.

Table 10.1 shows the four aspect_ratio_information codes currently defined for the aspect ratio. These apply to either the sample aspect ratio (SAR) of the source or the display aspect ratio (DAR). Square pels (SAR=1:1) are common in computer display monitors. The 3:4 display aspect ratio is used in traditional broadcast TV monitors. The 9:16 display aspect ratio is expected to be the HDTV standard. The 1:2.21 display ratio is a compromise aspect ratio for the 70mm CinemaScope and PanaVision formats used for widescreen movies and is not supported in the main profile (MP).

If the sequence_display_extension() function does not follow the sequence extension, then the reconstructed picture fills the active region of the display. Equation 10.2 gives the calculation for SAR from the DAR values:

$$SAR = \frac{DAR \times \text{horizontal\_size}}{\text{vertical\_size}} \tag{10.2}$$

Otherwise, Equation 10.6 defines SAR in terms of information from the sequence_display_extension() function. The relationship between SAR and DAR is discussed further in Section 10.2.5.

The four-bit frame_rate_code specifies a frame_rate_value (picture_rate in MPEG-1 terminology) from the list in Table 10.2. The solid circles indicate which values are valid for MP@ML.

A 12-bit bit_rate_extension shifted left 18 bits and added to the 18-bit bit_rate_value define the bit_rate measured in units of 400 bits/s, rounded up. The bit_rate is the upper bound of the compressed data

nominal frame rate	frame_rate_value	MP@ML	frame_rate_code
−	forbidden	.	0000
23.976	24÷1.001	●	0001
24	24	●	0010
25	25	●	0011
29.97	30÷1.001	●	0100
30	30	●	0101
50	50	.	0110
59.94	60÷1.001	.	0111
60	60	.	1000
−	reserved	.	1001
−	...	.	...
−	reserved	.	1111

Table 10.2: MPEG-2 frame_rate_code codes.

delivered to the input to the VBV. Zero is forbidden for bit_rate, but not for bit_rate_value since the preceding frame_rate_code has at most three trailing zeros out of four bits and the following marker_bit prevent start code emulation.

vbv_buffer_size is determined from the 10-bit vbv_buffer_size_value and eight msb from vbv_buffer_size_extension.

The constrained_parameters_flag is always '0' for ISO/IEC 13818-2 bitstreams.

The sequence header resets all quantization matrices back to the default values discussed in Chapter 5. User-defined quantization matrices can be downloaded in the sequence_header() or a quant_matrix_extension().

10.2.2   MPEG-2 sequence extension ●

Figure 10.3 gives the pseudocode for the sequence_extension() function. If the sequence header is repeated in a video sequence, the sequence extension is also repeated with all values unchanged. Many of its data elements are extensions of the data elements found in the sequence header and will not be discussed here. All extensions start with a 32-bit extension_start_code (0x000001B5). MPEG-2 extension start codes are followed by one of the four-bit extension_start_code_identifier listed in Table 10.3. The identifiers allowed in the main profile (MP) at any level are marked by a solid circle. If a reserved extension ID is decoded, the decoder is supposed to discard data until the next start code is reached; this allows it to decode bitstreams with compatible extensions that may be defined in the future.

MPEG-2 profiles and levels are discussed in Section 9.9 and 9.10. The

```
sequence_extension() {                      /* from ISO 13818-2 6.2.2.3     */
  extension_start_code(32);                 /* r/w 0x000001B5               */
  extension_start_code_identifier(4);       /* r/w '0001'                   */
  profile_and_level_indication(8);          /* r/w profile & level IDs      */
  progressive_sequence(1);                  /* r/w 1 if progressive else 0  */
  chroma_format(2);                         /* r/w '01' (4:2:0)             */
  horizontal_size_extension(2);             /* r/w width (2 msb)            */
  vertical_size_extension(2);               /* r/w height (2 msb)           */
  bit_rate_extension(12);                   /* r/w bit rate (12 msb)        */
  marker_bit(1);                            /* r/w '1'                      */
  vbv_buffer_size_extension(8);             /* r/w VBV buf. size (8 msb)    */
  low_delay(1);                             /* r/w 1 if low delay else 0    */
  frame_rate_extension_n(2);                /* r/w '00'                     */
  frame_rate_extension_d(5);                /* r/w '00000'                  */
  next_start_code();                        /* find next start code         */
}                                           /* end sequence_extension()     */
```

Figure 10.3: MPEG-2 sequence_extension() function.

Name	MP	extension_start_code_identifier
reserved	.	0000
Sequence extension ID	●	0001
Sequence display extension ID	●	0010
Quant matrix extension ID	●	0011
Copyright extension ID	●	0100
Sequence scalable extension ID	.	0101
reserved	.	0110
Picture display extension ID	●	0111
Picture coding extension ID	●	1000
Picture spatial scalable extension ID	.	1001
Picture temporal scalable extension ID	.	1010
reserved	.	1011
...	.	...
reserved	.	1111

Table 10.3: MPEG-2 extension_start_code_identifier codes.

Profile	MP	Profile identification codes
Reserved	.	000
High	.	001
Spatially scalable	.	010
SNR scalable	.	011
Main	●	100
Simple	●	101
Reserved	.	110
Reserved	.	111

Table 10.4: MPEG-2 profile identification.

Level	SP	MP	Level identification codes
Reserved	.	.	0000
...	.	.	...
Reserved	.	.	0011
High	.	●	0100
Reserved	.	.	0101
High 1440	.	●	0110
Reserved	.	.	0111
Main	●	●	1000
Reserved	.	.	1001
Low	●	●	1010
Reserved	.	.	1011
...	.	.	...
Reserved	.	.	1111

Table 10.5: MPEG-2 level identification.

profile and level of the bitstream are specified by the eight-bit **profile_and_level_indication** data element. This data element consists of three fields. The most significant bit is an escape bit, the next three bits give the profile (see Table 10.4, and the four least significant bits specify the level (see Table 10.5).

Decoders for a given profile and level must be able to decode bitstreams at the same level and any profile below them in Table 10.4. Similarly, they must be able to decode bitstreams that use the same profile and a lower level in Table 10.5. The one exception to these rules is that simple profile decoders at main level (SP@ML) must also be able to decode MP@LL bitstreams.

Undefined profiles and levels are reserved for future extensions to new profiles and levels. Future extensions that use an escape bit of '1' are not guaranteed to follow this subset decoding requirement.

format	blocks in macroblock	block_count	MP	chroma_format codes
reserved	–		.	00
4:2:0	Y1Y2 Cb5 Cr6 Y3Y4	6	●	01
4:2:2	Y1Y2 Cb5 Cr6 Y3Y4 Cb7 Cr8	8	.	10
4:4:4	Y1Y2 Cb5Cb9 Cr6Cr11 Y3Y4 Cb7Cb10 Cr8Cr12	12	.	11

Table 10.6: MPEG-2 chroma_format codes.

The progressive_sequence flag is set to one when the video sequence only contains progressive frame pictures. When it is zero, the video sequence may have both coded field pictures and frame pictures and the frame pictures may be interlaced or progressive.

Table 10.6 shows three defined chroma formats with an illustration using luminance and chrominance blocks. Only the 4:2:0 format is required for the main profile. The numbers indicate the order in which the blocks are coded. block_count is used in the macroblock() function to know how many blocks need to be coded.

The low_delay flag is set to '1' when the coded sequence does not include any B-pictures, the VBV description does not include frame reordering delay, and where big pictures can occur that take more than one normal picture time. In low delay operation the VBV buffer may underflow if the idealized decoder model is rigorously followed. When big pictures occur, pictures must be skipped and the temporal reference is incremented to reflect this skipping.

Low_delay must be set to '0' if B-pictures are coded or the frame reordering delay is included in the VBV description. No big pictures are allowed if this flag is '0'.

In the future, frame_rate_extension_n and frame_rate_extension_d will be used to calculate the frame rate. However, in all currently defined profiles they must be set to zero.[5]

10.2.3   MPEG-2 optional sequence extensions ●

MPEG-2 allows optional extension and user data to follow the video sequence, GOP, and picture headers and required extensions. Figure 10.4 shows a modified version of the pseudocode that can follow the video se-

---
[5]The MPEG-2 standard has more details.

```
extension_and_user_data_V() {         /* from ISO 13818-2 6.2.2.2  */
  while((nextbits(32)==extension_start_code) ||
        (nextbits(32)==user_data_start_code)) {
                                 /*while 0x000001B5 or 0x000001B2 */
    while (nextbits(32)==extension_start_code) {
                                 /* while 0x000001B5          */
      extension_start_code(32);      /* r/w 0x000001B5          */
      if (nextbits(4)=='‘Sequence display extension ID'’)
                                 /* if ‘0010’                */
        sequence_display_extension();/* r/w sequence display ext. */
      else                           /* else scalable extensions */
        ...;                         /* see standard            */
    }                                /* end while 0x000001B5    */
    if (nextbits(32)==user_data_start_code) /* if 0x000001B2    */
      user_data();                   /* r/w user data          */
  }                                  /* end while 0x000001B5 or...*/
}                                    /* end ext._and_user_data_V()*/
```

Figure 10.4: MPEG-2 extension_and user_data_V() function. If present, this follows the video sequence extension.

quence extension. The tag ending of this function, _V() indicates that the data follows the video sequence header.[6]

Extension and user data always start with either a extension_start_code (0x000001B5) or a user_data_start_code (0x000001B2). Referring to Figure 10.4, the only nonscalable extension defined by MPEG-2 is the sequence_display_extension(); the others are scalable and are indicated by the statement ...; in Figure 10.4.

10.2.4   MPEG-2 user data ●

Figure 10.5 shows the generic user_data() function. Since user data does not effect decoding, decoders can simply discard the user data bytes. To avoid accidental start code emulation, user data should not contain contiguous strings of 23 or more zero bits.

10.2.5   MPEG-2 sequence display extension ●

Figure 10.6 lists the sequence_display_extension() function. Since it affects display but not decodability, decoders may ignore the information in this extension. If the sequence display extension occurs after the first sequence header, MPEG-2 requires that it occur with unchanged values

---

[6]In ISO 13818-2 this is selected by function argument $i=0$.

```
user_data() {                        /* from ISO 13818-2 6.2.2.2.2  */
  user_start_code(32);               /* r/w 0x000001B2              */
  while (nextbits(24)!=0x000001 {    /* while not start code prefix */
    user_data(8);                    /* r/w a byte of user data     */
  }                                  /* end while not start code ...*/
  next_start_code();                 /* find next start code        */
}                                    /* end user_data()             */
```

Figure 10.5: MPEG-2 user_data() function.

```
sequence_display_extension() {       /* from ISO 13818-2 6.2.2.4     */
  extension_start_code_identifier(4); /* r/w '0010'                  */
  video_format(3);                   /* r/w video format            */
  colour_description(1);             /* r/w '1' if colors next       */
                                     /* r/w '0' if skip colors       */

  if (colour_description) {          /* if colors described next     */
    colour_primaries(8);             /*   source primaries           */
    transfer_characteristics(8);     /*   source transfer charact.   */
    matrix_coefficients(8);          /*   source matrix coefficients */
  }                                  /* end if colors next           */
  display_horizontal_size(14);       /* r/w display width            */
  marker_bit(1);                     /* r/w '1'                      */
  display_vertical_size(14);         /* r/w display height           */
  next_start_code();                 /* find next start code         */
}                                    /* end sequence_display_ext.()  */
```

Figure 10.6: MPEG-2 sequence_display_extension() function.

format	MP	video_format
component	●	000
PAL	●	001
NTSC	●	010
SECAM	●	011
MAC	●	100
Unspecified video format	●	101 (default)
reserved	.	110
reserved	.	111

Table 10.7: MPEG-2 video_format codes.

after every sequence header in the sequence. After the four-bit identifier ('0010'), one of six video formats can be specified as shown in Table 10.7. These are the source material formats prior to conversion to the YCbCr format used in the encoder.[7]

If detailed color primaries, transfer characteristics, and matrix coefficients are desired, the `colour_description` flag is set to '1'.

The chromaticity coordinates for the source primaries are given in Table 10.8. If `colour_primaries` are not explicitly set, then a default value of 1 is assumed. Additional information about color primaries and chromaticity coordinates can be found in Chapter 4.

The transfer characteristics are given in Table 10.9. A tutorial on color spaces and gamma correction is found in Chapter 4.

Table 10.10 lists the codes for the `matrix_coefficients`. In this table E'_Y, E'_R, E'_G, and E'_B range from 0 to 1. E'_{PB} and E'_{PR} range from -0.5 to $+0.5$. The formulas deriving the Y, Cb, and Cr are given below:

$$Y = (219E'_Y) + 16 \tag{10.3}$$

$$Cb = (224E'_{PB}) + 128 \tag{10.4}$$

$$Cr = (224E'_{PR}) + 128 \tag{10.5}$$

The display width and height can be sent in `display_horizontal_size` (in units of samples of the encoded frame) and `display_vertical_size` (in units of lines of the encoded frame). They may define a display size that is smaller or larger than the frame size. How these values are used is not standardized.

The relationships between SAR, DAR, and display size are as follows:

$$SAR = \frac{DAR \times \texttt{display\_horizontal\_size}}{\texttt{display\_vertical\_size}} \tag{10.6}$$

10.3   MPEG-2 GOP header ●

Figure 10.7 shows the syntax for the group of pictures header. The data elements have the same meaning as the MPEG-1 GOP (see Figure 8.7). The `closed_gop` flag is set to '1' if the pictures in the GOP do not depend on pictures outside the GOP. Thus, if B-pictures preceding the first I-picture in the GOP (in display order) depend only upon backward motion vectors, the `closed_gop` flag is set to '1'. It is set to '0' if forward motion vectors are used, because these vectors reference pictures outside the group.

The `broken_link` flag is always set to '0' in the encoder to indicate a conforming, decodable bitstream. The flag can be set to '1' in parts of a

---

[7]See Section 9.7 for background information about these video formats.

reference standard	primary	x	y	colour_primaries
forbidden				'0000 0000'
Rec. ITU-R BT.709	green	0.300	0.600	'0000 0001' (default)
	blue	0.150	0.060	
	red	0.640	0.330	
	white D65	0.3127	0.3290	
unspecified video	unknown character-istics			'0000 0010'
reserved				'0000 0011'
Rec. ITU-R BT.470-2 System M	green	0.21	0.71	'0000 0100'
	blue	0.14	0.08	
	red	0.67	0.33	
	white C	0.310	0.316	
Rec. ITU-R BT.470-2 System B, G	green	0.29	0.60	'0000 0101'
	blue	0.15	0.06	
	red	0.64	0.33	
	white D65	0.313	0.329	
SMPTE 170M	green	0.310	0.595	'0000 0110'
	blue	0.155	0.070	
	red	0.630	0.340	
	white D65	0.3127	0.3290	
SMPTE 240M (1987)	green	0.310	0.595	'0000 0111'
	blue	0.155	0.070	
	red	0.630	0.340	
	white D65	0.3127	0.3291	
reserved				'0000 1000'
...				...
reserved				'1111 1111'

Table 10.8: MPEG-2 colour_primaries codes.

```
group_of_pictures_header() {      /* from ISO 13818-2 6.2.2.6      */
   group_start_code(32);          /* r/w 0x000001B8                */
   time_code(25);                 /* r/w time code                */
   closed_gop(1);                 /* r/w nature of B-pictures      */
   broken_link(1);                /* r/w encoder: 0; editor: 0,1   */
   next_start_code();             /* find next start code          */
}                                 /* end group_of_pictures_header() */
```

Figure 10.7: MPEG-2 group_of_pictures_header() function.

reference standard	equations for transfer characteristics	transfer_ characteristics
forbidden		'0000 0000'
Rec. ITU-R BT.709	$V = (1.099L_c^0.45) - 0.099$ for $1 >= L_c >= 0.018$ $V = 4.500L_c for 0.018 > L_c$	'0000 0001' (default)
unspecified video	Image characteristics unknown	'0000 0010'
reserved		'0000 0011'
Rec. ITU-R BT.470-2 System M	Assumed display gamma 2.2	'0000 0100'
Rec. ITU-R BT.470-2 System B, G	Assumed display gamma 2.8	'0000 0101'
SMPTE 170M	$V = (1.099L_c^0.45) - 0.099$ for $1 >= L_c >= 0.018$ $V = 4.500L_c for 0.018 > L_c$	'0000 0110'
SMPTE 240M (1987)	$V = (1.1115L_c^0.45) - 0.1115$ for $1 >= L_c >= 0.0228$ $V = 4.0L_c for 0.0228 > L_c$	'0000 0111'
Linear	$V = L_c$	'0000 1000'
reserved ... reserved		'0000 1001' ... '1111 1111'

Table 10.9: MPEG-2 transfer_characteristics codes.

reference standard	equations to derive Y, Cb, and Cr from R, G, and B primaries	matrix_ coefficients
forbidden		'0000 0000'
Rec. ITU-R BT.709	$E'_Y = 0.7154E'_G + 0.0721E'_B + 0.2125E'_R$ $E'_{PB} = -0.386E'_G + 0.500E'_B - 0.155E'_R$ $E'_{PR} = -0.454E'_G + 0.046E'_B + 0.500E'_R$	'0000 0001' (default)
unspecified video	image characteristics are unknown	'0000 0010'
reserved		'0000 0011'
FCC	$E'_Y = 0.59E'_G + 0.11E'_B + 0.30E'_R$ $E'_{PB} = -0.331E'_G + 0.500E'_B - 0.169E'_R$ $E'_{PR} = -0.421E'_G + 0.079E'_B + 0.500E'_R$	'0000 0100'
Rec. ITU-R BT.470-2 System B, G	$E'_Y = 0.587E'_G + 0.114E'_B + 0.299E'_R$ $E'_{PB} = -0.331E'_G + 0.500E'_B - 0.169E'_R$ $E'_{PR} = -0.419E'_G + 0.081E'_B + 0.500E'_R$	'0000 0101'
SMPTE 170M	$E'_Y = 0.587E'_G + 0.114E'_B + 0.299E'_R$ $E'_{PB} = -0.331E'_G + 0.500E'_B - 0.169E'_R$ $E'_{PR} = -0.419E'_G + 0.081E'_B + 0.500E'_R$	'0000 0110'
SMPTE 240M (1987)	$E'_Y = 0.701E'_G + 0.087E'_B + 0.212E'_R$ $E'_{PB} = -0.384E'_G + 0.500E'_B - 0.116E'_R$ $E'_{PR} = -0.445E'_G + 0.055E'_B + 0.500E'_R$	'0000 0111'
reserved		'0000 1000'
...		...
reserved		'1111 1111'

Table 10.10: MPEG-2 matrix_coefficients codes.

```
extension_and_user_data_G() {           /* from ISO 13818-2 6.2.2.2  */
  while((nextbits(32)==extension_start_code) ||
        (nextbits(32)==user_data_start_code)) {
                                /*while 0x000001B5 or 0x000001B2 */
    if (nextbits(32)==user_data_start_code) /* if 0x000001B2      */
      user_data();                      /* r/w user data          */
  }                                     /* end while 0x000001B5 or...*/
}                                       /* end ext._and_user_data_G()*/
```

Figure 10.8: MPEG-2 extension_and user_data_G() function. If present, this follows the group of pictures header.

bitstream where an editor has removed or inserted data that make those parts undecodable. Since only 28 bits follow the group_start_code, the next_start_code() function must pad at least four bits.

In the current MPEG-2 profiles, the only extension and user data allowed after the GOP header is user data. This is shown in Figure 10.8.

10.4 MPEG-2 picture header and extensions ●

The syntax for the picture header is shown in Figure 10.9. It is quite similar to the header part of the MPEG-1 picture() function (see Figure 8.9). However, there are significant new restrictions and conventions. After the picture_start_code (0x00000100), the temporal_reference gives the display count, modulus 1024.

When low_delay is set to '0', MPEG-1 conventions still apply. The temporal reference is incremented by one for each frame in display order, and after a GOP header it is reset to zero. Both fields in a frame receive the same temporal reference number.

If the low_delay is set to '1' and a big picture occurs such that the picture is not complete in one picture time interval, the VBV buffer is reexamined N times ($N > 0$) and the temporal reference is incremented by $N + 1$ (modulo 1024).[8] If the big picture is followed immediately by a GOP header, then the temporal reference of the first coded picture in the GOP is set to N rather than zero. If the big picture is the first field of a frame, then the first field takes at least one picture time interval and the second field will not have the same temporal reference as the first field.

The picture_coding_type uses the same codes as MPEG-1 (see Table 8.5 except that D-pictures (code '100') "shall not be used." The 16-bit

---

[8]If $N=0$, the buffer is not reexamined. This means that the picture is removed and decoded normally after one picture time interval.

```
picture_header() {              /* from ISO 13818-2 6.2.3     */
  picture_start_code(32);       /* r/w 0x00000100            */
  temporal_reference(10);       /* r/w display count mod 1024 */
  picture_coding_type(3);       /* r/w field/frame I/P/B type */
  vbv_delay(16);                /* r/w VBV delay             */
  if (picture_coding_type==2)||(picture_coding_type==3) {
                                /*if P- or B-picture         */
    full_pel_forward_vector(1); /*   r/w '0' for MPEG-2      */
    forward_f_code(3);          /*   r/w '111' for MPEG-2    */
  }                             /* end if P-or B-picture     */
  if (picture_coding_type==3) { /* if B-picture              */
    full_pel_backward_vector(1);/*   r/w '0' for MPEG-2      */
    backward_f_code(3);         /*   r/w '111' for MPEG-2    */
  }                             /* end if B-picture          */
  while (nextbits(1)=='1') {    /* while extra picture info.  */
    extra_bit_picture(1);       /* r/w '1'                   */
    extra_information_picture(8); /* r/w byte of extra pic. info. */
  }                             /* end while extra picture info.*/
  extra_bit_picture(1);         /* r/w '0' to end extra info.  */
}                               /* end picture_header()      */
```

Figure 10.9: MPEG-2 `picture_header()` function.

vbv_delay is set to 0xFFFF for variable bit rate coding. If any vbv_delay
in a sequence is 0xFFFF, then all of them should be set to 0xFFFF.

The full_pel_forward_vector, forward_f_code, full_pel_backward_vector and backward_f_code are not used in MPEG-2. They are set to '0', '111', '0', and '111', respectively. Motion vector parameters are coded in the picture_coding_extension.

The syntax also defines a extra_bit_picture flag. If set to '1', a byte of picture information follows. However, since extra_information_picture data element is reserved for future revisions, extra_bit_picture should be set to '0'.

10.4.1   MPEG-2 picture coding extension ●

Figure 10.10 gives the pseudocode for the picture coding extension that is required immediately following the picture header in an MPEG-2 sequence. This extension contains MPEG-2 motion vector parameters, DC coefficient precision, field information, quantizer table selection, VLC table selection, zigzag scan pattern selection, and other parameters. It also may have optional information about a composite source picture.

After the extension_start_code (0x000001B5) comes the four-bit ex-

```
picture_coding_extension() { /* from ISO 13818-2 6.2.3.1          */
  extension_start_code(32);   /* r/w 0x000001B5                   */
  extension_start_code_identifier(4); /* r/w '1000'               */
  f_code[0][0](4);             /* r/w forward horizontal f_code    */
  f_code[0][1](4);             /* r/w forward vertical f_code      */
  f_code[1][0](4);             /* r/w backward horizontal f_code   */
  f_code[1][1](4);             /* r/w backward vertical f_code     */
  intra_dc_precision(2);       /* r/w DC precision for I-blocks    */
  picture_structure(2);        /* r/w top/bottom field or frame    */
  top_field_first(1);          /* r/w flag                         */
  frame_pred_frame_dct(1);     /* r/w '1' if frame pred/dct else '0'*/
  concealment_motion_vectors(1); /* r/w '1' if intra MV else '0'   */
  q_scale_type(1);             /* r/w '0' if q_scale[0] else '1'   */
  intra_vlc_format(1);         /* r/w '0' if DCT coef.Tbl.0 else '1'*/
  alternate_scan(1);           /* r/w '0' if zigzag scan[0] else '1'*/
  repeat_first_field(1);       /* r/w flag                         */
  chroma_420_type(1);          /* r/w flag                         */
  progressive_frame(1);        /* r/w '1' if prog. frame else '0'  */
  composite_display_flag(1);   /* r/w flag                         */
  if (composite_display_flag) { /* if composite video display      */
    v_axis(1);                 /* r/w v_axis info                  */
    field_sequence(3);         /* r/w field_sequence               */
    sub_carrier(1);            /* r/w sub_carrier                  */
    burst_amplitude(7);        /* r/w burst_amplitude              */
    sub_carrier_phase(8);      /* r/w sub_carrier_phase            */
  }                            /* end if composite video display   */
  next_start_code();           /* find next start code             */
}                              /* end picture_coding_extension()   */
```

Figure 10.10: MPEG-2 picture_coding_extension() function.

DC precision (bits)	reset value	DC range allowed	intra_dc_mult	MP	intra_dc_precision codes
8	128	0...255	8	•	00
9	256	0...511	9	•	01
10	512	0...1023	10	•	10
11	1024	0...2047	11	.	11

Table 10.11: MPEG-2 intra_dc_precision codes.

Picture structure	Abbreviation	MP	picture_structure
reserved	Res.	.	00
top field	T	•	01
bottom field	B	•	10
frame picture	frame	•	11

Table 10.12: MPEG-2 picture_structure codes.

tension_start_code_identifier ('1000'), followed by the f_codes. The f_code[0][0], f_code[0][1], f_code[1][0], and f_code[1][1] set the forward horizontal f_code, forward vertical f_code, backward horizontal f_code, and backward vertical f_code, respectively.

Intra_dc_precision selects a value for the DC precision of intra-coded blocks. Table 10.11 shows the four precision choices, the reset value for each precision, the full allowed range, the multipilier intra_dc_mult to dequantize intra DC coefficients, and the intra_dc_precision codes. Note that simple profile and main profile restrict the DC precision.

The picture_structure is shown in Table 10.12. The top field contains the top line in a frame. The bottom field contains the second and bottom line in a frame. The picture can also be coded as a complete frame. Since field pictures and frame pictures can be mixed in an interlaced video sequence, the top_field_first flag specifies the order. For progressive sequences, all pictures are coded frames.

Frame_pred_frame_dct is set to '1' if only frame prediction and frame DCTs are used. It is '0' for field pictures and '1' if progressive_frame is one. Table 10.13 shows the valid combinations of this data element with others related data elements.

Concealment_motion_vectors is set to '1' if motion vectors are coded with intra blocks. It is set to '0' if no motion vectors are coded with intra blocks.

The q_scale_type bit determines if quantiser_scale[0] or quantiser_scale[1] values in Table 10.18 are used to interpret quantiser_scale_code.

The intra_vlc_format bit selects whether Table 0 or Table 1 shown in Table 10.26 is used for encoding and decoding DCT coefficients.

Since MPEG-2 is intended for higher bit rates and allows field as well as frame compression, some assumptions about the best zigzag scanning order no longer hold true. If two fields are frozen at different instances of time with moving objects in them, intra-coded blocks containing both fields in the moving areas are likely to exhibit high vertical frequencies. At the higher bit rates, these frequencies are less likely to quantize to zero. Figure 10.11 shows the two zigzag scanning orders allowed in MPEG-2. Figure 10.11(a) is identical to the MPEG-1 scanning order. Figure 10.11(b) shows an alternative choice which picks up higher vertical frequencies sooner than the same horizontal frequencies. Figure 10.12 illustrates this bias toward vertical frequencies and should be compared with the uniform diagonal structure in Figure 2.10.

The alternate_scan bit selects scan[0] or scan[1] for the zigzag scan order of the DCT coefficients. The quantization matrices are downloaded in zigzag scanning order, but always in the scan[0] order used in MPEG-1.

The repeat_first_field bit is set to '0' for field pictures. For frame pictures its interpretation depends upon the values of progressive_sequence, progressive_frame and top_field_first. For the 2:3 pull-down technique used to convert 24 picture/s movies to 30 frame/s video, a frame may be displayed as three fields. Table 10.13 lists the various combinations of these parameters. Table 10.14 shows where the rules for combining these parameters were stated.

Table 10.15 lists the additional constraints placed on the usage of the data element repeat_first_field for simple and main profiles. The column MP@ML identifies restrictions for main profile at main level and simple profile at main level.[9]

The chroma_420_type bit is set equal to progressive_frame for 4:2:0 format. Otherwise, it is set to '0'.

When progressive_frame bit is '0', the two fields are interlaced and repeat_first_field is set to '0'. If progressive_frame is set to '1', the two fields are from a single progressive video frame and there is no time period between them. picture_structure must be a frame and frame_pred_frame_dct is '1'. Table 10.13 shows how this parameters combines with the other field and frame data elements.

The composite_display_flag bit indicates whether some composite video data elements are next in the bitstream. These elements are not used by the decoder. The composite display information elements can help

---

[9]See Section 10.4.4 for additional constraints for wide images.

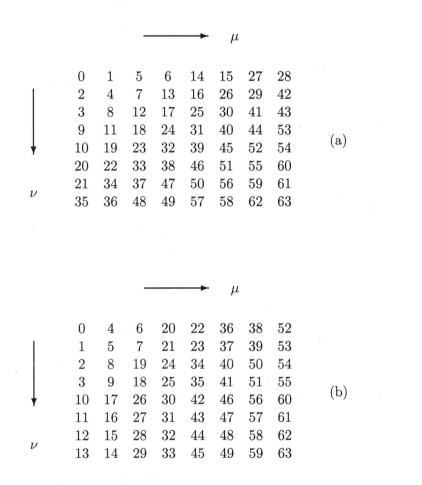

Figure 10.11: Zigzag ordering of DCT coefficients. (a) $\text{Scan}[0][\nu][\mu]$. This scanning order is also used in MPEG-1. (b) $\text{Scan}[1][\nu][\mu]$. This is an additional MPEG-2 choice.

pro-gres-sive_seq.	pro-gres-sive_frame	frame_pred_frame_dct	picture_structure	top_field_first	repeat_first_field	MP	decoder display outputs
0	0	0	00 Res.	x	x	.	reserved
0	0	0	01 T	0	0	●	T (1 field only)
0	0	0	01 T	0	1	.	illegal
0	0	0	01 T	1	x	.	illegal
0	0	0	10 B	0	0	●	B (1 field only)
0	0	0	10 B	0	1	.	illegal
0	0	0	10 B	1	x	.	illegal
0	0	0	11 Frame	0	0	●	BT (2 fields)
0	0	0	11 Frame	0	1	.	illegal
0	0	0	11 Frame	1	0	●	TB (2 fields)
0	0	0	11 Frame	1	1	.	illegal
0	0	1	00 Res.	x	x	.	reserved
0	0	1	01 T	x	x	.	illegal
0	0	1	10 B	x	x	.	illegal
0	0	1	11 Frame	0	0	●	BT (2 fields)
0	0	1	11 Frame	0	1	.	illegal
0	0	1	11 Frame	1	0	●	TB (2 fields)
0	0	1	11 Frame	1	1	.	illegal
0	1	0	xx	x	x	.	illegal
0	1	1	00 Res.	x	x	.	reserved
0	1	1	01 T	x	x	.	illegal
0	1	1	10 B	x	x	.	illegal
0	1	1	11 Frame	0	0	●	BT (2 fields)
0	1	1	11 Frame	0	1	●	BTB (3 fields)
0	1	1	11 Frame	1	0	●	TB (2 fields)
0	1	1	11 Frame	1	1	●	TBT (3 fields)
1	0	x	xx	x	x	.	illegal
1	1	0	xx	x	x	.	illegal
1	1	1	00 Res.	x	x	.	reserved
1	1	1	01 T	x	x	.	illegal
1	1	1	10 B	x	x	.	illegal
1	1	1	11 Frame	0	0	●	1 frame
1	1	1	11 Frame	0	1	.	2 frames
1	1	1	11 Frame	1	0	.	illegal
1	1	1	11 Frame	1	1	.	3 frames

Table 10.13: Field and frame parameter combinations. "x" represents both '0' and '1' combinations.

progressive_sequence	progressive_frame	frame_pred_frame_dct	picture_structure	top_field_first	repeat_first_field	rule
0 r	0		field			OK
0 r	1		field			OK
0 r	0		frame			OK
0 r	1		frame			OK
1 r			field			illegal
1 r	0		frame			illegal
1 r	1		frame			OK
	0 r				0	OK
	0 r				1	illegal
	1 r		field			illegal
	1 r	0	frame			illegal
	1 r	1	frame			OK
		0 r	field			OK
		1 r	field			illegal
	1	0 r				illegal
	1	1 r				OK
			00 r			reserved
			01 r			Top field
			10 r			Bottom field
			11 r			Frame
0			field	0 r		1 field
0			field	1 r		illegal
0			frame	0 r		BT
0			frame	1 r		TB
1			frame	0 r	0	1 frame
1			frame	0 r	1	2 frames
1			frame	1 r	0	illegal
1			frame	1 r	1	3 frames
			field		0 r	OK
			field		1 r	illegal
0	0		frame		0 r	2 fields
0	0		frame		1 r	illegal
0	1		frame	0	0 r	BT (2 fields)
0	1		frame	1	0 r	TB (2 fields)
0	1		frame	0	1 r	BTB (3 fields)
0	1		frame	1	1 r	TBT (3 fields)
1			frame		0 r	1 frame
1			frame	0	1 r	2 frames
1			frame	1	1 r	3 frames

Table 10.14: Rules for field and frame parameter combinations. The "r" indicates under which data element description the rule can be found.

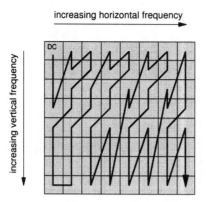

Figure 10.12: Zigzag scanning order of coefficients for MPEG-2 scan[1].

nominal frame rate	frame_ rate_ value	progres- sive_ sequence	MP@ ML	repeat_ first_ field
—	forbidden		.	
23.976	24÷1.001	0	●	0
23.976	24÷1.001	1	●	0
24	24	0	●	0
24	24	1	●	0
25	25	0	●	0,1
25	25	1	●	0
29.97	30÷1.001	0	●	0,1
29.97	30÷1.001	1	●	0
30	30	0	●	0,1
30	30	1	●	0
50	50	0	.	0,1
50	50	1	.	0
59.94	60÷1.001	0	.	0,1
59.94	60÷1.001	1	.	0,1
60	60	0	.	0,1
60	60	1	.	0,1
—	reserved		.	
—	...		.	
—	reserved		.	

Table 10.15: Constraints on repeat_first_field in MPEG-2 SP and MP.

field	frame	field_sequence code
1	1	000
2	1	001
3	2	010
4	2	011
5	3	100
6	3	101
7	4	110
8	4	111

Table 10.16: MPEG-2 field_sequence codes.

an MPEG decoder's postprocessing stage understand the nature of the original source video signal to synchronize the field phase of the MPEG decoder display with the composite source signal's original phase.

The v_axis bit is set to '1' if the signal before encoding came from a PAL system and had a positive sign. Otherwise it is set to '0'.[10]

The field_sequence records the field number in an eight field sequence from PAL or in a four-field sequence from NTSC systems. Table 10.16 gives the definitions.

The sub_carrier bit is set to '1' if the subcarrier/line frequency relationship is correct otherwise it is set to '0'. This is a composite signal attribute, and has nothing to do with the YCbCr component world of MPEG. In composite signals, color data is modulated on a subcarrier of the luminance carrier. The color channel is interleaved in such a way that it is spectrally orthogonal to the luminance carrier (like harmonics of a Fourier series are orthogonal to each other, yet in the sampled signal, they are summed together). The *Television Engineering Handbook* [Ben85] gives an extensive description of PAL and NTSC. *Digital Pictures* [NH95] also provides a brief description of the purpose of composite signals.

The burst_amplitude defines for NTSC and PAL signals the seven least significant bits of the subcarrier burst amplitude. This synchronization pulse is encoded onto each composite scan line.

The subcarrier can change phase over the course of time. The sub_carrier_phase is an eight-bit number from 0 to 255. Equation 10.7 converts it to phase.

$$phase = (360 \deg \div 256) * \texttt{sub\_carrier\_phase} \qquad (10.7)$$

---

[10]v_axis is strictly to convey phase information about the source PAL signal, prior to conversion to YCbCr. Section D.2.4 provides a hint at the purpose.

```
extension_and_user_data_P() {          /* from ISO 13818-2 6.2.2.2  */
  while((nextbits(32)==extension_start_code) ||
        (nextbits(32)==user_data_start_code)) {
                              /*while 0x000001B5 or 0x000001B2 */
    while (nextbits(32)==extension_start_code) {
                              /* while 0x000001B5            */
      extension_start_code(32);    /* r/w 0x000001B5            */
      if (nextbits(4)==''Quant matrix extension ID'')
                              /* if '0011'                 */
        quant_matrix_extension();   /* r/w quant matrix extension*/
      elseif (nextbits(4)==''Copyright extension ID'')
                              /* if '0100'                 */
        copyright_extension();     /* r/w copyright extension   */
      elseif (nextbits(4)==''Picture display extension ID'')
                              /* if '0111'                 */
        picture_display_extension(); /* r/w picture display ext.  */
      else                     /* else scalable extensions  */
        ...                    /* see standard              */
    }                         /* end while 0x000001B5      */
    if (nextbits(32)==user_data_start_code) /* if 0x000001B2      */
      user_data();              /* r/w user data             */
  }                           /* end while 0x000001B5 or...*/
}                             /* end ext._and_user_data_P()*/
```

Figure 10.13: MPEG-2 extension_and user_data_P() function. If present, this follows the picture coding extension.

10.4.2   MPEG-2 picture extensions and user data ●

The extension_and_user_data_P() function calls the optional picture extensions, quant_matrix_extension(), copyright_extension(), and picture_display_extension(), or the user_data() function.

10.4.3   MPEG-2 quant matrix extension ●

The quant_matrix_extension() allows new quantization matrices to be set. Each sequence header resets these matrices to their default values given in Chapter 14. The Quant matrix extension ID ('0011') is the extension_start_code_identifier. The next six lines can set new quantization matrices for the intra and nonintra blocks. This is the same as the six code lines in the sequence header following the constrained_parameters_flag. These are the only two quantization matrices allowed for 4:2:0 data. For the 4:2:2 and 4:4:4 formats, separate chrominance matrices for intra and nonintra blocks can also be downloaded.

```
quant_matrix_extension() {                /* from ISO 13818-2 6.2.3.2 */
  extension_start_code_identifier(4); /* r/w '0011'                */
  load_intra_quantiser_matrix(1);     /* r/w intra q matrix flag   */
  if (load_intra_quantiser_matrix)    /* if flag set               */
    intra_quantiser_matrix[1..64];    /*   r/w 64 8-bit values     */
  load_non_intra_quantiser_matrix(1); /* r/w nonintra q mat. flag  */
  if (load_non_intra_quantiser_matrix)/* if flag set               */
    non_intra_quantiser_matrix[1..64];/*   r/w 64 8-bit values     */
  load_chroma_intra_quantiser_matrix(1);  /* r/w '0' for MP        */
  if (load_chroma_intra_quantiser_matrix) /* if flag set           */
    ...;                              /* see standard              */
  load_chroma_non_intra_quantiser_matrix(1);  /* r/w '0' for MP    */
  if (load_chroma_non_intra_quantiser_matrix) /* if flag set       */
    ...;                              /* see standard              */
  next_start_code();                  /* find next start code      */
}                                     /* end quant_matrix_ext.()   */
```

Figure 10.14: MPEG-2 `quant_matrix_extension()` function.

10.4.4   MPEG-2 picture display extension ●

Since the display process is not defined by the standard, the information in the `picture_display_extension()` function shown in Figure 10.15 can be ignored by decoders. This extension allows the position of the display area (defined in the `sequence_display_extension`) to be moved for each picture. This is useful for selecting the most interesting region from a 9:16 wide video screen for a 3:4 narrow screen on a picture by picture basis. A `picture_extension` is only allowed if a `sequence_display_extension` followed the last sequence header. If a given picture has no `picture_extension`, then the most recent frame center offsets are used. A sequence header resets the frame offsets to zero.

The picture display extension ID for the `picture_display_extension()` function is '0111'. Following this come the 16-bit data elements `frame_centre_horizontal_offset` and `frame_centre_vertical_offset`, each terminated by a `marker_bit` to avoid start code emulation. These code the `number_of_frame_centre_offset` values.

Table 10.17 shows how to calculate the number of frame offsets. Vertical offsets are in units of 1/16th of the vertical line-to-line spacing in the frame; horizontal offsets are in units of 1/16 pel. If the display area is shifted down or to the right relative to the reconstructed frame, the offset is positive.

```
picture_display_extension() {           /* from ISO 13818-2 6.2.3.3 */
  extension_start_code_identifier(4); /* r/w '0111'                */
  for (i=0; i<number_of_frame_centre_offsets; i++) {
                                        /* for each offset pair     */
    frame_centre_horizontal_offset(16); /* r/w horizontal offset */
    marker_bit(1);                      /* r/w '1'                  */
    frame_centre_vertical_offset(16); /* r/w vertical offset      */
    marker_bit(1);                      /* r/w '1'                  */
  }                                     /* end for loop             */
  next_start_code();                    /* find next start code     */
}                                       /* end picture_display_ext.()*/
```

Figure 10.15: MPEG-2 `picture_display_extension()` function.

progressive_ sequence	picture_ structure	top_ field_ first	repeat_ first_ field	number of frame offsets
0	field			1
0	frame		0	2
0	frame		1	3
1			0	1
1		0	1	2
1		1	1	3

Table 10.17: Number of frame center offsets in MPEG-2.

```
copyright_extension() {                 /* from ISO 13818-2 6.2.3.6 */
  extension_start_code_identifier(4); /* r/w '0100'                */
  copyright_flag(1);                    /* r/w                      */
  copyright_identifier(8);              /* r/w                      */
  original_or_copy(1);                  /* r/w                      */
  reserved(7);                          /* r/w '0000000'            */
  marker_bit(1);                        /* r/w '1'                  */
  copyright_number_1(22);               /* r/w copyright # (22 msb) */
  marker_bit(1);                        /* r/w '1'                  */
  copyright_number_2(22);               /* r/w copyright # (22 mid.)*/
  marker_bit(1);                        /* r/w '1'                  */
  copyright_number_3(22);               /* r/w copyright # (22 lsb) */
  next_start_code();                    /* find next start code     */
}                                       /* end copyright_extension()*/
```

Figure 10.16: MPEG-2 `copyright_extension()` function.

```
picture_data() {                      /* from ISO 13818-2 6.2.3.7 */
  do {                                /* do slices                */
    slice();                          /*   r/w   slices           */
  } while (nextbits(32)==slice_start_code)
                      /* while 0x00000101 - 0x000001AF, do slices */
  next_start_code();                  /* find next start code     */
}                                     /* end picture_data()       */
```

Figure 10.17: MPEG-2 picture_data() function.

10.4.5 MPEG-2 copyright extension ●

The copyright_extension() function communicates copyright information for a picture. The copyright extension ID ('0100') is the extension_start_code_identifier for this function. The copyright_flag is set to '1' when pictures up to the next copyright_extension() function are copyrighted. A '0' value means the pictures may or may not be copyrighted. The eight-bit copyright_identifier identifies the copyright registration authority.[11] copyright_identifier should be zero when copyright_flag is '0'.

The original_or_copy bit is set to '1' if the material is original and '0' if it is a copy. The next seven bits are reserved for future extensions and are set to '0000000'. Marker bits break up the 64-bit copyright number to prevent start code emulation. The 20-bit copyright_number_1 are the most significant bits and the 22-bit copyright_number_3 are the least significant bits of copyright number. If copyright_flag is '1' but the identification number of the copyrighted work is not available, the copyright number is set to zero. if copyright_flag is '0', copyright number is always zero.

10.4.6 MPEG-2 picture data function ●

The picture_data() function calls the slice() function as long as the next start code is one of the slice_start_codes. They are 32-bit codes ranging from 0x00000101 to 0x000001AF. The final byte of these start codes is called the slice_vertical_position. If the slice_vertical_position_extension is present, then the slice_vertical_position is reduced from a range of 1 to 175 to a range of 1 to 127 (i.e., the vertical position is a 10-bit value with the seven least significant bits obtained from the slice_vertical_position, and the three most significant bits from the slice_vertical_position_extension).

---

[11]ISO/IEC JTC1/SC29 (MPEG's parent committee) designated the registration authority.

```
slice(){                             /* from ISO 13818-2 6.2.4      */
  slice_start_code(32);              /* r/w 0x00000101-0x000001AF   */
  if (vertical_size>2800)            /* if more than 175 slices     */
    slice_vertical_position_extension(3);/* extend slice positions*/
  if (scalable extensions)           /* if scalable extensions      */
    ...;                             /*   see standard              */
  quantiser_scale_code(5);           /* r/w quantizer scale         */
  if (nextbits(1)=='1'{              /* if '1', extra slice info.   */
    intra_slice_flag(1);             /* r/w 1 if next 2 lines else 0 */
    intra_slice(1);                  /* r/w 1 if all intra mb else 0 */
    reserved_bits(7);                /* r/w '0000 000'              */
    while (nextbits(1)=='1'{         /* while '1', extra slice info. */
      extra_bit_slice(1);            /* r/w '1'                     */
      extra_information_slice(8);    /* r/w byte of extra slice info.*/
    }                                /* end while '1', extra slice...*/
  }                                  /* end if '1', extra slice info.*/
  extra_bit_slice(1);                /* r/w '0' to end extra info.  */
  do {                               /* do mb                       */
    macroblock();                    /*   process a macroblock      */
  } while (nextbits(23)!=0)          /* while not 23 zeros, do mb   */
  next_start_code();                 /* find next start code        */
}                                    /* end slice() function        */
```

Figure 10.18: MPEG-2 slice() function.

10.5   MPEG-2 slice header ●

The slice_start_code has values in the range from (0x00000101 through
0x000001AF). The least significant byte, the slice_vertical_position,
gives the vertical position of the first macroblock in the slice. In general,
more than one slice can start in the same row and every slice must contain
at least one macroblock. However, in restricted environments slices must
cover the image area and are restricted to be a single macroblock row. Main
profile is a restricted environment.

Macroblocks are not skipped in I-pictures, and in any type of picture the
first and last macroblock of each slice may not be skipped. In B-pictures,
macroblocks may not be skipped immediately after I-blocks.

The product of 16 lines per block row × 175 block rows is 2800 lines.
If the image vertical_size is greater than 2800 lines, a three-bit slice_
vertical_position_extension is used to extend the range of the slice vertical position. However the least significant byte of the slice code is then
restricted to 1 through 128. Note that since the main profile at high level
(MP@HL) has a maximum of 1920 lines, the extension is not needed for
main profile. The mb_row parameter starts at zero, and thus is the slice_

```
mb_row = slice_vertical_position; /* last byte of slice start code*/
mb_row = mb_row - 1;              /* mb row starts with 0          */
if vertical_size > 2800          /* if vertical size too large    */
  mb_row = (slice_vertical_position_extension << 7) + mb_row;
                                 /* extend mb_row with extension */
```

Figure 10.19: MPEG-2 `mb_row` calculation.

`vertical_position` decremented by one. Figure 10.19 shows how to calculate the macroblock row.

The `sequence_scalable_extension()` function (sequence scalable extension ID '0101') may be present in the video sequence if MPEG-2 scalable extensions are used. Scalable extensions and data partitioning are beyond the scope of this chapter.

The five-bit `quantiser_scale_code` is interpreted in two different ways in Table 10.18, depending on the value of the `quantiser_scale_type` data element in the picture header. The data element `quantiser_scale_code` can also be set in the macroblock header.

The `intra_slice_flag` bit signals whether or not the `intra_slice` bit, `reserved_bits` and extra slice information are present. The `intra_slice` data element is set to '1' to indicates that all of the macroblocks are intra in the slice. It is assumed to be '0' unless explicitly set. This information is helpful for DSM applications in fastforward playback and fastreverse playback. The next seven bits, `reserved_bits` should be set to zero until ITU-T — ISO/IEC defines future extensions.

When `extra_bit_slice` is set to '1', eight bits of `extra_information_slice` follow. A zero value in `extra_bit_slice` stops this process. For now, `extra_bit_slice` shall be '0'. Future extensions may use '1'. Decoders are required to remove and discard the `extra_information_slice` bytes if `extra_bit_slice` is set to '1'.

Until the next start code occurs, macroblocks are coded in the bitstream.

10.6   MPEG-2 macroblock header ●

If the `macroblock_address_increment` is greater than 1, macroblocks are skipped. Except for the macroblock stuffing code, the MPEG-1 codes given in Table 8.6 are used for `macroblock_address_increment`. Macroblock stuffing is not allowed in MPEG-2, except in MPEG-1 compatibility mode. Zero stuffing is allowed at slice start code boundaries.

The `macroblock_modes()` function communicates the `macroblock_type` data element and extra motion vector related information. It is described

quantiser_scale[0]	quantiser_scale[1]	quantiser_scale_code	
decimal	decimal	decimal	binary
forbidden	forbidden	0	00000
2	1	1	00001
4	2	2	00010
6	3	3	00011
8	4	4	00100
10	5	5	00101
12	6	6	00110
14	7	7	00111
16	8	8	01000
18	10	9	01001
20	12	10	01010
22	14	11	01011
24	16	12	01100
26	18	13	01101
28	20	14	01110
30	22	15	01111
32	24	16	10000
34	28	17	10001
36	32	18	10010
38	36	19	10011
40	40	20	10100
42	44	21	10101
44	48	22	10110
46	52	23	10111
48	56	24	11000
50	64	25	11001
52	72	26	11010
54	80	27	11011
56	88	28	11100
58	96	29	11101
60	104	30	11110
62	112	31	11111

Table 10.18: The two MPEG-2 mappings between quantiser_scale_code and quantiser_scale.

```
macroblock(){                              /* from ISO 13818-2 6.2.5   */
  while (nextbits(11)=='00000001000')      /* while macroblock escape  */
    macroblock_escape(11);                 /*   r/w '0000 0001 000'    */
  macroblock_address_increment(1-11);      /* r/w mb address increment */
  macroblock_modes();                      /* r/w macroblock modes     */
  if (macroblock_quant)                    /* if q scale change        */
    quantiser_scale_code(5);               /*   r/w quantiser scale    */
  if ((macroblock_motion_forward)||
      (macroblock_intra && concealment_motion_vectors))
                                           /* if forward MVs           */
    motion_vectors(s0);                    /*   r/w forward MVs        */
  if (macroblock_motion_backward)          /* if backward MVs          */
    motion_vectors(s1);                    /*   r/w backward MVs */
  if (macroblock_intra && concealment_motion_vectors)
                                           /* if possible start code   */
    marker_bit(1);                         /*   r/w '1'                */
  if (macro_block_pattern)                 /* if any blocks coded      */
    coded_block_pattern();                 /*   r/w coded block pattern*/
  for (i=0; i<block_count; i++){           /* for all blocks in mb     */
    block(i);                              /* r/w block data           */
  }                                        /* end for all blocks in mb */
}                                          /* end macroblock() function*/
```

Figure 10.20: MPEG-2 macroblock() function.

```
macroblock_modes() {                    /* from ISO 13818-2 6.2.5.1 */
  macroblock_type(1-9);                 /* r/w macroblock type      */
  if spatial_temporal_weight_code_flag==1 /* never true for MP      */
    ...;                                /* see standard for scalable*/
  if (macroblock_motion_forward || macroblock_motion_backward) {
                                        /* if motion vectors        */
    if (picture_structure=="frame") { /* if frame ('11')          */
      if (frame_pred_frame_dct==0)    /*   if field possible      */
        frame_motion_type(2);         /*     r/w frame motion type*/
    }                                 /*   end if field possible  */
    else                              /* else field ('01', '10')  */
      field_motion_type(2);           /*   r/w field motion type  */
  }                                   /* end if motion vectors    */
  if ((picture_structure=="frame") && (frame_pred_frame_dct==0) &&
      (macroblock_intra||macroblock_pattern))
                                      /* if ambiguity about type  */
    dct_type(1);                      /*   r/w dct type           */
}                                     /* end macroblock_modes()   */
```

Figure 10.21: MPEG-2 macroblock_modes() function.

in more detail later in this section. If the macroblock_quant flag is 1 as derived from macroblock_type, then quantiser_scale_code is changed.

If forward motion vectors are needed (i.e., a value of 1 for macroblock_motion_forward is derived from macroblock_type or an intra-coded block with concealment motion vectors is being coded), then the forward motion vectors are coded in the motion_vectors(s0) function. The parameter s0 has a value of 0. This form is used to make it clear that motion_vector(s0) is a function with a parameter and not a data element with zero bits.

If macroblock_motion_backward is 1 (derived from macroblock_type), then backward motion vectors are coded in the motion_vectors(s1) function. In this case, s1 has the value of 1. If the forward motion vectors are concealment vectors, then a marker_bit is needed to prevent accidental start code emulation. If macroblock_pattern indicates that some blocks are coded, then the coded_block_pattern() function indicates which of the six blocks in the macroblock need additional data. Those blocks are coded in the block(i) function.

10.6.1 Macroblock modes ●

The macroblock_type variable length codes for I-, P-, and B-pictures (without scalability) are found in MPEG-1 Table 8.7. The spatial_temporal_weight_code_flag, a new parameter determined from macroblock_type,

frame_motion_type	field_motion_type	MP	code
reserved	reserved	.	00
field-based prediction	field-based prediction	●	01
frame-based prediction	16×8 MC	●	10
dual-prime	dual-prime	●	11

Table 10.19: MPEG-2 frame_motion_type and field_motion_type codes.

frame_motion_type	dmv	mv_format	motion_vector_count
reserved	—	—	—
field-based prediction	0	field	2
frame-based prediction	0	frame	1
dual-prime	1	field	1

Table 10.20: MPEG-2 variables set by frame_motion_type.

is always '0' for nonscalable I-, P-, and B-pictures. When the spatial_temporal_weight_code_flag is 0, spatial_temporal_weight_class is set to 0 and not sent in the bitstream.

The data element frame_motion_type specifies whether field, frame, or dual prime motion vector prediction is used for macroblocks in frame-coded pictures. Its codes are found in Table 10.19. For frame pictures where frame_pred_frame_dct is 1 or intra macroblocks with concealment motion vectors, frame_motion_type is not coded and is assumed to be frame-based prediction.

The field_motion_type codes are also found in Table 10.19. Since a field has half the vertical lines of a frame, a field macroblock can be coded with two sets of motion vectors. Each set of motion vectors are used for either the top or the bottom half of the macroblock. This 16×8 MC prediction for fields shares the same code as the frame-based prediction for frames.

Table 10.20 shows three parameters derived from the frame_motion_type. Table 10.21 shows the same variables derived from field_motion_type. The dmv flag indicates that dual prime is the prediction type and that the small dmvector needs to be coded. The mv_format clarifies the field versus frame format. Two sets of motion vectors are needed for frame prediction in frame-coded pictures and for 16×8 MC prediction in field-coded pictures.

The DCT coefficients for field pictures are always organized as independent fields. If picture_structure is frame and frame_pred_frame_dct is

field_motion_type	dmv	mv_format	motion_vector_count
reserved	–	–	–
field-based prediction	0	field	1
16×8 MC	0	field	2
dual-prime	1	field	1

Table 10.21: MPEG-2 variables set by field_motion_type.

picture_structure	frame_pred_frame_dct	dct_type	code
field	x	field '1'	not coded
frame	0	field	'1'
frame	1	frame	'0'
frame	1	frame '0'	not coded

Table 10.22: MPEG-2 dct_type codes.

'1' (frame), then the DCT coefficients are clearly coded as a frame. If a frame picture is coded with frame_pred_frame_dct set to zero (field) and the macroblock is coded (either an intra-coded macroblock or macroblock_pattern is nonzero), the DCT is done either on fields or on frames, depending on the data element dct_type. Table 10.22 shows the code for dct_type, including cases where it is not coded.

10.6.2 Motion vectors ●

The motion_vectors(s) function codes forward (s=0) and backward (s=1) motion vectors. Depending upon motion_vector_count, one or two sets of motion vectors are needed. For one set of motion vectors, the motion_vertical_field_select[0][s] flag needs to be coded only if the mv_format indicates field and the prediction is not dual prime. This data element selects the top (0) reference field or bottom (1) reference field. For two sets of motion vectors, the prediction is already known to be field; the motion_vertical_field_select[0][s] (field selection for the first set of motion vectors) and motion_vertical_field_select[1][s] (field selection for the second set of motion vectors) precede each set of motion vectors.

The motion_vector(r,s) function is called by motion_vectors(s) to code the first (r=0) and second (r=1) sets of forward (s=0) and backward (s=1) motion vectors. The third index, t, in motion_code[r][s][t] and motion_residual[r][s][t] is for horizontal (0) and vertical (1) motion.

```
motion_vectors(s) {                       /* from ISO 13818-2 6.2.5.2  */
  if (motion_vector_count==1) {           /* if only 1 set of MVs       */
    if ((mv_format==field)&&(dmv!=1))/* if field & not dual prime */
      motion_vertical_field_select[0][s](1);
                                          /* r/w 1st ver. field select */
    motion_vector(0,s);                   /* r/w 1 set of MVs        .  */
  }                                       /* end if only 1 set of MVs  */
  else {                                  /* else 2 sets of MVs         */
    motion_vertical_field_select[0][s](1);
                                          /* r/w 1st ver. field select */
    motion_vector(0,s);                   /* r/w 1st set of MVs         */
    motion_vertical_field_select[1][s](1);
                                          /* r/w 2nd ver. field select */
    motion_vector(1,s);                   /* r/w 2nd set of MVs         */
  }                                       /* end else 2 sets of MVs     */
}                                         /* end motion_vectors(s)      */
```

Figure 10.22: MPEG-2 `motion_vectors(s)` function.

```
motion_vector(r,s) {                   /* from ISO 13818-2 6.2.5.2.1   */
  motion_code[r][s][0](1-11);          /* r/w motion code horizontal   */
  if ((f_code[s][0]!=1)&&(motion_code[r][s][0]!=0))
                                       /* if horizontal remainder needed */
    motion_residual[r][s][0](1-8);     /* r/w horizontal remainder     */
  if (dmv==1)                          /* if dual prime                */
    dmvector[0](1-2);                  /*   r/w dmvector horizontal     */
                                       /* now do the same for vertical */
  motion_code[r][s][1](1-11);          /* r/w motion code vertical     */
  if ((f_code[s][1]!=1)&&(motion_code[r][s][1]!=0))
                                       /* if vertical remainder needed */
    motion_residual[r][s][1](1-8);     /* r/w vertical remainder       */
  if (dmv==1)                          /* if dual prime                */
    dmvector[1](1-2);                  /*   r/w dmvector vertical       */
}                                      /* end motion_vector() function */
```

Figure 10.23: MPEG-2 `motion_vector(r,s)` function. r has the values of 0, 1, 2, and 3, for the first, second, third, and fourth sets of motion vectors. s has the values 0 (forward) and 1 (backward).

dmvector[t] value	VLC code
+1	10
0	0
−1	11

Table 10.23: MPEG-2 dmvector[t] codes.

```
coded_pattern() {            /* from ISO 13818-2 6.2.5.3      */
  coded_block_pattern_420;   /* r/w VLC code for 1st 6 blocks */
  if (chroma_format==4:2:2)  /* if '10' then 2 more blocks    */
    ...;                     /* see standard                  */
  if (chroma_format==4:4:4)  /* if '11' then 6 more blocks    */
    ...;                     /* see standard                  */
}                            /* end coded_block_pattern()     */
```

Figure 10.24: MPEG-2 coded_pattern() function.

The motion_code[r][s][0] (horizontal) uses the same variable-length codes as MPEG-1 (see Table 8.9). However, the forward/backward and horizontal/vertical have been removed from the motion_code name and replaced with array indices. As long as f_code[s][0] is not one and motion_code[r][s][0] is not zero, a motion_residual[r][s][0] is coded in f_code minus 1 bits. If the motion compensation is dual prime, then one or two more bits code dmvector[0] (horizontal). The variable dmv will always be zero if there is a second motion vector ($r=1$). Since dual prime is only allowed for P-pictures, dmv will be zero for backward motion vectors ($s=1$) too. Therefore, it only needs one index to distinguish between horizontal and vertical displacements. Table 10.23 gives the variable-length codes for dmvector[t].

The horizontal motion vector ($t=0$) is coded in the top half of the function. The vertical motion vector ($t=1$) is coded with identical-looking code in the bottom half of the function.

10.6.3   MPEG-2 coded block pattern ●

The coded_pattern() function extends the original six blocks for 4:2:0 format chrominance to 8 or 12 blocks for 4:2:2 and 4:4:4 format chrominance. The code_block_pattern_420 indicates whether the first six blocks are skipped or coded. Its codes are identical to MPEG-1 Table 8.8 except that the forbidden cpb=0 entry is replaced with the code '0000 0000 1' for

```
block(i) {                              /* from ISO 13818-2 6.2.6  */
  if (pattern_coded[i]) {               /* if ith block coded      */
    if (macroblock_intra) {             /* if intra-coded          */
      if (i<4){                         /* if luminance block      */
        dct_dc_size_luminance(2-9);     /*   r/w size              */
        if (dct_dc_size_luminance!=0)   /*   if size not zero      */
          dct_dc_differential(1-11);    /*     r/w differential DC */
      }                                 /* end if luminance block  */
      else {                            /* else chrom. block       */
        dct_dc_size_chrominance(2-10);  /*   r/w size              */
        if (dct_dc_size_chrominance!=0)/*   if size not zero      */
          dct_dc_differential(1-11);    /*     r/w differential DC */
      }                                 /* end else chrom. block   */
    }                                   /* end if intra-coded      */
    else {                              /* else nonintra mb        */
      First DCT coefficient;            /* r/w 1st run/level       */
    }                                   /* end else nonintra       */
    while (nextbits(2or4)!=End of block)/* while not end-of-block */
      Subsequent DCT coefficients;      /*   r/w run/level         */
    end-of-block(2or4);                 /* r/w '01' or '0110'      */
  }                                     /* end if ith block coded  */
}                                       /* end block(i) function   */
```

Figure 10.25: MPEG-2 block(i) function.

MPEG-2 non-4:2:0 chrominance structures. The main profile is restricted to 4:2:0 formats.

10.7 MPEG-2 block(i) function ●

Figure 10.25 gives pseudocode for the block(i) function. The block(i) function only codes coefficients for blocks with pattern_code[i] nonzero. If the block is intra coded (determined by macroblock_intra set to one), then for the first four macroblocks (i less than 4) the number of bits needed to code the luminance DC coefficient, dct_dc_size_luminance, is coded using Table 10.24. If the number of bits is greater than zero, the dct_dc_differential is sent with that many bits. For chrominance blocks the number of bits needed to code the chrominance DC coefficient, dct_dc_size_chrominance, is coded using Table 10.25. If dct_dct_size_chrominance is nonzero, that many bits is used to code dct_dc_differential.

If the block is not intra-coded, then the DC coefficient is coded with the rest of the AC coefficients using run-level VLC codes from table 0 listed in Table 10.26. The first nonzero coefficient has a different VLC code since the

Luminance DC difference	MPEG-1	MP	dct_dc_size_ luminance
0	●	●	100
1	●	●	00
2	●	●	01
3	●	●	101
4	●	●	110
5	●	●	1110
6	●	●	1111 0
7	●	●	1111 10
8	●	●	1111 110
9	.	●	1111 1110
10	.	●	1111 1111 0
11	.	.	1111 1111 1

Table 10.24: MPEG-1 and MPEG-2 dct_dc_size_luminance codes.

chrominance DC difference	MPEG-1	MP	dct_dc_size_ chrominance
0	●	●	00
1	●	●	01
2	●	●	10
3	●	●	110
4	●	●	1110
5	●	●	1111 0
6	●	●	1111 10
7	●	●	1111 110
8	●	●	1111 1110
9	.	●	1111 1111 0
10	.	●	1111 1111 10
11	.	.	1111 1111 11

Table 10.25: MPEG-1 and MPEG-2 dct_dc_size_chrominance codes.

end-of-block code is not allowed at that point.

For subsequent coefficients the `intra_vlc_table` flag set in the picture coding extension selects between table 0 and table 1 for intra-coded blocks. Nonintra-coded blocks always use table 0. Until the end-of-block is reached (i.e., the rest of the coefficients are all zero), the run-level pairs are coded. The luminance and chrominance blocks use the same VLC tables.

Two VLC tables for run-level codes are listed in Table 10.26. Run-level table 0 is used in MPEG-1; table 1 is appropriate for the higher bitrates and resolutions addressed in some MPEG-2 profiles. Note that the entries in table 1 are identical those in table 0 for more than half of the run-level pairs.

If the run-level is not in the tables, it is coded with the escape code followed by a fixed-length six-bit code for the run. The level is coded as a 12 bit signed integer. Thus, the levels range from -2047 to +2047 (but zero is forbidden). The value -2048 ('1000 0000 0000') is not listed in the table [12].

10.8   MPEG-2 MP@ML and SP@ML additional restrictions ●

Most of the restrictions for main profile at main level have been identified in the data elements code tables. However, there are some additional constraints. Table 10.27 gives the maximum number of bits in a macroblock for the chrominance formats. At most two macroblocks in a macroblock row may exceed these values. The bits for the macroblock are counted from the first `macroblock_escape` or `macroblock_address_increment` to the last bit of the end-of-block code (or to the `coded_block_pattern()` if all blocks are skipped). The slice information is not counted as part of the macroblock.

In addition, Figure 10.26 gives additional constraints for MP@ML and SP@ML.

10.9   MPEG-2 MP@ML video data elements ●

Table 10.28 lists all data elements for MPEG-2 main profile at main level in alphabetical order in the form that they appear in the pseudocode. If a data element appears with explicit array indices, it is listed in the table in that manner (e.g., `f_code[0][1]`). Derived values are indented a few spaces. A solid circle indicates whether that data element was set (or used) in the video sequence header and extensions (v), group of pictures header

---

[12] A technical corrigendum suggested that the -2048 value be reserved, but the outcome was uncertain when this book went to press.

run/level	VLC table 0	VLC table 1
0/1	1s (first)	10s
0/1	11s (next)	10s
0/2	0100 s	110s
0/3	0010 1s	0111 s
0/4	0000 110s	1110 0s
0/5	0010 0110 s	1110 1s
0/6	0010 0001 s	0001 01s
0/7	0000 0010 10s	0001 00s
0/8	0000 0001 1101 s	1111 011s
0/9	0000 0001 1000 s	1111 100s
0/10	0000 0001 0011 s	0010 0011 s
0/11	0000 0001 0000 s	0010 0010 s
0/12	0000 0000 1101 0s	1111 1010 s
0/13	0000 0000 1100 1s	1111 1011 s
0/14	0000 0000 1100 0s	1111 1110 s
0/15	0000 0000 1011 1s	1111 1111 s
0/16	0000 0000 0111 11s	0000 0000 0111 11s
0/17	0000 0000 0111 10s	0000 0000 0111 10s
0/18	0000 0000 0111 01s	0000 0000 0111 01s
0/19	0000 0000 0111 00s	0000 0000 0111 00s
0/20	0000 0000 0110 11s	0000 0000 0110 11s
0/21	0000 0000 0110 10s	0000 0000 0110 10s
0/22	0000 0000 0110 01s	0000 0000 0110 01s
0/23	0000 0000 0110 00s	0000 0000 0110 00s
0/24	0000 0000 0101 11s	0000 0000 0101 11s
0/25	0000 0000 0101 10s	0000 0000 0101 10s
0/26	0000 0000 0101 01s	0000 0000 0101 01s
0/27	0000 0000 0101 00s	0000 0000 0101 00s
0/28	0000 0000 0100 11s	0000 0000 0100 11s
0/29	0000 0000 0100 10s	0000 0000 0100 10s
0/30	0000 0000 0100 01s	0000 0000 0100 01s
0/31	0000 0000 0100 00s	0000 0000 0100 00s
0/32	0000 0000 0011 000s	0000 0000 0011 000s
0/33	0000 0000 0010 111s	0000 0000 0010 111s
0/34	0000 0000 0010 110s	0000 0000 0010 110s
0/35	0000 0000 0010 101s	0000 0000 0010 101s
0/36	0000 0000 0010 100s	0000 0000 0010 100s
0/37	0000 0000 0010 011s	0000 0000 0010 011s
0/38	0000 0000 0010 010s	0000 0000 0010 010s
0/39	0000 0000 0010 001s	0000 0000 0010 001s
0/40	0000 0000 0010 000s	0000 0000 0010 000s

Table 10.26: (a) MPEG-2 variable length codes for AC coefficients. The sign bit 's' is '0' for postive and '1' for negative.

run/level	VLC table 0	VLC table 1
1/1	011s	010s
1/2	0001 10s	0011 0s
1/3	0010 0101 s	1111 001s
1/4	0000 0011 00s	0010 0111 s
1/5	0000 0001 1011 s	0010 0000 s
1/6	0000 0000 1011 0s	0000 0000 1011 0s
1/7	0000 0000 1010 1s	0000 0000 1010 1s
1/8	0000 0000 0011 111s	0000 0000 0011 111s
1/9	0000 0000 0011 110s	0000 0000 0011 110s
1/10	0000 0000 0011 101s	0000 0000 0011 101s
1/11	0000 0000 0011 100s	0000 0000 0011 100s
1/12	0000 0000 0011 011s	0000 0000 0011 011s
1/13	0000 0000 0011 010s	0000 0000 0011 010s
1/14	0000 0000 0011 001s	0000 0000 0011 001s
1/15	0000 0000 0001 0011 s	0000 0000 0001 0011 s
1/16	0000 0000 0001 0010 s	0000 0000 0001 0010 s
1/17	0000 0000 0001 0001 s	0000 0000 0001 0001 s
1/18	0000 0000 0001 0000 s	0000 0000 0001 0000 s
2/1	0101 s	0010 1s
2/2	0000 100s	0000 111s
2/3	0000 0010 11s	1111 1100 s
2/4	0000 0001 0100 s	0000 0011 00s
2/5	0000 0000 1010 0s	0000 0000 1010 0s
3/1	0011 1s	0011 1s
3/2	0010 0100 s	0010 0110 s
3/3	0000 0001 1100 s	0000 0001 1100 s
3/4	0000 0000 1001 1s	0000 0000 1001 1s
4/1	0011 0s	0001 10s
4/2	0000 0011 11s	1111 1101 s
4/3	0000 0001 0010 s	0000 0001 0010 s
5/1	0001 11s	0001 11s
5/2	0000 0010 01s	0000 0010 0s
5/3	0000 0000 1001 0s	0000 0000 1001 0s
6/1	0001 01s	0000 110s
6/2	0000 0001 1110 s	0000 0001 1110 s
6/3	0000 0000 0001 0100 s	0000 0000 0001 0100 s
7/1	0001 00s	0000 100s
7/2	0000 0001 0101 s	0000 0001 0101 s
8/1	0000 111s	0000 101s
8/2	0000 0001 0001 s	0000 0001 0001 s

Table 10.26: (b) Continuation of MPEG-2 variable length codes for AC coefficients.

run/level	VLC table 0	VLC table 1
9/1	0000 101s	1111 000s
9/2	0000 0000 1000 1s	0000 0000 1000 1s
10/1	0010 0111 s	1111 010s
10/2	0000 0000 1000 0s	0000 0000 1000 0s
11/1	0010 0011 s	0010 0001 s
11/2	0000 0000 0001 1010 s	0000 0000 0001 1010 s
12/1	0010 0010 s	0010 0101 s
12/2	0000 0000 0001 1001 s	0000 0000 0001 1001 s
13/1	0010 0000 s	0010 0100 s
13/2	0000 0000 0001 1000 s	0000 0000 0001 1000 s
14/1	0000 0011 10s	0000 0010 1s
14/2	0000 0000 0001 0111 s	0000 0000 0001 0111 s
15/1	0000 0011 01s	0000 0011 1s
15/2	0000 0000 0001 0110 s	0000 0000 0001 0110 s
16/1	0000 0010 00s	0000 0011 01s
16/2	0000 0000 0001 0101 s	0000 0000 0001 0101 s
17/1	0000 0001 1111 s	0000 0001 1111 s
18/1	0000 0001 1010 s	0000 0001 1010 s
19/1	0000 0001 1001 s	0000 0001 1001 s
20/1	0000 0001 0111 s	0000 0001 0111 s
21/1	0000 0001 0110 s	0000 0001 0110 s
22/1	0000 0000 1111 1s	0000 0000 1111 1s
23/1	0000 0000 1111 0s	0000 0000 1111 0s
24/1	0000 0000 1110 1s	0000 0000 1110 1s
25/1	0000 0000 1110 0s	0000 0000 1110 0s
26/1	0000 0000 1101 1s	0000 0000 1101 1s
27/1	0000 0000 0001 1111 s	0000 0000 0001 1111 s
28/1	0000 0000 0001 1110 s	0000 0000 0001 1110 s
29/1	0000 0000 0001 1101 s	0000 0000 0001 1101 s
30/1	0000 0000 0001 1100 s	0000 0000 0001 1100 s
31/1	0000 0000 0001 1011 s	0000 0000 0001 1011 s
End_of_block	10	0110
Escape	0000 01	0000 01

Table 10.26: (c) Continuation of MPEG-2 variable length codes for AC coefficients.

chroma_ format	MP	maximum bits in a macroblock
4:2:0	●	4608
4:2:2	.	6144
4:4:4	.	9216

Table 10.27: MPEG-2 restrictions on the maximum bits in a macroblock.

```
if ((vertical_size > 480) || (frame_rate=='0011')) {
                                /* if wide image or 25 Hz    */
      if (picture_coding_type==3)   /*   and if B-picture       */
        repeat_first_field=0;       /*  can't repeat first field */
}                                 /* end if wide image or 25Hz */
if (vertical_size > 480)          /* if wide image             */
   frame_rate= '0011';            /*   force frame rate 25 Hz  */
```

Figure 10.26: MPEG-2 additional restrictions on MP@ML and SP@ML.

(g), picture header and extensions (p), slice code (s), macroblock code (m), or block code (b). An open circle shows where a derived value is set or implicitly used. MPEG-2 syntax adds a new data type (dt), signed integer, most significant bit first (simsbf). Table 10.29 shows this expanded set of data types.

The number of bits (# of bits) column lists the maximum number of bits ever allowed by the syntax. The value range has been restricted to the minimum and maximum required for MP@ML compliant decoders. Decoders may choose to implement more levels, optional scalable and advanced functions data elements and expanded ranges.

A similar table for simple profile at main level (SP@ML) decoders would restrict the backward f_codes (f_code[1][0] and f_code[1][1]) to 15, the derived macroblock_motion_backward to 0, and not allow backward motion vectors.

video data element name	set v g p s m b	used v g p s m b	dt	# of bits	value range
alternate_scan	· · ● · · ·	· · · · · ·	U	1	0,1
aspect_ratio_information	● · · · · ·	· · · · · ·	U	4	1,2,3
backward_f_code	· · ● · · ·	· · · · · ·	U	3	'111'
bit_rate_extension	● · · · · ·	· · · · · ·	U	12	0,...,4095
bit_rate_value	● · · · · ·	· · · · · ·	U	18	$1..2^{18}-1$
broken_link	· ● · · · ·	· · · · · ·	U	1	0,1
burst_amplitude	· · ● · · ·	· · · · · ·	U	7	0,...,127
chroma_format	● · · · · ·	· · · · · ●	U	2	'01' (4:2:0)
block_count	○ · · · · ·	· · · · · ●			6
chroma_420_type	· · ● · · ·	· · · · · ·	U	1	0,1
closed_gop	· ● · · · ·	· · · · · ·	U	1	0,1
coded_block_pattern_420	· · · · ● ·	· · · · · ·	V	3-9	1,...,63
pattern_code[0]	· · · · ○ ·	· · · · · ●			0,1
...	· · · · ○ ·	· · · · · ●			0,1
pattern_code[5]	· · · · ○ ·	· · · · · ●			0,1
colour_description	● · · · · ·	● · · · · ·	U	1	0,1
colour_primaries	● · · · · ·	· · · · · ·	U	8	1,2,4,5,6,7
composite_display_flag	· · ● · · ·	· · ● · · ·	U	1	0,1
concealment_motion_ vectors	· · ● · · ·	· · · · · ●	U	1	0,1
constrained_parameters_ flag	● · · · · ·	· · · · · ·	B	1	'0'
copyright_identifier	· · ● · · ·	· · · · · ·	U	8	0,...,255
copyright_number_1	· · ● · · ·	· · · · · ·	U	20	$0,...,2^{20}-1$
copyright_number_2	· · ● · · ·	· · · · · ·	U	22	$0,...,2^{22}-1$
copyright_number_3	· · ● · · ·	· · · · · ·	U	22	$0,...,2^{22}-1$
copyright_flag	· · ● · · ·	· · · · · ·	B	1	0,1
dct_dc_differential	· · · · · ●	· · · · · ·	U	1-11	$-2^{10},...,0,...,2^{10}-1$
dct_dc_size_chrominance	· · · · ● ·	· · · · · ●	V	2-10	0,...,10
dct_dc_size_luminance	· · · · ● ·	· · · · · ●	V	2-9	0,...,10
dct_type	· · · · ● ·	· · · · · ·	U	1	0,1
display_horizontal_size	● · · · · ·	· · · · · ·	U	14	1,...,16376
display_vertical_size	● · · · · ·	· · · · · ·	U	14	1,...,16376
dmvector[0] (h.)	· · · · ● ·	· · · · · ·	V	1-2	-1,0,1
dmvector[1] (v.)	· · · · ● ·	· · · · · ·	V	1-2	-1,0,1
end of block	· · · · · ●	· · · · · ●	V	2,4	'10','0110'

Table 10.28: (a) Summary of data elements in MPEG-2 main profile video.

video data element name	set						used						dt	# of bits	value range
	v	g	p	s	m	b	v	g	p	s	m	b			
extension_start_code	●		●				●	●					B	32	0x000001B5
extension_start_code_identifier	●		●										U	4	1,...,4,7,8
Copyright Extension ID			○						●				-	4	'0100'
Picture Coding Ext. ID			○						●				-	4	'1000'
Picture Display Ext. ID			○						●				-	4	'0111'
Quant Matrix Ext. ID			○						●				-	4	'0011'
Sequence Display Ext. ID	○						●						-	4	'0010'
Sequence Extension ID	○						●						-	4	'0001'
extra_bit_picture			●						○				U	1	'0'
extra_bit_slice				●						○			U	1	'0'
extra_information_picture			●										U	8	reserved
extra_information_slice				●									U	8	reserved
f_code[0][0] (forward h.)			●										U	4	1,...,8
f_code[0][1] (forward v.)			●										U	4	1,...,5
f_code[1][0] (backward h.)			●										U	4	1,...,8,15
f_code[1][1] (backward v.)			●										U	4	1,...,5,15
field_motion_type					●								U	2	1,2,3
dmv					○							●			0,1
motion_vector_count					○							●			1,2
mv_format					○							●			field
field_sequence			●										U	3	0,...,7
first DCT coefficient						●							V	2-24	
forward_f_code			●										B	3	'111'
frame_centre_horizontal_offset			●										S	16	$-2^{15},\ldots,2^{15}-1$
frame_centre_vertical_offset			●										S	16	$-2^{15},\ldots,2^{15}-1$
frame_motion_type					●								U	2	1,2,3
dmv					○							●			0,1
motion_vector_count					○							●			1,2
mv_format					○							●			field,frame
frame_pred_frame_dct			●									●	U	1	0,1
frame_rate_code	●												U	4	1,...,5
frame_rate_extension_d	●												U	5	'00000'
frame_rate_extension_n	●												U	2	'00'
full_pel_backward_vector			●										B	1	'0'
full_pel_forward_vector			●										B	1	'0'
group_start_code		●					●						B	32	0x000001B8

Table 10.28: (b) Summary of data elements in MPEG-2 main profile video.

video data element name	set v g p s m b	used v g p s m b	dt	# of bits	value range
horizontal_size_extension	●	. . . . . .	U	2	0
horizontal_size_value	●	. . . . . .	U	12	1,...,720
intra_dc_precision	. . ● . . .	. . . . . .	U	2	8,9,10
intra_quantiser_matrix[0]	● . ● . . .	. . . . . .	U	8	8
...	● . ● . . .	. . . . . .	U	8	1,...,255
intra_quantiser_matrix[63]	● . ● . . .	. . . . . .	U	8	1,...,255
intra_slice	. . . ● . .	. . . . . .	U	1	0,1
intra_slice_flag	. . . ● . .	. . . ○ . .	B	1	0,1
intra_vlc_format	. . . ● . .	. . . . . .	U	1	0,1
load_chroma_intra_quantiser_ matrix	. . . ● . .	. . . ● . .	U	1	'0'
load_chroma_non_intra_ quantiser_matrix	. . . ● . .	. . . ● . .	U	1	'0'
load_intra_quantiser_matrix	● . ● . . .	● . ● . . .	U	1	0,1
load_non_intra_quantiser_matrix	● . ● . . .	● . ● . . .	U	1	0,1
low_delay	●	. . . . . .	U	1	0,1
macroblock_address_increment .	. . . . ● .	. . . . . .	V	1-11	1,...,33
macroblock_escape	. . . . ● .	. . . . ○ .	B	11	33
macroblock_type	. . . . ● .	. . . . . .	V	1-9	
macroblock_intra	. . . . ○ .	. . . . ● ●	-	1	0,1
macroblock_motion_backward .	. . . . ○ .	. . . . ● .	-	1	0,1
macroblock_motion_forward .	. . . . ○ .	. . . . ● .	-	1	0,1
macroblock_pattern	. . . . ○ .	. . . . ● .	-	1	0,1
macroblock_quant	. . . . ○ .	. . . . ● .	-	1	0,1
spatial_temporal_ weight_code_flag	. . . . ○ .	. . . . ● .	-	1	'0'
spatial_temporal_ weight_code_class	. . . . ○ .	. . . . ● .	-		'0'
marker_bit	● ○ ● . ● .	. . . . . .	B	1	'1'
motion_code[r][s][0] (h.)	. . . . ● .	. . . . . .	V	1-11	-16, ...,16
motion_code[r][s][1] (v.)	. . . . ● .	. . . . . .	V	1-11	-16, ...,16
motion_residual[r][s][0] (h.)	. . . . ● .	. . . . . .	U	1-8	0,...,127
motion_residual[r][s][1] (v.)	. . . . ● .	. . . . . .	U	1-8	0,...,127
matrix_coefficients	●	. . . . . .	U	8	1,2,4,5,6,7
motion_vertical_ field_select[0][s] (1st set) ...	. . . . ● .	. . . . . .	U	1	0,1

Table 10.28: (c) Summary of data elements in MPEG-2 main profile video.

video data element name	set v g p s m b	used v g p s m b	dt	# of bits	value range
motion_vertical _field_select[1][s] (2nd set)	. . . . ● .	. . . . . .	U	1	0,1
non_intra_quantiser_matrix[0]	● . ● . . .	. . . . . .	U	8	1,. . .,255
. . .	● . ● . . .	. . . . . .	U	8	1,. . .,255
non_intra_quantiser_matrix[63]	● . ● . . .	. . . . . .	U	8	1,. . .,255
original_or_copy	. . ● . . .	. . . . . .	B	1	0,1
picture_coding_type	. . ● . . .	. . . ● . ●●	U	3	1,2,3
picture_start_code	. . ● . . .	●	B	32	0x00000100
picture_structure	. . ● . . .	. . . . . ●	U	2	1,2,3
field	. . ○ . . .	. . . . . ●			1,2
frame	. . ○ . . .	. . . . . ●			3
profile_and_level_indication	●		U	8	
escape bit	○		U	1	0
profile	○		U	3	4,5
level	○		U	4	8,10
progressive_frame	. . ● . . .		U	1	0,1
progressive_sequence	●		U	1	0,1
quantiser_scale_code	. . ●● . .		U	5	1,. . .,31
q_scale_type	. . ● . . .		U	1	0,1
repeat_first_field	. . ● . . .		U	1	0,1
reserved	. . ● . . .		U	7	'0000000'
reserved_bits	. . . ● . .		U	7	'0000000'
sequence_end_code	●	●	B	32	0x000001B7
sequence_header_code	●	●	B	32	0x000001B3
slice_start_code 1	. . ● . . .	. . . ● . .	B	32	0x00000101
. . .	. . ● . . .	. . . ● . .	B	32	0x000001xx
slice_start_code 175	. . ● . . .	. . . ● . .	B	32	0x000001AF
slice_vertical_position	. . ● . . .		U	8	1,. . .,175
slice_vertical_position_ext.	. . ● . . .		U	3	0
sub_carrier	. . ● . . .		U	1	0,1
sub_carrier_phase	. . ● . . .		U	8	0,. . .,255
subsequent DCT coefficients	. . . . ● .	. . . . . ●	V	3-24	
temporal_reference	. . ● . . .		U	10	0,. . .,1023
time_code	. ●		B	25	
drop_frame_flag	. ○		U	1	0,1
time_code_hours	. ○		U	5	0,. . .,23

Table 10.28: (d) Summary of data elements in MPEG-2 main profile video.

video data element name	set v g p s m b	used v g p s m b	dt	# of bits	value range
time_code_minutes	. o	. . . . . .	U	6	0,...,59
marker_bit	. o	. . . . . .	B	1	'1'
time_code_seconds	. o	. . . . . .	U	6	0,...,59
time_code_pictures ...	. o	. . . . . .	U	6	0,...,59
top_field_first	. . ● . . .	. . . . . .	U	1	0,1
transfer_characteristics ..	●	. . . . . .	U	8	1,2,4,5,6,7,8
user_data	● ● ● . . .	. . . . . .	-	8	0,...,255
user_data_start_code	● ● ● . . .	● ● ● . . .	B	32	0x000001B2
v_axis	. . ● . . .	. . . . . .	U	1	0,1
vbv_buffer_size_extension ..	●	. . . . . .	U	8	0,...,255
vbv_buffer_size_value	●	. . . . . .	U	10	0,...,1023
vbv_delay	. . ● . . .	. . . . . .	U	16	0,...,0xFFFF
vertical_size_extension ..	●	. . . . . .	U	2	0
vertical_size_value	●	. . . . . .	U	12	2,4,...,576
vertical_size	o	. . . ● . .			2,4,...,576
video_format	●	. . . . . .	U	3	0,...,5

Table 10.28: (e) Summary of data elements in MPEG-2 main profile video.

data element type	original pseudo- code mnemonic	summary chart
bit string, left bit first	bslbf	B
signed integer, most significant bit first	simsbf	S
unsigned integer, most significant bit first	uimsbf	U
variable-length code, left bit first	vlclbf	V
fixed 1-bit	'1'	'1'
fixed 0-bit	'0'	'0'

Table 10.29: Data types in the MPEG-2 video bitstream.

11

Motion Compensation

A key element in MPEG inter compression is motion compensation. In inter compression the pels in a region of a reference picture (or pictures) are used to predict pels in a region of the current picture. Differences between the reference picture and the current picture are then coded to whatever accuracy is affordable at the desired bitrate. If there is motion, some parts of the image used for prediction may be shifted relative to the current image, and motion compensation is used to minimize the effects of these shifts.

In this chapter we look at motion compensation from the decoder's point of view. The motion vector data elements have already been decoded from the bitstream as described in Chapter 8, and now must be converted into horizontal and vertical motion displacements.

Chapter 12 describes how the motion displacements are used in the pel reconstruction process. Motion estimation, the process by which the encoder selects these motion vectors, is discussed in Chapter 13.

11.1   Motion in sequences of pictures ○

The pictures in moving picture sequences often have large regions in which very small changes occur from one picture to the next in the sequence. The picture-to-picture correlation in these regions is very high and any reasonable video compression scheme should take advantage of this. Obviously, there are usually some changes from picture to picture, or we would only have to send the first picture of the sequence. Changes are caused by a number of different mechanisms, among them noise, scene lighting, and motion.

If the only difference from one scene to the next is random noise, perhaps from video amplifiers, the most logical choice for prediction of a given region would be a region with the same spatial coordinates in the reference picture. Since we are dealing with lossy compression in which distortions too small to be visible may be ignored, small differences due to noise can often be ignored and the region can be copied from the reference picture without modification. Only a few bits are needed to tell the decoder that there are no changes in these regions.

Relative change from one picture to the next can be caused by camera motion such as panning or vibration, and even by instabilities in video synchronization. In this case the motion is often describable by a single motion vector, a horizontal and vertical displacement of the image from one picture to the next. More generally, however, relative motion also occurs, either because of movement of objects within the scene, motion of the camera, or changes in scene lighting that might cause shadows to shift. The general case is therefore best served by individual motion vectors for a multiplicity of regions within the picture.

Relative motion within a picture gives rise to many interesting technical problems. For example, consider the relatively simple case of rigid body motion of an object in a picture: The motion of a rigid body has six degrees of freedom, three for spatial translations and three for rotational motions. Deformable bodies are less easily analyzed; multiple objects, camera motion, and scene lighting changes introduce further complications. Although there is a significant body of literature on modeling of motion, analysis of unconstrained scenes captured under unconstrained conditions is well beyond the current state of the art.

In a video sequence, the scene is projected through a lens onto a two-dimensional array of sensors, and captured at discrete time intervals. Real motion becomes *apparent motion*, in which the eye perceives motion as it views the sequence of still pictures. In each still picture the motion can be described by a two-dimensional array of motion vectors that give displacements relative to a reference picture in the sequence. If good values of these displacements are available, pels in the reference picture can be used to predict the current picture. Figure 11.1 illustrates this for MPEG P-pictures.

To paraphrase a famous quotation, no model of scene motion can accurately predict all of the regions all of the time. The complexities of the motion, and problems such as uncovering background for which there is no prior knowledge make this virtually impossible. The real role of motion compensation is to improve the coding efficiency by improving the prediction from the reference picture. Prediction is always followed by further information, even if this information only serves to tell the decoder that the prediction requires no corrections.

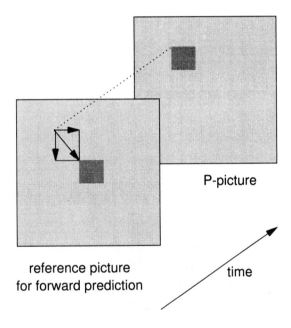

P-picture

reference picture
for forward prediction   time

Figure 11.1: P-picture motion vector displacements. Positive displacements in the reference picture are to the right and down, relative to the macroblock being coded.

11.2   Motion compensation in MPEG-1 ○

Since pictures are rectangular arrays of pels, it is convenient to describe motion vectors in terms of horizontal and vertical displacements. Hopefully, the displacements are those that give the best match between a given region in the current picture and the corresponding displaced region in the reference picture. The main technical issues in using motion vectors are the precision of the motion vectors, the size of the regions assigned to a single motion vector, and the criteria used to select the best motion vector value. Note that the decoder simply decodes the motion vectors; the processing to determine the motion vector values is done entirely in the encoder.

The size of the region assigned to a single motion vector determines the number of motion vectors that are needed to describe the motion. A trade-off must be made between the accuracy in predicting complex motion in the picture and the expense of transmitting the motion vectors. In addition, if statistical methods are used to select the optimal motion vectors, the size of the region used to generate the match criterion will affect the variance in the criterion.[1] Smaller regions require more complex estimation techniques incorporating techniques such as noise smoothing. Indeed, if the regions approach a single pel in size, the techniques developed for pel-recursive motion estimation [MPG85] become more and more relevant.

In MPEG the size of the region is a macroblock and the same values for the motion displacements are used for every pel in the macroblock. An MPEG-1 macroblock has four luminance blocks and two chrominance blocks; while the same motion displacement information is used for all blocks, the actual displacement must be scaled to reflect the differing resolutions for luminance and chrominance.

Precision of the motion vectors is another important variable. Intuitively, motion vector displacements should be to at least integer pel precision, but higher precision may be desirable in many instances. The issue here is accuracy of prediction versus the cost of transmitting higher-precision displacements. In MPEG-1 the precision of the motion vectors may be either full pel or half-pel, and is selected by bits in the picture header. Girod [Gir93] presents arguments suggesting that quarter-pel precision is nearly optimum. However, the optimality is broad, and half-pel precision appears to be almost as good.

If fractional-pel motion vectors are used, there is the question of how the prediction is made when the displacement is a fractional pel value. Bilinear interpolation is perhaps the simplest filter to use, but more complex filters using a larger surrounding region are known to be more accurate. In this

---

[1]Typically, the regions consist of a uniform grid of rectangular blocks of pels and the search for the best match is called *block matching*.

case, the issue is simplicity versus performance, and MPEG opts for the simplicity of bilinear interpolation.

The size of the region to which the motion vector applies is also important. In MPEG-1 a single set of displacements is applied to the entire macroblock, a choice which is convenient but not necessarily optimal; in MPEG-2 a finer granularity is also allowed.[2]

Because relatively uniform motion over large areas of the picture is quite common, motion vectors of adjacent regions are usually highly correlated. This is true even when the regions are as large as the 16×16 pel macroblocks of MPEG. For this reason MPEG motion vectors are coded differentially, the prediction being derived from the motion vector for the preceding macroblock.

A number of techniques may be used to estimate the motion vectors. For most of the MPEG developmental testing, a full search of the predicting image was used to find the best motion vector, the criterion for best match being the minimization of either the *mean absolute distortion* (MAD) or the *mean square error* (MSE.[3] Other best-match criteria and other search strategies that are computationally more efficient than the full search may be used. It is also possible to use optical flow techniques to derive the motion vectors, and one might even do trial encodings to find the match which gives the minimum number of coded bits. Pragmatically, this last approach, albeit computationally intensive, is guaranteed to give the best coding performance.[4]

11.3   P-picture motion-compensated prediction ◐

P-pictures use forward motion-compensated prediction, so named because the pel predictions are projected forward in time from an earlier I- or P-picture in the sequence. Each macroblock has a motion vector associated with it. Since motion vectors tend to be highly correlated from macroblock to macroblock, the horizontal or vertical motion vector displacement, MD, is predicted from the corresponding motion vector displacement of the preceding macroblock in the slice, and only the difference, dMD, is coded. Thus, if PMD is the previous macroblock motion vector displacement,

$$dMD = MD - PMD \qquad (11.1)$$

---

[2] Another possibility is to fit a surface to the motion displacement information, thereby allowing for individual interpolated displacements for each pel in the macroblock. Unless the coding efficiency gains were very large, it would be difficult to justify this additional complexity.

[3] From a performance point of view, these two criteria are similar, but reasons are given in Chapter 13 for preferring MAD.

[4] But not the best rate-distortion.

This equation holds for both horizontal and vertical displacements.

The following rules apply to motion vector coding in P-pictures:

The predicted motion displacements are reset to zero at the start of a slice.

The predicted motion displacements are reset to zero after a macroblock is intra coded.

The predicted motion displacements are reset to zero when the macroblock is skipped.

The predicted motion displacements are reset to zero when `macroblock_motion_forward` is zero.

Because of subsampling of chrominance relative to luminance the relationship of MD to pel displacements depends on which component is being coded.

Within a given picture the displacements are either to full pel or half-pel precision, as defined in the picture header.

Motion displacement values that fall outside of the reference picture are not permitted.

By definition, a positive horizontal displacement means that the area in the reference picture is to the right, relative to the current macroblock position. Similarly, a positive vertical displacement means that the area in the reference picture is below, relative to the current macroblock position. Figure 11.1 illustrates these conventions. The reference picture precedes the current picture in time, and the motion vectors are therefore *forward* motion vectors. The sign conventions are natural for hardware implementation, in that positive is forward in time for normal raster scan video signals.

11.4   B-picture motion-compensated prediction ◑

B-pictures may use either forward or backward motion-compensated prediction or both. In backward motion-compensated prediction the reference picture occurs later in the sequence. Figure 11.2 illustrates the sign conventions for backward motion vectors. Note that the sign conventions for displacements are always relative to the current picture, independent of the direction in time.

Forward motion-compensated prediction in B-pictures is done much the same as in P-pictures, except for different rules for resetting the predicted value of the motion vector.

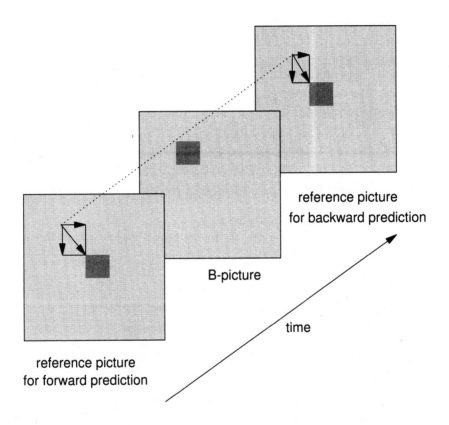

Figure 11.2: B-picture motion vector displacements. In both reference pictures positive displacements are to the right and down, relative to the macroblock being coded.

Backward motion-compensated prediction in B-pictures is done in exactly the same manner as forward motion-compensated prediction in B-pictures, except that the reference picture is an I- or P-picture that occurs later in the sequence. Among other things, backward motion-compensated prediction provides a means of predicting uncovered background.

In B-pictures the following rules apply:

> The predicted motion displacements are reset to zero at the start of a slice.

> The predicted displacements are reset to zero after a macroblock is intra coded.

> In skipped macroblocks the motion displacements are predicted from the preceding macroblock in the slice.

> When `motion_vector_forward` is zero, the forward motion displacements are predicted from the preceding macroblock in the slice.

> When `motion_vector_backward` is zero, the backward motion displacements are predicted from the preceding macroblock in the slice.

> The motion-compensated prediction is obtained from either an I- or a P-picture that occurs earlier in coded order in the sequence; B-pictures may not be used as a predictive reference.

> Because of subsampling of chrominance relative to luminance the relationship of MD to pel displacements depends on which component is being coded.

> Within a given picture the displacements are either to full pel or half pel precision, as defined in the picture header.

> Motion displacement values that fall outside of the reference picture are not permitted.

Both forward and backward motion-compensated prediction may be used simultaneously to give interpolated motion-compensated prediction. The prediction, `pel`, is a simple average of the pel values from the forward (`pel_for`) and backward (`pel_back`) motion-compensated reference pictures. Thus, for the `[i][j]`th element:

$$\mathtt{pel[i][j]} = (\mathtt{pel\_for[i][j]} + \mathtt{pel\_back[i][j]})//2 \qquad (11.2)$$

where the $//$ means that the quotient is rounded up to the nearest integer when the pel sum is odd.

header parameter	set in header	number of bits	displacement parameter
`full_pel_forward_vector`	p	1	precision
`forward_f_code`	p	3	range
`full_pel_backward_vector`	p	1	precision
`backward_f_code`	p	3	range
`motion_horizontal_forward_code`	mb	VLC	principal
`motion_horizontal_forward_r`	mb	`forward_r_size`	residual
`motion_vertical_forward_code`	mb	VLC	principal
`motion_vertical_forward_r`	mb	`forward_r_size`	residual

Table 11.1: Header parameters for motion vector computation: "p" stands for picture header; "mb" for macroblock header.

A possible reason for using interpolated motion-compensated prediction is to average the noise in the two reference pictures, thereby improving the prediction. This is, perhaps, one of the reasons why MPEG does not temporally weight the average in favor of the reference picture that is closer in time to the current picture.

11.5   Generic motion displacements ❶

The precision and range of motion vectors are set by several parameters in the picture header (see Section 8.5). For P-pictures the one-bit parameter `full_pel_forward_vector` sets the precision (1=full pel, 0=half pel), and `forward_f_code` sets the range. For B-pictures, the forward parameters are followed by an equivalent set of parameters for backward motion vector displacements.

In addition to the picture header parameters, differential motion vector displacements are transmitted for each macroblock, if the macroblock type is appropriate (see Section 8.7). Table 11.1 lists the various motion displacement parameters in the picture and macroblock headers.

Since forward and backward displacements are handled identically, we use the generic parameters listed in Table 11.2 in our discussion in the next section on the decomposition of the motion vector displacements into principal and residual parts.

11.6   Displacement principal and residual ❶

To code a motion displacement, MPEG decomposes it into two parts, a principal part and a residual. In this section we treat all motion displacements

generic name	full data element names	range
`full_pel_vector`	`full_pel_forward_vector` `full_pel_backward_vector`	0,1
`f_code`	`forward_f_code` `backward_f_code`	1,...,7
`r_size`	`forward_r_size` `backward_r_size`	0,...,6
`f`	`forward_f` `backward_f`	1,2,4,8,16,32,64

`motion_code`	`motion_horizontal_forward_code`
	`motion_vertical_forward_code`
	`motion_horizontal_backward_code`
	`motion_vertical_backward_code`
		-16,...,+16

`motion_r`	`motion_horizontal_forward_r`
	`motion_vertical_forward_r`
	`motion_horizontal_backward_r`
	`motion_vertical_backward_r`
		0,...,(f-1)

`r`	`compliment_horizontal_forward_r`
	`compliment_vertical_forward_r`
	`compliment_horizontal_backward_r`
	`compliment_vertical_backward_r`
		0,...,(f-1)

Table 11.2: Generic names for motion displacement parameters.

as signed integers. The interpretation of these integers in terms of full or half pel displacements comes later.

The principal part, which we designate by dMD_p, is a signed integer given by the product of motion_code and a scaling factor, f:

$$dMD_p = \text{motion\_code} \times \text{f} \tag{11.3}$$

where (by definition) motion_code is an integer in the range $\{-16, \ldots, +16\}$ coded in the bitstream by means of a VLC, and the scaling factor, f, is a power of two given by:

$$\text{f} = 2^{\text{r\_size}} \tag{11.4}$$

The exponent, r_size, is an integer, and is related to the picture header parameter, f_code, by:[5]

$$\text{r\_size} = \text{f\_code} - 1 \tag{11.5}$$

Since f_code is a three-bit parameter with values in the range $\{1 \ldots 7\}$, r_size has values in the range $\{0 \ldots 6\}$, and f therefore is a power of 2 in the range $\{1, 2, 4, \ldots 64\}$.

The residual, r, is by definition, a positive integer, but is defined a little differently than one might expect:

$$\text{r} = |dMD_p| - |\text{dMD}| \tag{11.6}$$

With this definition, $|dMD_p|$ must be $\geq |\text{dMD}|$. The value of r is coded in the bitstream by concatenating the ones-complement of r, motion_r, to the VLC for motion_code. From the definition of the ones-complement operator, motion_r is related to r by:

$$\text{motion\_r} = (\text{f} - 1) - \text{r} \tag{11.7}$$

where $\text{f} - 1$ creates a bit string of length r_size with all bits set.

The value of f must be chosen such that for the largest positive or negative differential displacement to be coded in the picture

$$-(16 \times \text{f}) \leq \text{dMD} < (16 \times \text{f}) \tag{11.8}$$

The smallest f that satisfies these criteria should be used. The reason for the positive bound magnitude excluding the value $(16 \times \text{f})$ will be discussed in Section 11.7.

Note that the values of dMD used in Equation 11.8 may or may not be the largest actual displacements. For example, if the constrained_parameters_

---

[5]Defining f_code such that a value of 0 is illegal is done to prevent accidental start code emulation.

f_code	r_size	f	f - 1	r	motion_r
1	0	1	0	-	0
2	1	2	1	0,1	1,0
3	2	4	3	0-3	3,...,0
4	3	8	7	0-7	7,...,0
5	4	16	15	0-15	15,...,0
6	5	32	31	0-31	31,...,0
7	6	64	63	0-63	63,...,0

Table 11.3: Range of motion_r as a function of f_code.

flag is set in the video sequence header, f may not exceed 8. Encoders may also choose to handle very large displacements suboptimally in order to obtain better coding efficiency for small displacements.

Given the value of f, the motion_code for the differential displacement of a macroblock is given by:

$$\text{motion\_code} = \frac{\text{dMD} + \text{Sign(dMD)} \times (\text{f} - 1)}{\text{f}} \qquad (11.9)$$

where the rounding is such that motion_code $\times$ f $\geq$ dMD.

The way f and r are defined make motion_code values of zero special. Independent of the value of f, when motion_code=0, there is no residual. There is also no residual when f=1. In this respect the coding of motion vectors is done exactly as in H.261. Indeed, compatibility with H.261 was a key influence in shaping the MPEG motion vector coding. MPEG, however, allows for much larger motion vectors, reflecting the more diverse nature of the video sources that must be coded.

Table 11.3 tabulates the differential motion displacement parameters, illustrating how the values are interlinked.

The decoder must reconstruct the value of dMD from motion_code, f, and r. If we solve for dMD using Equations 11.6 and 11.3, we get:

$$\text{dMD} = \text{motion\_code} \times \text{f} - \text{Sign(motion\_code)} \times \text{r} \qquad (11.10)$$

This is added to PMD to get the current motion displacement. Note, however that the modulo calculation described in the following section must be used to keep the motion displacements within the prescribed range.

11.7   Wraparound of motion displacements ◑

Normally, a differencing operation such as in Equation 11.1 requires that dMD be one bit greater in precision than MD and PMD. However, a remapping

motion_code	f	motion_code × f	r	dMD	PMD	MD
+1	4	4	2	+2	+63	-63
+1	4	4	3	+1	+63	-64
0	4	0	0	0	+63	63
-1	4	-4	3	-1	+63	62
...	...	...	...	...	...	...
+2	4	8	3	+5	10	15
+1	4	4	0	+4	10	14
+1	4	4	1	+3	10	13
+1	4	4	2	+2	10	12
+1	4	4	3	+1	10	11
0	4	0	0	0	10	10
-1	4	-4	3	-1	10	9
-1	4	-4	2	-2	10	8
-1	4	-4	1	-3	10	7
-1	4	-4	0	-4	10	6
-2	4	-8	3	-5	10	5
...	...	...	...	...	...	...
+1	4	4	2	+2	-64	-62
+1	4	4	3	+1	-64	-63
0	4	0	0	0	-64	-64
-1	4	-4	3	-1	-64	+63

Table 11.4: Examples of wraparound modulo calculation for r_size = 2. For this case, f is 4.

displacement	motion vector		
	temporal direction	spatial direction	pel precision
right_for	forward	horizontal	full
right_half_for	forward	horizontal	half
right_back	backward	horizontal	full
right_half_back	backward	horizontal	half
down_for	forward	vertical	full
down_half_for	forward	vertical	half
down_back	backward	vertical	full
down_half_back	backward	vertical	half

Table 11.5: Derived motion vector displacements.

of values by modulo $32 \times f$ (with the zero point shifted by $16 \times f$) is used to keep the precisions the same. The remapping calculation is as follows:

$$\text{MD} = [(\text{PMD} + \text{dMD} + 16 \times f)\%(32 \times f)] - 16 \times f \qquad (11.11)$$

where % signifies the modulo operation. If the original arithmetic does not overflow the modulo range, the remapping has no effect. A few numerical examples in Table 11.4 illustrate this.

A rule that is equivalent to Equation 11.11 is as follows: Whenever the value of MD is below the minimum allowed value, $32 \times f$ is added to it; whenever the value of MD is greater than the maximum allowed value, $32 \times f$ is subtracted from it.

A careful reader may have noticed a slight problem at this point. If the range is truly from $-16 \times f$ to $+16 \times f$, why is the value PMD $= +63$ used in Table 11.4 for the maximum positive PMD? In fact, neither differential motion displacements nor motion displacements exactly equal to $16 \times f$ are allowed (See Equation 11.8.). Therefore, for the example in Table 11.4, motion vector displacements from -64 to +63 are allowed. A displacement of +64 is not allowed, because it is exactly equal to $16 \times f$. A motion_code value of +16 is legal, however, provided that the residual is not zero.

In general, MD and dMD values are restricted to the range min to max, where min $= -16 \times f$ and max $= (16 \times f) - 1$.

11.8   Coding and decoding of motion vectors ◑

Depending on whether the computation is for B- or P-pictures, the motion vector calculation produces up to four displacements. As tabulated in Table 11.5, these four displacements correspond to two temporal prediction directions (forward and backward in time) and two displacement directions (horizontal and vertical). In addition, there are two displacement precisions, full pel and half pel. Furthermore, because the resolution of chrominance blocks is lower, the interpretation depends on whether the block in question is luminance or chrominance.

The parameters used to derive displacements in the decoder come from the picture header and the macroblock header. Table 11.1 lists these parameters. Note that `picture_type` and `macroblock_type` determine when these parameters are present in the bitstream.

The MPEG standard provides four separate, interleaved, pseudocode procedures for computing the four spatial and temporal combinations of motion displacements. However, the procedures that determine the motion displacements are identical for all four cases, and we have therefore replaced them with a single pseudocode function call. Just as in the preceding sections, the names given by MPEG to variables are shortened to the simpler generic names listed in Table 11.2 by deleting forward/backward and horizontal/vertical from the compound names used in the standard.

11.8.1   Decoding motion displacements ◑

Depending on which bits are set in `macroblock_type`, the macroblock may contain information about both forward and backward motion vectors, either forward or backward alone, or no motion vector information at all.

If any motion vector information is decoded (see Section 8.7 in Chapter 8), the actual motion displacements must be calculated. A generic procedure for doing this, `motion_displacement()` is shown in Figure 11.3.[6] The parameters required for the call to this procedure are the motion displacement prediction, `PMD`, and the decoded parameters, `motion_code` and `motion_r`. The end result of the calculation is the new motion displacement, `MD`. Note that for all conditions where `motion_r` is not present in the bitstream, the first test in this procedure will set `motion_r` to zero.

At the end of this procedure, `MD` is simply a signed integer representing either full pel or half-pel displacement units. This is specified in the picture header for the motion displacement being calculated.

11.8.2   Forward motion vectors ◑

Given that the macroblock header has been decoded, a decoder must calculate motion vector displacements for each temporal and spatial direction coded for the macroblock. Figure 11.4 contains the pseudocode function, `motion_forward()`, that sets up parameters for the call to `motion_displacement` for forward motion displacements.

---

[6]This procedure is similar to the procedure defined in MPEG-2.

```
motion_displacement() {
   if ((f==1)||(motion_code==0)) { /* if minimum scale or dMD=0   */
      dMD=motion_code            /* no motion_r, so set to motion_code*/
   else {                        /* else residual part is sent        */
      dMD=1+f*(ABS(motion_code)-1);/* scale VLC coded value of dMD */
      r=motion_r;               /* compute residual of dMD           */
      dMD=dMD + r;              /* compute |dMD| from scale, residual*/
      if (motion_code < 0)      /* if motion_code negative, then     */
         dMD=-dMD;              /*    dMD is negative                */
   }                            /* end else residual part is sent    */
   MD=PMD+dMD;                  /* compute MD from dMD, previous MD   */
   range=32*f;                  /* compute range                     */
   if (MD>max)                  /* if greater than max MD allowed    */
      MD=MD-range;              /*    remap to value within range    */
   if (MD<min)                  /* if less than minimum MD allowed   */
      MD=MD+range;              /*    remap to value within range    */
}                               /* end motion_displacement() function*/
```

Figure 11.3: Pseudocode for calculating a motion displacement.

```
motion_forward() {                    /* calculate forward MD values    */
   /* calculate horizontal motion displacement                          */
   PMD=recon_right_for_prev;    /* set horizontal MD prediction   */
   f=horizontal_forward_f;      /* set horizontal f value         */
   motion_r=horizontal_forward_motion_r;   /* set motion_r        */
   motion_displacement();       /* calc. horizontal MD and PMD    */
   recon_right_for_prev=MD;     /* save prediction                */
   recon_right_for=MD;          /* set horizontal motion displ.   */

   /* calculate vertical motion displacement                            */
   PMD=recon_down_for_prev;     /* set vertical MD prediction     */
   f=vertical_forward_f;        /* set vertical f value           */
   motion_r=vertical_forward_motion_r;   /* set motion_r           */
   motion_displacement();       /* calc. vertical MD and PMD      */
   recon_down_for_prev=MD;      /* save prediction                */
   recon_down_for=MD;           /* set horizontal motion displ.   */
}
```

Figure 11.4: Pseudocode for calculating forward motion displacements.

```
motion_backward() {              /* calculate backward MD values  */
   /* calculate horizontal motion displacement                   */
   PMD=recon_right_back_prev;    /* set horizontal MD prediction  */
   f=horizontal_backward_f;      /* set horizontal f value        */
   motion_r=horizontal_backward_motion_r;   /* set motion_r       */
   motion_displacement();        /* calc. horizontal MD and PMD   */
   recon_right_back_prev=MD;     /* save prediction               */
   recon_right_back=MD;          /* set horizontal motion displ.  */

   /* calculate vertical motion displacement                      */
   PMD=recon_down_back_prev;     /* set vertical MD prediction     */
   f=vertical_backward_f;        /* set vertical f value          */
   motion_r=vertical_backward_motion_r;   /* set motion_r         */
   motion_displacement();        /* calc. vertical MD and PMD     */
   recon_down_back_prev=MD;      /* save prediction               */
   recon_down_back=MD;           /* set vertical motion displ.    */
}
```

Figure 11.5: Pseudocode for calculating backward motion displacements.

11.8.3   Backward motion vectors ◑

Figure 11.5, which applies to backward motion displacements, has exactly the same structure as Figure 11.4.

11.8.4   P- and B-picture motion vectors ◑

At this point we have laid the groundwork for the processing of P-picture macroblocks (Figure 11.6) and B-picture macroblocks (Figure 11.7). Note that if macroblock_intra is set, the bits signaling forward and backward motion vector coding must be clear. Note also that the rules for when to clear the motion displacement predictions are different for B- and P-pictures.

11.8.5   Calculating displacements for Y and C ◑

Because the resolution of chrominance is a factor of 2 lower both horizontally and vertically, the results of the motion vector computation are also dependent on whether the block in question is luminance or chrominance. Defining YMD as the motion displacement for luminance,

```
YHF=MD & 1;              /* set half-pel flag for luminance  */
YMD=MD >> 1;             /* whole pel motion displacement    */
```

where if YHF is set, interpolation of the prediction will be needed.

```
motion_displacement_P() {          /* calculate P-picture MDs    */
   if (macroblock_motion_forward) {  /* if motion vectors coded    */
      motion_forward();
   } else {                          /* no forward motion vectors  */
      recon_right_for=0;             /* clear motion displacement  */
      recon_down_for=0;              /* clear motion displacement  */
      recon_right_for_prev=0;        /* clear prediction           */
      recon_down_for_prev=0;         /* clear prediction           */
   }
   if (full_pel_forward_vector) {    /* double values if full pel */
      recon_right_for=recon_right_for<<1;
      recon_down_for=recon_down_for<<1;
   }
}
```

Figure 11.6: Motion vector coding for P-pictures.

Similarly, for chrominance we define CMD as the motion displacement and CHF as the half-pel flag signaling that interpolation is needed. Then, for chrominance:

```
CHF=(MD/2) & 1;           /* set half-pel flag for chrominance */
CMD=(MD/2) >> 1;          /* whole pel motion displacement     */
```

For horizontal forward or backward motion vectors, the displacement to the right from the macroblock in the current picture to the predictive area in the reference picture is:

right_half_for=YHF or right_half_for=CHF
right_for=YMD   or   right_for=CMD

Note that by convention, displacements to the right are positive, whereas displacements to the left are negative.

For vertical forward or backward motion vectors, the displacement down from the current picture to the reference picture is:

down_half_for=YHF or down_half_for=CHF
down_for=YMD   or   down_for=CMD

Displacements down are positive and displacements up are negative.

11.9   Strategies for determining motion vectors ◗

One of the many difficult tasks the encoder must perform is determining the motion vectors that are transmitted to the decoder. Chapter 13 will

```
motion_displacement_B() {          /* calculate B-picture MDs    */
   if (macroblock_intra) {         /* if intra-coded macroblock  */
      recon_right_for_prev=0;      /* clear motion predictions   */
      recon_down_for_prev=0;
      recon_right_back_prev=0;
      recon_down_back_prev=0;
   }
   if (macroblock_motion_forward) /* if motion vectors coded     */
      motion_forward();
   } else {                        /* take prediction as-is       */
      recon_right_for=recon_right_for_prev; /* pass pred. through */
      recon_down_for=recon_down_for_prev;
   }
   if (macroblock_motion_backward){  /* if motion vectors coded   */
      motion_backward();
   } else {                          /* take prediction as-is     */
      recon_right_back=recon_right_back_prev;/* pass pred. through*/
      recon_down_back=recon_down_back_prev;
   }
   if (full_pel_forward_vector) { /* double values if full pel    */
      recon_right_for=recon_right_for<<1;
      recon_down_for=recon_down_for<<1;
   }
   if (full_pel_backward_vector){ /* double values if full pel    */
      recon_right_back=recon_right_back<<1;
      recon_down_back=recon_down_back<<1;
   }
}
```

Figure 11.7: Motion vector coding for B-pictures.

discuss the strategies for doing this. Among the topics considered there is the criterion for determining a best motion vector value, the precision of motion vectors, the range of vector values, and fast search strategies.

The MPEG syntax already provides a number of constraints on the problem, however: The motion vector precision can only be half-pel or full pel. The range of the motion vectors is constrained to ±128 pels if the constrained parameter bit is set. The motion vector is constant over the macroblock and the macroblock is always 16×16 pels.

11.10   Motion compensation in MPEG-2 ◑

Table 10.19 in Chapter 10 lists the frame and field motion types for MPEG-2. As is seen in that table, motion types can be classified as frame or field, and motion-compensated predictions are made from the full frame or separately from individual fields.

Aside from the direct analogs of MPEG-1 applied to the full macroblock or to the two 16×8 sections of the macroblock (for the two interlaced fields) a new class of motion-compensated prediction called dual-prime is introduced. Dual-prime motion compensation applies only to interlaced video and is a way of averaging predictions from two adjacent fields of opposite parity when coding a given field or frame.

11.10.1   MPEG-2 motion vectors ●

MPEG-2 uses different names in its motion vector calculation. Table 11.6 lists the MPEG-2 and MPEG-1 full names along side our generic names. If there is a one-to-one map, the MPEG-2 full name immediately precedes each MPEG-1 full name. Otherwise, it precedes the set of MPEG-1 full names.

Figure 11.8 shows the MPEG-2 motion vector calculation. The parameters r, s, and t uniquely identify the motion vector being calculated. The r index keeps track of whether this is the first (0), second (1), third (2), or fourth (3) set of motion vectors. The s index is 0 for forward and 1 for backward motion vectors. The t index is 0 for horizontal and 1 for vertical motion vector components.

Frame prediction in frame pictures and field prediction in field pictures need only one set of motion vectors ($r=0$). Field prediction in frame pictures and 16×8 MC prediction in field pictures need two sets of motion vectors ($r=0,1$). The derived motion vectors in dual prime prediction create extra motion vectors ($r=2,3$). These derived motion vectors are created after the calculation in Figure 11.8; therefore, in this calculation of the motion displacement, PMV[r][s][t], the r index only has values of 0 and 1.

generic name	full name	MPEG-2	MPEG-1
dMD	delta	●	.
	down_little or down_big	.	●
	right_little or right_big	.	●
dmd	dmvector	●	.
dmd_h	dmvector[0]	●	.
dmd_v	dmvector[1]	●	.
f	f	●	.
	backward_f	.	●
	forward_f	.	●
f_code	f_code[0][t]	●	.
	forward_f_code	.	●
	f_code[1][t]	●	.
	backward_f_code	.	●
MD	vector'[r][s][0]	●	.
	recon_right_for	.	●
	vector'[r][s][1]	●	.
	recon_down_for	.	●
max	high	●	.
	max	.	●
min	low	●	.
	min	.	●
motion_code	motion_code[r][0][0]	●	.
	motion_horizontal_forward_code	.	●
	motion_code[r][0][1]	●	.
	motion_vertical_forward_code	.	●
	motion_code[r][1][0]	●	.
	motion_horizontal_backward_code	.	●
	motion_code[r][1][1]	●	.
	motion_vertical_backward_code	.	●
PMD	prediction	●	.
	PMV[r][s][0]	●	.
	recon_right_for_prev	.	●
	PMV[r][s][1]	●	.
	recon_down_for_prev	.	●

Table 11.6: (a) Generic names for MPEG-1 and MPEG-2 motion vector terms.

generic name	full name	MPEG-2	MPEG-1
r	residual[r][0][0]	●	.
	motion_horizontal_forward_r	.	●
	residual[r][0][1]	●	.
	motion_vertical_forward_r	.	●
	residual[r][1][0]	●	.
	motion_horizontal_backward_r	.	●
	residual[r][1][1]	●	.
	motion_vertical_backward_r	.	●
range	range	●	.
	32*backward_f	.	●
	32*forward_f	.	●
r_size	r_size	●	.
	backward_r_size	.	●
	forward_r_size	.	●
YHF	half_flag[0]	●	.
CHF	right_half_for	.	●
	half_flag[1]	●	.
	down_half_for	.	●
YMD	int_vec[0]	●	.
CMD	right_for	.	●
	int_vec[1]	●	.
	down_for	.	●

Table 11.6: (b) Generic names for MPEG-1 and MPEG-2 motion vector terms.

```
motion_vector_calc(r,s,t) {        /* from ISO/IEC 13818-2 7.6.3.1 */
  r_size=f_code[s][t] - 1;         /* number of remainder bits      */
  f=1 << r_size;                   /* principle step size           */
  high=(16*f) -1;                  /* maximum MD allowed            */
  low=(-16)*f;                     /* minimum MD allowed            */
  range=32*f;                      /* allowed range                 */
  if ((f==1)||                     /* if f is 1 or motion_code is   */
      (motion_code[r][s][t]==0))   /*   zero then no residual bits  */
    delta=motion_code[r][s][t];    /*   dMD = motion_code           */
  else {                           /* else residual bits sent       */
    delta=1+f*(Abs(motion_code[r][s][t])-1);
                                   /* dMD=1+f*(abs(motion_code)-1)  */
    delta=delta+motion_residual[r][s][t]; /* dMD=dMD+r              */
    if (motion_code[r][s][t]<0)    /* if negative                   */
      delta=-delta;                /*    dMD=-dMD                   */
  }                                /* end else residual bits sent   */
  prediction=PMV[r][s][t];         /* set PMD                       */
  if ((mv_format=="field") &&      /* if field prediction           */
      (t==1) &&                    /*    of vertical component      */
      (picture_structure=="frame"))/*    in frame picture then      */
    prediction=prediction DIV 2;   /*    halve vertical PMD         */
  vector'[r][s][t]=prediction + delta; /* MD=PMD+dMD                */
  if (vector'[r][s][t] < low)      /* if MD less than min           */
    vector'[r][s][t]=vector'[r][s][t] + range; /* add 32*f          */
  if (vector'[r][s][t] > high)     /* if MD greater than max        */
    vector'[r][s][t]=vector'[r][s][t] - range; /* subtract 32*f     */
  if ((mv_format=="field") &&      /* if field prediction           */
      (t==1) &&                    /*    of vertical component      */
      (picture_structure=="frame"))/*    in frame picture           */
    PMV[r][s][t]=vector'[r][s][t]*2; /* PMD=MD*2                    */
  else                             /* else no need to scale         */
    PMV[r][s][t]=vector'[r][s][t];/*    PMD=MD                      */
}                                  /* end motion_vector_calc()      */
```

Figure 11.8: MPEG-2 motion vector calculation.

This code closely resembles the `motion_displacement()` function until the `mv_format=='field'`... test. MPEG-1 supports neither interlaced video nor fields, and if the motion vectors are operating on fields of frame pictures, the vertical prediction displacement has to be cut in half to match to the field dimensions. (It is restored to frame dimensions before it is used as a prediction for the next macroblock.)

11.10.2   MPEG-2 dual-prime prediction ●

In MPEG-2 main profile dual-prime motion-compensated prediction is only used for P-pictures with no B-pictures between the P-picture and its reference picture. The constraint that there are no intervening B-pictures between the reference picture and the picture being coded guarantees a fixed temporal relationship between the reference fields and the predicted fields. Once the motion vector for a macroblock in a field of given parity is known relative to a reference field of the same parity, it is extrapolated or interpolated to obtain a prediction of the motion vector for the opposite parity reference field. This prediction is adjusted by adding a small shift to account for the half-pel vertical offset between the two fields. Then, the small horizontal and vertical corrections $(+1, 0, -1)$ coded in the bitstream are added. In calculating the pel values of the prediction the motion-compensated predictions from the two reference fields are averaged, a process that tends to reduce the noise in the data.

If a field picture is being coded, the coded motion vector always applies to the reference field of the same parity. Obtaining the motion vector for the opposite parity field between the two fields always involves interpolation.

If a frame picture is being coded, the coded motion vector still applies to the fields of the same parity, but a single vector is used for both fields. Obtaining motion vectors for the opposite parity fields now involves both interpolation and extrapolation, but the arithmetic is in other respects the same.

Two parameters, m and e, are used in the calculation of the motion vector for the opposite parity field. m is determined by whether the prediction of the second motion vector is obtained by interpolation (m=1) or extrapolation (m=3). e is determined by the direction of the half-pel shift needed to align the reference field to the field being coded.

Table 11.7 summarizes the values of m and e for valid combinations of `picture_structure`, `top_field_first` and the parity of the reference (for the derived motion vector) and predicted fields. Note that the reference and predicted fields always have opposite parity. If the coded horizontal and vertical motion vector displacements are `MD_h` and `MD_v` respectively, and the differential corrections are `dmd_h` and `dmd_v`, the calculated displacements for the opposite parity field (denoted by a ') are given by:

picture_ structure	top_field_ first	parity_ ref	parity_ pred	scale m	shift e
T '01'	-	B '1'	T '0'	1	−1
B '10'	-	T '0'	B '1'	1	+1
frame '11'	no '0'	T '0'	B '1'	1	+1
frame '11'	no '0'	B '1'	T '0'	3	−1
frame '11'	yes '1'	T '0'	B '1'	3	+1
frame '11'	yes '1'	B '1'	T '0'	1	−1

Table 11.7: MPEG-2 dual prime scaling factor m and line shift e.

generic name	full name	MPEG-2	MPEG-1
	p[y][x] (final prediction)	●	.
P[y][x]	pel_pred[y][x]	●	.
	pel[i][j]	.	●
	pel_pred_backward[y][x]	●	.
	pel_back[i][j]	.	●
	pel_pred_forward[y][x]	●	.
	pel_for[i][j]	.	●
Ref[y][x]	pel_ref[y][x]	●	.
	pel_future[i][j]	.	●
	pel_past[i][j]	.	●

Table 11.8: Generic names for MPEG-1 and MPEG-2 prediction terms.

```
MD_h'= (m*MD_h)//2 + dmd_h;      /* scale, halve, add dmd         */
MD_v'= (m*MD_v)//2 + dmd_v + e; /* scale, halve, add dmd, shift */
```

11.10.3 Motion-compensated predictions ●

To form a prediction P[y][x] for each sample in the prediction block, samples in the reference image are selected directly or created by interpolating neighbors. Figure 11.9 shows the motion-compensated prediction calculation function described in our generic terms for the four combinations of the horizontal (HF_h) and the vertical (HF_v) half-sample flag values. If neither flag is set (i.e., HF_h and HF_v both zero), then the prediction is just the sample that is MD_h to the right and MD_h down in the reference image from the point (y,x) where the prediction will be used, i.e. Ref[y+MD_v][x+MD_h]. If the vertical half-sample flag is set and the horizontal flag is not set, then the prediction comes from an interpolation (average) of the previous reference sample (Ref[y+MD_v][x+MD_h]) and the sample one line below it

```
mvpred_calc(y,x) {                          /* see ISO 13818-2 7.6.4            */
  if (!HF_h) && (!HF_v)                      /* if none are half pel            */
    P[y][x]=Ref[y+MD_v][x+MD_h];             /*   prediction offset by MD       */
  if (!HF_h) && (HF_v)                       /* if only vertical half pel       */
    P[y][x]=(Ref[y+MD_v][x+MD_h]             /*   then at x+MD_h average         */
      + Ref[y+MD_v+1][x+MD_h])//2;           /*   vertical neighbors            */
  if (HF_h) && (!HF_v)                       /* if only horizontal half pel     */
    P[y][x]=(Ref[y+MD_v][x+MD_h]             /*   then at y+MD_v average         */
      + (Ref[y+MD_v][x+MD_h+1])//2;          /*   horizontal neighbors          */
  if (HF_h) && (HF_v)                        /* if both axis are half pel       */
    P[y][x]=(Ref[y+MD_v][x+MD_h]             /*   the sum four neighbors        */
      + Ref[y+MD_v][x+MD_h+1]                /*   and integer divide by 4       */
      + Ref[y+MD_v+1][x+MD_h]                /*   with half values rounded      */
      + Ref[y+MD_v+1][x+MD_h+1])//4;         /*   up to nearest integer         */
}                                            /* end mvpred_calc(x,y)            */
```

Figure 11.9: MPEG-2 motion vector prediction calculation.

(Ref[y+MD_v+1][x+MD_h]). If only the horizontal half-sample flag is set, then a horizontal interpolation is needed. If both flags are set, then four neighbors are summed and divided by four. The type of division (//) used rounds the half-point up to the next larger integer.

12

Pel Reconstruction

Recent chapters have dealt with the MPEG-1 system syntax, video syntax, and motion-compensated prediction. All of these are necessary precursors to the pel reconstruction process that actually creates the decoded pictures. To reconstruct the pel values, the coefficients in each block must be dequantized and the inverse DCT (IDCT) taken. Then, for nonintra blocks in P- and B-pictures, the motion-compensated prediction must be added to the results of the IDCT. The pel reconstruction process in MPEG-1 and MPEG-2 is the central topic of this chapter.

The IDCT is not rigidly standardized in MPEG; rather, the standard requires that the IDCT meet certain tolerances. The intent here is to prevent the buildup of excessive differences between encoder and decoder IDCT outputs when a sequence has many P-pictures pictures between each I-picture. The predictions of P-pictures are daisy-chained, and small differences between implementations then accumulate. *IDCT mismatch*, the term for this effect, is a subtle problem. This chapter presents an analysis of the rather different procedures used in MPEG-1 and MPEG-2 for reducing IDCT mismatch.

12.1 Intra-coded block dequantization ◑

Figure 12.1 gives the `invQ_intra()` function, pseudocode for the dequantization (i.e., inverse quantization) for intra-coded blocks. (A mathematical description of quantization and dequantization is given in Section 3.6.2.) The flowchart corresponding to this procedure is shown in Figure 12.2. The pseudocode converts from zigzag scan order to row-column indexing almost immediately, whereas the flowchart does that conversion as nearly the last step. Both forms are valid.

Following the pseudocode in Figure 12.1 for all of the coefficients in the

```
invQ_intra() {                    /* from ISO/IEC 11172-2 2.4.4.1     */
  /* reconstruct coefficients for blocks in intra-macroblock          */
  for (m=0; m<8; m++) {           /* for horizontal index 0 to 7      */
    for (n=0; n<8; n++) {         /* for vertical index 0 to 7        */
      i=scan[m][n];               /* zigzag order index               */
      dct_recon[m][n]=
              (2*dct_zz[i]*quantizer_scale*intra_quant[m][n])/16;
                                  /* dequantize DCT coefficient       */
      if ((dct_recon[m][n]&1)==0)     /* if even value, oddify        */
         dct_recon[m][n]=dct_recon[m][n]-Sign(dct_recon[m][n]);
                                  /* -1 if positive, +1 if neg.*/
      if (dct_recon[m][n]>2047)       /* if too large and positive */
         dct_recon[m][n]=2047;        /*    clamp to max range        */
      if (dct_recon[m][n]<-2048)      /* if too large and negative */
         dct_recon[m][n]=-2048;       /*    clamp to min range        */
    }                             /* end for vertical index 0 to 7    */
  }                               /* end for horizontal index 0 to 7  */
  dct_recon[0][0]=dct_zz[0]*8/* overrule calculation for DC coef. */
}                                 /* end invQ_intra() function        */
```

Figure 12.1: InvQ_intra() function.

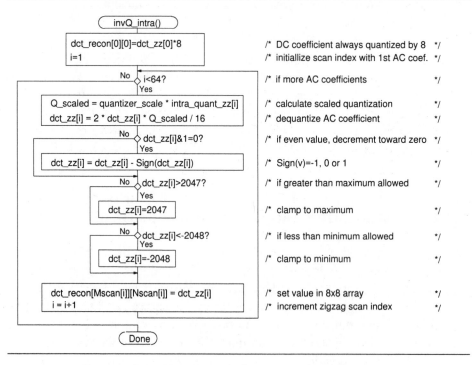

Figure 12.2: Flowchart for invQ_intra() function.

8×8 block, the zigzag-order index, i, for the mth column and nth row is determined from the `scan[m][n]` array. Then the coefficient at horizontal index m and vertical index n is dequantized by multiplying the decoded coefficient at zigzag index i, `dct_zz[i]`, by the `quantizer_scale` and the intra quantization matrix term, `intra_quant`, for that coefficient. Note that in the intra-coded block dequantization process (see Section 3.6.2) there is a multiplication by 2 and a truncation after division by 16.[1] Then, if the dequantized coefficient `dct_recon[m][n]` is even, it is forced to the adjacent odd integer value that is closer to zero. The process of guaranteeing that all dequantized AC coefficients are odd is known as *oddification*. Later sections in this chapter will explain in greater detail why "oddification" is part of the reconstruction process. The dequantized coefficients are clamped if necessary to the maximum (2047) and minimum (−2048) bounds.

In the last step in Figure 12.1, the dequantized DC coefficient, `dct_recon[0][0]`, is simply eight times the quantized DC coefficient, overruling the results from the loop above. Note that this recalculation of the DC coefficient removes oddification and clamping for the DC term.

12.2   DC prediction in intra macroblocks ◑

For intra-coded macroblocks all six blocks are always coded. The MPEG standard presents individual pseudocode for the coefficient reconstruction of the first luminance block, the subsequent luminance blocks, the Cr block, and the Cb block. In this section we have revised those steps and created function names to emphasize the commonality of everything except the DC prediction during dequantization. The function `invQ_intra()` performs the coefficient dequantization of individual blocks and was explained in the previous section. Note that the decoded value of the DC coefficient is a differential value, whereas the actual reconstructed dequantized value is the sum of this differential value and the prediction.

Figure 12.3 gives a function for reconstructing the coefficients for the first luminance block in an intra-coded macroblock. It calls the `invQ_intra()` function to dequantize the coefficients in the block. If the previous macroblock was skipped, the DC coefficient prediction is 1024 (8×128); otherwise, the prediction, `dct_dc_y_past` is obtained from the last luminance block in the previous macroblock. The prediction is added to the differential dequantized DC coefficient. The prediction for the DC coefficient for the next block luminance block is the reconstructed DC coefficient in this block.

---

[1]This dequantization process is exactly the same as truncation after division by 8 without multiplication by 2. We assume that this presentation was chosen to stress the similarity with the nonintra coded dequantization process where the multiplication by 2 is needed for the rounding operations.

```
invQ_intra_Y1() {                /* see ISO/IEC 11172-2 2.4.4.1      */
  /* reconstruct coefficients for 1st Y block in intra-macroblock */
  invQ_intra()                   /* compute 8x8 invQ for 1st Y block */
  if ((macroblock_address - past_intra_address)>1)
                                 /* if macroblocks skipped          */
     dct_recon[0][0]=(128*8)+dct_recon[0][0];
                                 /* add to default prediction       */
  else                           /* else previous block is prediction */
     dct_recon[0][0]=dct_dc_y_past+dct_recon[0][0];
                                 /* add to prediction               */
  dct_dc_y_past=dct_recon[0][0];/* set prediction for next Y block*/
}                                /* end invQ_intra_Y1() function    */
```

Figure 12.3: InvQ_intra_Y1() function.

```
invQ_intra_Ynext() {             /* see ISO/IEC 11172-2 2.4.4.1      */
  /* reconstruct coefficients for next Y block in intra-macroblock*/
  invQ_intra()                   /* compute 8x8 invQ for Y block     */
  dct_recon[0][0]=dct_dc_y_past+dct_recon[0][0];
                                 /* add to prediction               */
  dct_dc_y_past=dct_recon[0][0]; /*set prediction for next Y block*/
}                                /* end invQ_intra_Ynext() function  */
```

Figure 12.4: InvQ_intra_Ynext() function.

Figure 12.4 shows the invQ_intra_Ynext() function, a procedure for dequantizing the remaining luminance blocks in an intra-coded macroblock. After calling the dequantization function, invQ_intra(), the DC coefficient from the previous luminance block is added to the differential dequantized DC coefficient to give the reconstructed DC coefficient, dct_recon[0][0]. This becomes the DC prediction for the next block.

Figure 12.5 lists invQ_intra_Cb(), a function that performs the coefficient reconstruction for the Cb block of the macroblock. It is identical to the invQ_intra_Y1() function except that if the previous block was not skipped, the prediction has been saved from the previous Cb block in dct_dc_cb_past and the final DC coefficient is saved for the next Cb block.

An examination of the pseudocode in Figure 12.6 for dequantizing the Cr block will show that it is identical to Figures 12.3 and 12.5 with one exception: The prediction from the previous Cr block is saved in dct_dc_cr_past and the DC coefficient is saved in the same variable.

```
invQ_intra_Cb() {              /* see ISO/IEC 11172-2 2.4.4.1      */
   /* reconstruct coefficients for Cb block in intra-macroblock    */
   invQ_intra()                /* compute 8x8 invQ for Cb block     */
   if ((macroblock_address - past_intra_address)>1)
                               /* if macroblocks skipped            */
      dct_recon[0][0]=(128*8)+dct_recon[0][0];
                               /* add to default prediction         */
   else                        /* else previous block is prediction */
      dct_recon[0][0]=dct_dc_cb_past+dct_recon[0][0];
                               /* add to prediction                 */
   dct_dc_cb_past=dct_recon[0][0];
                               /* set prediction for next Cb block  */
}                              /* end invQ_intra_Cb() function      */
```

Figure 12.5: InvQ_intra_Cb() function.

```
invQ_intra_Cr() {              /* see ISO/IEC 11172-2 2.4.4.1      */
   /* reconstruct coefficients for Cr block in intra-macroblock    */
   invQ_intra()                /* compute 8x8 invQ for Cr block     */
   if ((macroblock_address - past_intra_address)>1)
                               /* if macroblocks skipped            */
      dct_recon[0][0]=(128*8)+dct_recon[0][0];
                               /* add to default prediction         */
   else                        /* else previous block is prediction */
      dct_recon[0][0]=dct_dc_cr_past+dct_recon[0][0];
                               /* add to prediction                 */
   dct_dc_cr_past=dct_recon[0][0];
                               /* set prediction for next Cr block  */
}                              /* end invQ_intra_Cr() function      */
```

Figure 12.6: InvQ_intra_Cr() function.

```
invQ_non_intra()                /* from ISO/IEC 11172-2 2.4.4.2-3   */
  /* dequantize coefficients in predictive-coded macroblock        */
  for (m=0; m<8; m++) {          /* for horizontal index 0 to 7     */
    for (n=0; n<8; n++) {        /* for vertical index 0 to 7       */
      i=scan[m][n];              /* zigzag order index              */
      dct_recon[m][n]=(((2*dct_zz[i])+Sign(dct_zz[i]))*
                       quantizer_scale*non_intra_quant[m][n])/16;
                               /* round/scale/dequantize DCT coef. */
      if ((dct_recon[m][n]&1)==0)    /* if even value, oddify       */
         dct_recon[m][n]=dct_recon[m][n]-Sign(dct_recon[m][n]);
                                 /* -1 if positive, +1 if neg.*/
      if (dct_recon[m][n]>2047)      /* if too large and positive */
         dct_recon[m][n]=2047;       /*    clamp to max range     */
      if (dct_recon[m][n]<-2048)     /* if too large and negative */
         dct_recon[m][n]=-2048;      /*    clamp to min range     */
      if (dct_zz[i]==0)         /* if quantized coef is zero      */
         dct_recon[m][n]=0;     /*   force dequantized coef to zero */
    }                           /* end for vertical index 0 to 7   */
  }                             /* end for horizontal index 0 to 7 */
}                               /* end invQ_non_intra() function   */
```

Figure 12.7: InvQ_non_intra() function.

12.3   Nonintra-coded block dequantization ◑

Since there is no DC coefficient prediction, the nonintra block dequantization procedure does not distinguish between luminance and chrominance blocks or between DC and AC coefficients. Figure 12.7 gives pseudocode for the dequantization of nonintra blocks, i.e., predictive-coded blocks. For all coefficients in the 8×8 block, the zigzag scan index, i, is determined and the decoded coefficient, dct_zz[i], is dequantized. The quantizer_scale scales the appropriate term from the nonintra quantizer matrix. Unlike the intra-coded dequantization process, the Sign() function is used to move the doubled value a half-quantization interval away from zero before the scaling and division by 16 with truncation toward zero.

The oddification and clamping is the same as for intra-coded blocks. A test is made to guarantee that if the quantized coefficient is zero, the final reconstructed coefficient remains zero.

Macroblocks can be skipped completely in predictive-coded pictures. In that case dct_recon[m][n]=0 for all m,n in skipped macroblocks and when pattern_code[i]==0.

12.4   IDCT mismatch ◐

Good compression of motion video sequences generally depends upon inter coding to achieve the desired performance. Since inter-coded blocks are added to previously reconstructed pels or data predicted from such pels, any differences between the IDCT used in the encoder and the IDCT used in the decoder will propagate until a block is intra coded. (see Figures 5.9 and 5.10 in Chapter 5). The difference between the encoder IDCT reconstructions and the decoder IDCT reconstructions is called IDCT mismatch error. These discrepancies accumulate and, if not controlled, can become visible, annoying distortions. This section discusses the IDCT mismatch problem and the techniques used to control it.

The FDCT and IDCT calculations are theoretically perfectly matched. An FDCT followed by an IDCT returns exactly the original input data, if done to infinite precision. Encoder quantization and variations in the FDCT implementations create deviations from this perfect FDCT model. Once the quantized coefficients are calculated, however, the Huffman encoding, transmission, and eventual Huffman decoding are lossless processes. Thus, all IDCT calculations start from the same quantized coefficients that the encoder produced.[2] Differences between encoder FDCTs may make a big difference in image quality, but have no effect on IDCT mismatch.

There are two major causes of IDCT mismatch. The first is the finite precision nature of practical IDCT implementations. The second is that the ideal IDCT will, under certain circumstances, provide outputs precisely at $N + 0.5$, where N is an integer. This makes the output particularly sensitive to small differences between the finite precision arithmetic procedures and rounding conventions.

12.4.1   IEEE Standard Specification for IDCT ◐

The MPEG standard could have fully specified the IDCT architecture and precision and thus eliminated the IDCT mismatch problem. Instead the committee chose to follow the example set by the CCITT SGXV committee in developing its H.261 teleconferencing standard. The numerical characteristics of the 8×8 IDCT are required to meet tight criteria designed to keep the additional noise introduced by mismatch between the encoder IDCT and the decoder IDCT below the level of noticeable quality degradation. Occasional forced intra coding is required to keep any errors from propagating for too many pictures.

The MPEG documents specify IEEE Std 1180-1990, *IEEE Standard Specifications for the Implementations of the 8×8 Inverse Discrete Cosine*

---

[2]For the purposes of this chapter we will discount the possibility of transmission errors or implementation errors resulting in errors in the decoded coefficients.

Transform [IEE91]. The abstract of this document indicates that it is expected to be useful "in visual telephony and similar applications where the 8×8 IDCT results are used in a reconstruction loop." The foreword indicates that the specifications were developed in the CCITT committee and sponsored by the IEEE Circuits and Systems Standards Committee as an IEEE standard at the international committee's request.

This standard gives a procedure for the accuracy measurement. Specific code is used to generate random numbers and to create integer samples in the range $-L$ to $+H$. The data is then rearranged into 8×8 blocks. Data representing 10,000 blocks are generated for ($L = 256$, $H = 255$), ($L = H = 5$), and ($L = H = 300$). For each 8×8 block the FDCT is executed using at least 64-bit floating point accuracy. The transform coefficients are rounded to the nearest integer and clamped to a range of -2048 to 2047.

Using the clamped transform coefficients as input, the IDCT is executed for each block using at least 64-bit floating point accuracy. The output is clipped to -256 to 255 to create the reference IDCT reconstructed pel $f_k(x, y)$, where $x, y = 0, \ldots, 7$ and $k = 1, \ldots, 10,000$. This creates the reference data for the test.

Then, using the proposed IDCT implementation, or an exact simulation thereof, the IDCT is executed and the output clamped to -256 to 255 to generate the test IDCT reconstructed pel $\hat{f}_k(y, x)$. For each of the 10,000 blocks and for each of the three ranges, calculate the peak, mean, and mean square error between the reference and test data for each sample location in the block. Then repeat the process is with opposite sign for every original input pel.

The measured error ($e_k(y, x)$) between the reference data and the test data has to meet certain requirements:

At any pel location (y, x) in the 8×8 blocks, the pel [3] peak error $ppe(y, x)$ magnitude should be less than or equal to 1.

The pel mean square error, $pmse(y, x)$, at location (y, x) in the 8×8 blocks should be less than or equal to 0.06.

The overall mean square error, $omse(y, x)$, should be less than or equal to 0.02.

The pel mean error, $pme(y, x)$, magnitude at location (y, x) in the 8×8 blocks should be less than or equal to 0.015.

The overall mean error, $ome(y, x)$, magnitude should be less than or equal to 0.0015.

---

[3]We substitute "pel" for "pixel" to stay consistent with MPEG notation.

For all zero input, the test IDCT should generate all zero output.

The equations for these error terms are given in Equations 12.1 to 12.6.

$$e_k(y,x) = \hat{f}_k(y,x) - f_k(y,x) \tag{12.1}$$

$$ppe(y,x) = max(|e_1(y,x)|, ..., |e_{10,000}(y,x)|) \tag{12.2}$$

$$pmse(y,x) = \frac{\displaystyle\sum_{k=1}^{10,000} e_k^2(y,x)}{10,000} \tag{12.3}$$

$$omse(y,x) = \frac{\displaystyle\sum_{x=0}^{7}\sum_{y=0}^{7}\sum_{k=1}^{10,000} e_k^2(y,x)}{64,000} \tag{12.4}$$

$$pme(y,x) = \frac{\displaystyle\sum_{k=1}^{10,000} e_k(y,x)}{10,000} \tag{12.5}$$

$$ome(y,x) = \frac{\displaystyle\sum_{x=0}^{7}\sum_{y=0}^{7}\sum_{k=1}^{10,000} e_k(y,x)}{64,000} \tag{12.6}$$

Having specified the required accuracy, this IEEE standard goes further to specify a solution to the problem when the output of the IDCT is exactly $N+0.5$, where N is an integer. Assuming that only one coefficient is nonzero, the 2-D IDCT in Equation 12.7 gives:

$$f(x,y) = \frac{C(\mu)}{2}\frac{C(\nu)}{2}F(\mu,\nu)\cos\left[(2x+1)\mu\pi/16\right]\cos\left[(2y+1)\nu\pi/16\right] \tag{12.7}$$

where the constant, $C(\mu)$ (and $C(\nu)$) is given by:

$$C(\mu) = 1/\sqrt{2} \quad \text{if } \mu = 0$$
$$C(\mu) = 1 \quad\quad \text{if } \mu > 0$$

For μ and ν equal to zero or four, the equation reduces to

$$f(x,y) = (\pm 1/8)F(\mu,\nu) \tag{12.8}$$

If $F(\mu,\nu)$ is equal to $(8m+4)$, the rounding of the 0.5 can give m or $m+1$ depending upon tiny differences in precision. The solution proposed in the IEEE standard is to avoid even values for $F(\mu,\nu)$. Hence, the oddification rules used in MPEG-1. Section 12.8 explores the cases that lead to mismatch in greater detail.

12.5   Limiting error propagation ◐

To guarantee that IDCT mismatch errors do not propagate forever, macroblocks must be intra-coded at least once every 132 P-pictures, a process called *forced updating*. In practice, forced updating is rare; I-pictures are commonly 10 to 20 pictures apart in order to simply image editing, fast forward playback, fast reverse playback, and semirandom access.

12.6   H.261 and H.263 oddification ◐

Oddification is not unique to MPEG-1. The CCITT SGXV specialist group XV extensively studied this problem as it developed the H.261 teleconferencing standard. [ITU93] Instead of a matrix of 64 quantization values, that standard uses the same quantization value $QUANT$ for all coefficients in the macroblock except intra DC coefficients. The following equations specify the reconstruction levels (REC) from the quantized coefficient $LEVEL$ transmitted in the H.261 bitstream.

$$
\begin{aligned}
REC &= \quad QUANT(2 \times LEVEL + 1) && \text{if } LEVEL > 0, QUANT \text{ odd} \\
REC &= \quad QUANT(2 \times LEVEL - 1) && \text{if } LEVEL < 0, QUANT \text{ odd} \\
REC &= \quad QUANT(2 \times LEVEL + 1) - 1 && \text{if } LEVEL > 0, QUANT \text{ even} \\
REC &= \quad QUANT(2 \times LEVEL - 1) + 1 && \text{if } LEVEL < 0, QUANT \text{ even} \\
REC &= \quad 0 && \text{if } LEVEL = 0 \\
REC &= \quad 8 \times LEVEL && \text{if intra DC coefficient}
\end{aligned}
$$

$$(12.9)$$

The reconstruction values, REC, are clipped to a range of -2048 and 2047 even if this undoes the oddification.

$QUANT$ ranges in size from 1 to 31 and the stepsize is twice $QUANT$. The ± 1 immediately following $LEVEL$ introduces a dead zone around zero for all coefficients except the intra DC coefficient; in MPEG-1 only the nonintra coefficients have a dead zone. The -1 (for $LEVEL$ positive and $QUANT$ even) and the $+1$ (for $LEVEL$ negative and $QUANT$ even) are the oddification. In H.261 the parity of $QUANT$ is sufficient to determine if oddification is needed. The H.263 teleconferencing standard[ITU95] adopted the same reconstruction equations, although it expresses them differently.

12.7   MPEG-2 pel reconstruction ◐

MPEG-2 uses different names in its pel reconstruction calculation. Table 12.1 lists the MPEG-1 and MPEG-2 full names next to a descriptive phrase. If there is an equivalency, the MPEG-1 and MPEG-2 names are on

Description	MPEG-1 name	MPEG-2 name
zigzag scan order (1-D)	i	n
zigzag scan order (2-D)	scan[m][n]	scan[0][ν][μ] scan[1][ν][μ]
DC dequan. coeff.	dct_recon[0][0]	F[0][0]
dequantized oddified clamped toggled	dct_recon[m][n] dct_recon[m][n] dct_recon[m][n]	F"[ν][μ] F'[ν][μ] F[ν][μ]
Y DC predictor Cb DC predictor Cr DC predictor	dct_dc_y_past dct_dc_cb_past dct_dc_cr_past	dc_dct_pred[0] dc_dct_pred[1] dc_dct_pred[2]
DC coefficient	dct_zz[0]	QFS[0]
VLD output (1-D)	dct_zz[i]	QFS[n]
VLD output (2-D)	F(m,n)	QF[ν][μ]
quantizer scale	quantizer_scale	quantiser_scale[0] quantiser_scale[1]
VLD output DC	dct_dc_size_luminance dct_dc_size_chrominance	dc_dct_size_luminance dc_dct_size_chrominance
DC extra bits	dct_dc_differential	dc_dct_differential
quantization matrix (2-D)	intra_quant[m][n] non_intra_quant[m][n]	W[0][μ][ν] W[1][μ][ν] W[2][μ][ν] W[3][μ][ν]
reconstructed sample	pel	d[y][x]
output of IDCT	f(x,y)	f[y][x] (9-bit)
MC prediction	pel[i][j]	p[y][x]
MC prediction reference samples	pel_past[i][j] pel_future[i][j] pel_for[i][j] pel_back[i][j]	pel_ref[y][x] pel_ref[y][x] pel_ref[y][x] pel_ref[y][x]

Table 12.1: MPEG-1 and MPEG-2 names for pel reconstrution terms.

```
invQ_intra() {                    /* for ISO/IEC 13818-2 7.4.5          */
   /* reconstruct coefficients for blocks in intra-macroblock          */
   sum=0;                         /* clear toggle sum                   */
   for (m=0; m<8; m++) {          /* for horizontal index 0 to 7        */
      for (n=0; n<8; n++) {       /* for vertical index 0 to 7          */
         if (n==0) && (m==0) {    /* if intra DC                        */
            F[0][0]=intra_dc_mult*QF[0][0]; /* scale by 8 or 4 or 2     */
         } else {                 /* else intra AC coefficient          */
            F[m][n]=(2*QF[m][n]*quantiser_scale*W[0][m][n])/32;
         }                        /* scale and dequantize DCT coef.     */
         if (F[m][n]>2047) {      /* if too large                       */
            F[m][n]=2047;         /*    clamp to max range              */
         }                        /* end - if too large                 */
         if (F[m][n]<-2048) {     /* if too small                       */
            F[m][n]=-2048;        /*    clamp to min range              */
         }                        /* end - if too small                 */
         sum=sum+F[m][n];         /* sum all coefficients               */
      }                           /* end for vertical index 0 to 7      */
   }                              /* end for horizontal index 0 to 7    */
   if ((sum&1)==0) {              /* if even sum                        */
      if ((F[7][7]&1)!=0) {       /* if odd                             */
         F[7][7]=F[7][7]-1;       /*    toggle last bit                 */
      } else {                    /* else even                          */
         F[7][7]=F[7][7]+1;       /*    toggle last bit                 */
      }
   }                              /* end - if even sum                  */
}                                 /* end invQ_intra() function          */
```

Figure 12.8: MPEG2 invQ_intra() function.

the same line. Otherwise, separate lines are used. In the MPEG-2 pseudocode in the next few sections we will cast the reconstruction pseudocode in the MPEG-1 formalism as much as possible. Therefore, many of the MPEG-2 names in Table 12.1 will not be used here. However, the table may be useful when comparing our treatment to the actual MPEG-2 standard.

12.7.1 MPEG-2 intra-coded block reconstruction ◗

Figure 12.8 shows pseudocode for the MPEG-2 invQ_intra() function. The code has been kept as close as possible to Figure 12.1 to make comparison with MPEG-1 easier, and in the process, the name distinction for each stage (dequantizing, clamping, and toggling) has been dropped. The toggle sum, the sum of all of the dequantized coefficients, is cleared before reconstructing the DCT coefficients. m and n are used here for the horizontal and vertical indices.

The intra DC coefficient (m and n are zero) is handled as a special case. The quantized DC coefficient QF[0][0] is multiplied by the appropriate value of intra_dc_mult given in Table 10.11 for the selected DC coefficient precision. The AC quantized coefficients QF[m][n] are doubled, multiplied by the quantizer scale and the intra quantizer matrix coefficient W[0][m][n].[4] The integer division by 32 with truncation toward zero appears to be different from the procedure followed in MPEG-1. However, quantiser_scale has a factor of 2 contained in it if q_scale_type selects quantiser_scale[0] in Table 10.18 and a similar, though not so obvious, scaling is contained in quantiser_scale[1].

The dequantized coefficients are immediately checked to see if they are out of range and clamped if necessary. Then the clamped value is added to the toggle sum. If the final toggle sum of the 64 coefficients is even, the coefficient in the lower right corner (m and n are seven) has its least significant bit toggled so that the dequantized coefficient sum is odd. The intra DC term is added to the DC prediction before the IDCT is computed.

12.7.2 MPEG-2 nonintra-coded block reconstruction ◗

Figure 12.9 gives the pseudocode for the nonintra coded blocks. In nonintra coding the DC term is a difference and is coded as if it was simply another AC coefficient. The dequantization step includes adding the sign of the quantized coefficient QF[m][n] to its doubled value (see Section 3.6.2. This introduces a dead zone around zero. Note that quantization matrix 1 (W[1][m][n]) is used. The clamping and toggling procedures are the same as for intra-coded blocks.[5]

12.8 IDCT mismatch control ●

The IEEE Std 1180 recommends that even parity IDCT coefficients be forced to odd values as the means of controlling the IDCT mismatch problem (see Section 12.4.1). MPEG-1, H.261, and H.263 all use this oddification technique. The IEEE solution assumes that the primary problem comes from single nonzero IDCT coefficients in a block. Higher order combinations were found in CCITT studies to be unimportant. However, when pushing to MPEG-2 higher qualities and higher bit rates, these assumptions are no longer valid. The two papers discussed in the following sections explored

---

[4]For 4:2:0 picture format the intra-coded blocks always use quantizer matrix 0. Quantizer matrix 2 may be used for the chrominance blocks for 4:2:2 and 4:4:4 picture formats.

[5]For 4:2:0 picture format the nonintra-coded blocks always use quantizer matrix 1. Quantizer matrix 3 may be used for the chrominance blocks for 4:2:2 and 4:4:4 picture formats.

```
invQ_non_intra() {              /* for ISO/IEC 13818-2 7.4.5       */
  /* reconstruct coefficients for blocks in nonintra-macroblock    */
  sum=0;                        /* clear toggle sum                */
  for (m=0; m<8; m++) {         /* for horizontal index 0 to 7     */
    for (n=0; n<8; n++) {       /* for vertical index 0 to 7       */
      F[m][n]=(2*QF[m][n]+Sign(QF[m][n]))*quantiser_scale
                                *W[1][m][n])/32;
                                /* scale and dequantize DCT coef.  */
      if (F[m][n]>2047) {       /* if too large                    */
        F[m][n]=2047;           /*   clamp to max range            */
      }                         /* end - if too large              */
      if (F[m][n]<-2048) {      /* if too small                    */
        F[m][n]=-2048;          /*   clamp to min range            */
      }                         /* end - if too small              */
      sum=sum+F[m][n];          /* sum all coefficients            */
    }                           /* end for vertical index 0 to 7   */
  }                             /* end for horizontal index 0 to 7 */
  if ((sum&1)==0) {             /* if even sum                     */
    if ((F[7][7]&1)!=0) {       /* if odd                          */
      F[7][7]=F[7][7]-1;        /*   toggle last bit               */
    } else {                    /* else even                       */
      F[7][7]=F[7][7]+1;        /*   toggle last bit               */
    }
  }                             /* end - if even sum               */
}                               /* end invQ_non_intra() function   */
```

Figure 12.9: MPEG2 invQ_non_intra() function.

the implications for MPEG-2 and led to a different solution. In this section we will review these papers and go into greater detail about the mismatch problem.

12.8.1   Review of Katayama's paper ●

In a document submitted to MPEG, Katayama [Kat93] makes the interesting observation that the IDCT given in Equation 12.10 can be recast into the form given in Equation 12.12.

$$f(x, y) = \sum_{\mu=0}^{7} \sum_{\nu=0}^{7} \frac{C(\mu)}{2} \frac{C(\nu)}{2} F(\mu, \nu) \cos\left[(2x+1)\mu\pi/16\right] \cos\left[(2y+1)\nu\pi/16\right]$$

(12.10)

where:

x and y are spatial indices;

μ and ν are DCT frequency indices;

$F(\mu, \nu)$ is the DCT coefficient for frequencies μ ,ν;

The constant $C(a)$ (where a is either μ or ν) is given by:

$$\begin{aligned} C(a) &= 1/\sqrt{2} \quad \text{for } a = 0 \\ &= 1 \qquad\quad a > 0 \end{aligned}$$

(12.11)

The IDCT can be written as:

$$f(x, y) = \sum_{\mu=0}^{7} \sum_{\nu=0}^{7} C_{x,y,\mu,\nu} F(\mu, \nu)$$

(12.12)

where

$$C_{x,y,\mu,\nu} = (1/4)C(\mu)C(\nu) \cos\left[(2x+1)\mu\pi/16\right] \cos\left[(2y+1)\nu\pi/16\right] \quad (12.13)$$

Because $(2x+1)\mu$ is zero only when μ is zero, the $C_{x,y,\mu,\nu}$ can be reduced to a simpler format. Define $k = (2x+1)\mu$ and $l = (2y+1)\nu$. Then,

$$\begin{aligned} C_{x,y,\mu,\nu} &= C_{(2x+1)\mu} C_{(2y+1)\nu} \\ C_{x,y,\mu,\nu} &= C_k C_l \end{aligned}$$

(12.14)

where C_k is given by:

$$\begin{aligned} C_k &= 1/\sqrt{2} \qquad\quad \text{if } k = 0 \\ C_k &= \cos\left(k\pi/16\right) \quad \text{if } k > 0 \end{aligned}$$

(12.15)

Because of the IDCT normalization requirements,

$$C_0 = (1/2)\sqrt{2} \qquad (12.16)$$

From trignometric identities, the cosine terms C_k can be shown to have the following irrational forms:

$$
\begin{aligned}
C_1 &= (1/2)\sqrt{2 + \sqrt{2 + \sqrt{2}}} \\[2mm]
C_2 &= (1/2)\sqrt{2 + \sqrt{2}} \\[2mm]
C_3 &= (1/2)\sqrt{2 + \sqrt{2 - \sqrt{2}}} \\[2mm]
C_4 &= (1/2)\sqrt{2} \\[2mm]
C_5 &= (1/2)\sqrt{2 - \sqrt{2 - \sqrt{2}}} \\[2mm]
C_6 &= (1/2)\sqrt{2 - \sqrt{2}} \\[2mm]
C_7 &= (1/2)\sqrt{2 - \sqrt{2 + \sqrt{2}}}
\end{aligned}
\qquad (12.17)
$$

The IDCT always has products $C_k C_l$, and products of square roots are not necessarily irrational numbers. Indeed, the products $C_0 C_0$, $C_0 C_4$, $C_4 C_0$, and $C_4 C_4$ are 1/2, and in this case, the IDCT has a normalization of 1/8. Consequently, DCT coefficients of 4, 12, 20, ... of either sign will be exactly 0.5 before conversion to integer format. Slight differences between implementations in the integer arithmetic procedures in the IDCT can cause the output to round up or down, and this is the cause of IDCT mismatch from rounding.

MPEG-1 oddification attempts to cure this by forcing the dequantized IDCT coefficients to odd values. If only one coefficient is likely to occur at an amplitude that triggers mismatch, oddification will prevent it.

However, oddification does not prevent mismatch for combinations of coefficients. In general, whenever linear combinations of $C_k C_l$ products can produce a rational number, mismatch can occur at particular amplitudes. For example, consider the combination

$$IDCT = \left(\frac{1}{4}\right)[C_2 C_2 F(2,2) + C_6 C_6 F(6,6)] \qquad (12.18)$$

Since

$$C_2 C_2 = \left(\frac{1}{4}\right)(2 + \sqrt{2}) \qquad (12.19)$$

	0	1	2	3	4	5	6	7
0	$+C_0$	$+C_0$	$+C_0$	$+C_0$	$+C_0$	$+C_0$	$+C_0$	$+C_0$
1	$+C_1$	$+C_3$	$+C_5$	$+C_7$	$-C_7$	$-C_5$	$-C_3$	$-C_1$
2	$+C_2$	$+C_6$	$-C_6$	$-C_2$	$-C_2$	$-C_6$	$+C_6$	$+C_2$
3	$+C_3$	$-C_7$	$-C_1$	$-C_5$	$+C_5$	$+C_1$	$+C_7$	$-C_3$
4	$+C_4$	$-C_4$	$-C_4$	$+C_4$	$+C_4$	$-C_4$	$-C_4$	$+C_4$
5	$+C_5$	$-C_1$	$+C_7$	$+C_3$	$-C_3$	$-C_7$	$+C_1$	$-C_5$
6	$+C_6$	$-C_2$	$+C_2$	$-C_6$	$-C_6$	$+C_2$	$-C_2$	$+C_6$
7	$+C_7$	$-C_5$	$+C_3$	$-C_1$	$+C_1$	$-C_3$	$+C_5$	$-C_7$

Table 12.2: Array of C_k values.

and

$$C_6 C_6 = \left(\frac{1}{4}\right)(2 - \sqrt{2}) \qquad (12.20)$$

This can be recast as two terms:

$$IDCT = \left(\frac{1}{8}\right)[F(2,2) + F(6,6)] + \left(\frac{1}{16}\right)\sqrt{2}\,[F(2,2) - F(6,6)] \quad (12.21)$$

Mismatch cannot occur when the coefficient amplitudes are not matched, because the $\sqrt{2}$ terms do not cancel. In addition, however, even when the amplitudes are matched, only those amplitude values that give an IDCT output with a fractional part equal to 0.5 can produce mismatch. This occurs when the matched amplitudes are each equal to $4m + 2$, where m is an integer.

There are only 8 possible values for the magnitude of subscript k in Equation 12.15 and $|k| = 0$ is a special case in which the weighting is given by $1/\sqrt{2}$ instead of 1. Because of this, the values C_k are repeated many times in the 8x8 IDCT array. Table 12.2 shows this array for μ down the left and x across the top.

The cross-product of two cosine values gives the array of values in Table 12.3 for $k, l = 0 \ldots 7$. Note that $\cos(0)$ is replaced by $1/\sqrt{2}$ to reflect the $C(\mu)$ and $C(\nu)$ values for a zero cosine argument. Four cross-products are exactly 0.5, and these are the values where IDCT mismatch problems are severe, because any one of these coefficients can by itself trigger mismatch between two IDCTs.

12.8.2   Review of Yagasaki's paper ●

Yagasaki [Yag93] notes that the MPEG-1 solution of oddification is insufficient for MPEG-2 because MPEG-2 operates at higher bit rates. The

	0	1	2	3	4	5	6	7
0	0.500	0.694	0.653	0.588	0.500	0.393	0.271	0.138
1	0.694	0.962	0.906	0.815	0.694	0.545	0.375	0.191
2	0.653	0.906	0.854	0.768	0.653	0.513	0.354	0.180
3	0.588	0.815	0.768	0.691	0.588	0.462	0.318	0.162
4	0.500	0.694	0.653	0.588	0.500	0.393	0.271	0.138
5	0.393	0.545	0.513	0.462	0.393	0.309	0.213	0.108
6	0.271	0.375	0.354	0.318	0.271	0.213	0.146	0.075
7	0.138	0.191	0.180	0.162	0.138	0.108	0.075	0.038

Table 12.3: Array of $C_k C_l$ values.

MPEG-1 oddification can not prevent mismatch where there is more than one coefficient with μ and ν equal to 0 or 4.:

$$f(x,y) = (1/8)\left(F(0,0) \pm F(0,4) \pm F(4,0) \pm F(4,4)\right) \qquad (12.22)$$

$f(x,y)$ will be exactly 0.5 when the coefficients sum to a multiple of four but not a multiple of eight.

Other combinations can also occur that give mismatch after MPEG-1 oddification. Two coefficients with the same value at special positions can cause a mismatch if they sum to a multiple of four and not a multiple of eight. Two examples are $F(3,1) = F(1,3)$ and $F(1,5) = F(5,1)$.

This paper proposed a new method of oddification. All of the reconstructed coefficients are summed and if the sum is odd, nothing further is done. (Note that none of the mismatch cases has an odd sum.) If the sum is even, only one coefficient needs to be modified. The highest frequency coefficient, $F(7,7)$ is one that gives little perturbation and is also the least visible. Yagasaki proposed to increase $F(7,7)$ by one unit if the coefficient sum was even or zero.

The paper noted that an exclusive OR between the least significant bits in all coefficients was sufficient to determine if the sum was odd or even.

The paper gave simulation results for 60 frames from the flower garden sequence for a GOP of 15 frames, an I- or P-frame every three pictures, and mquant=0.5, 1 (fixed).[6] For intra DC precision of 8 bits and mquant=1, the mismatch decreased from 3,232 mismatches without oddification to 1,916 mismatches with MPEG-1 oddification, and to zero mismatches with the proposed method. For intra DC precision of 8 bits and mquant=0.5, the

---

[6]Using the smallest values from the two MPEG-2 quantiser_scale tables with the default nonintra quantization matrix (fixed 16) gives 1 and 0.5 for mquant.

mismatch decreased from 41,262 mismatches without oddification to 9,576 mismatches with MPEG-1 oddification, and to 868 mismatches with the proposed method. Results also showed that for all bit rates, the proposed method has improved signal-to-noise ratio compared to the MPEG-1 oddification method.

This proposal is close to the MPEG-2 method of toggling the least significant bit of $F(7,7)$ if the sum of all of the coefficients is even.

13

Motion estimation

Motion compensation is a normative part of MPEG, whereas motion estimation, the topic of this chapter, is not. Motion compensation refers to the use of motion displacements in the coding and decoding of the sequence; motion estimation refers to the determination of the motion displacements.

In the encoder the difference between source picture and prediction is coded; in the decoder this difference is decoded and added to the prediction to get the decoded output. Both encoder and decoder use the same motion displacements in determining where to obtain the prediction. However, the encoder must *estimate* the displacements before encoding them in the bitstream; the decoder merely decodes them. Chapter 11 has already described the syntax for the coding of motion displacements.

Motion estimation is absolutely necessary in any complete implementation of an MPEG encoder, but the procedures for motion estimation are not normative. What is normative is the rules and constraints such as one vector per macroblock, precisions of full pel or half-pel, the range of motion vector sizes, and the restriction to simple translational motions. However, these are motion compensation rules, not motion estimation rules.

Motion estimation techniques for simple translational motion fall into two main classes, pel-recursion and block matching. Pel-recursion techniques are used primarily for systems where the motion vectors can vary from pel to pel. Block matching techniques are used primarily where a single motion vector is applied to a block of pels. The treatment in this chapter is oriented almost exclusively to block matching techniques, for these apply more directly to MPEG.

When determining the optimal motion displacement of the prediction, a full search over every possible motion displacement is guaranteed to produce the best possible value. This assumes, however, that the criterion for best is known, and it also assumes that the computational resources needed for a full search are available. Motion estimation is one of the most demanding

283

aspects of MPEG, from the point of view of computational complexity.

If computational constraints apply, there are a number of fast search algorithms available. Generally, these algorithms use approximations that decrease the quality of the search, sometimes finding displacement values that are not the best. In this chapter we discuss the criteria for determining the best motion displacements, and review the literature on fast search techniques. We also provide an quantitative analysis of the relative merits of the various fast search algorithms.[1]

13.1   Criterion for block matching ◑

A search for a best motion vector value requires, of course, a criterion for what is best. In general, displacements are chosen that either maximize correlation or minimize error between a macroblock and a corresponding array of pel values (possibly interpolated to fractional pel positions) in the reference picture. Correlation calculations are computationally expensive, and error measures such as mean square error (MSE) and mean absolute distortion (MAD) are more commonly used. MAD is perhaps the simplest and most commonly used measure of best match[MPG85].

MAD is defined for a 16×16 pel macroblock as follows: Let $MAD(x, y)$ be the MAD between a 16×16 array of pels of intensities $V_n(x + i, y + j)$ at macroblock position (x, y) in source picture n, and a corresponding 16×16 array of pels of intensity $V_m(x + dx + i, y + dy + j)$ at macroblock position $(x + dx, y + dy)$ in reference picture m:

$$MAD(x, y) = (1/256) \sum_{i=0}^{15} \sum_{j=0}^{15} |V_n(x + i, y + j) - V_m(x + dx + i, y + dy + j)|$$

(13.1)

The 16×16 array in picture m is displaced horizontally by dx and vertically by dy. By convention, x, y refers to the upper left corner of the macroblock, indices i, j refer to values to the right and down, and displacements dx, dy are positive when to the right and down.

Note that the *sum of absolute distortions* (SAD), is sometimes used in place of MAD. It is equivalent, but omitting the normalization by 256 allows an integer representation that is usually computationally more efficient.

Another measure that could be used, but has not been (as far as the authors know), is minimization of the bitstream. This requires, in principle, trial codings for each possible vector — a prohibitive computational require-

---

[1]This chapter draws heavily from two technical reports published by W. Pennebaker [Pen96b, Pen96e]. The material is reproduced in this chapter with permission.

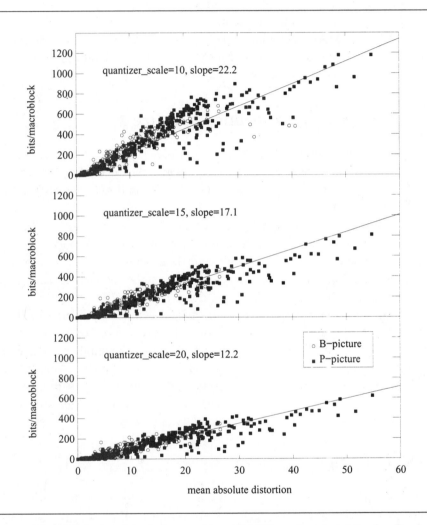

Figure 13.1: Relationship between luma MAD and coded bits for P-picture and B-picture macroblocks.

ment for most applications — and we are not aware of any implementations that use this technique.

One does not have to work this hard, however. If a linear relationship exists between coded bits and a given distortion measure, minimization of the distortion measure is equivalent to minimization of coded bits. Therefore, identifying a search criterion is really simply a matter of finding a distortion measure that is proportional to coded bits. Without belaboring the point, it turns out that MAD has exactly this property, albeit with some scatter in the data. The relationship between MAD and coded bits is quite linear for MPEG coding techniques, as can be seen in Figure 13.1[Pen96e].

13.2   The search reference picture ◑

When coding the difference between a given picture and a reference picture in the sequence, the reference picture is always the reconstructed version. Otherwise, very large differences between encoder and decoder outputs will develop very quickly.

What should the reference be, however, in determining motion displacements? Given that the objective is to minimize the differences that are coded, should the reference be the reconstructed picture or the original source picture? Or, will the quantization noise in the reconstructed pictures interfere with the motion displacement determination?

Figure 13.2 shows that the same linearity between bits and MAD that was found for source references (Figure 13.1) also exists for reconstructed references. In fact, the scatter appears to be slightly smaller with reconstructed references. Thus, although virtually all of the data presented in this chapter is for source picture references, it is entirely possible that reconstructed references can provide lower rate-distortion. This interesting possibility remains to be explored fully.

13.3   How large should the search range be? ◑

The complexity of motion vector estimation is a monotonic function of the search range. For some of the fast search search techniques discussed later in this chapter, the complexity scales linearly with the search range; for full search, the complexity is proportional to the square of the search range. Therefore, it is important to establish reasonable bounds for the search.

Histograms of the frequency of occurrence of motion displacement magnitudes can provide some guidance. Figure 13.3 gives histograms of the motion displacement magnitudes representative of the flower garden sequence. The linear increase in distribution width with temporal distance is as expected. For this experiment the data were obtained using a full search over $\pm16dT$

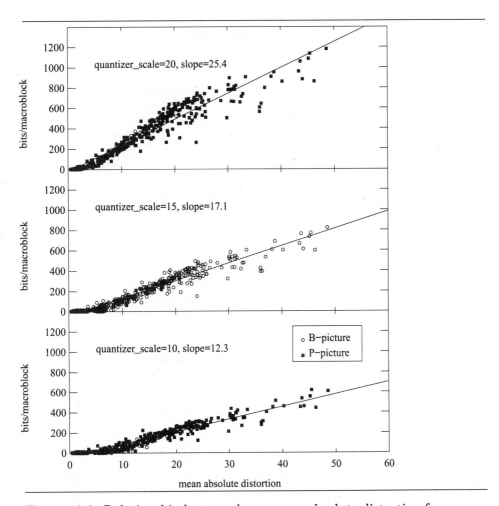

Figure 13.2: Relationship between luma mean absolute distortion for a macroblock and coded bits when reconstructed pictures are used as references.

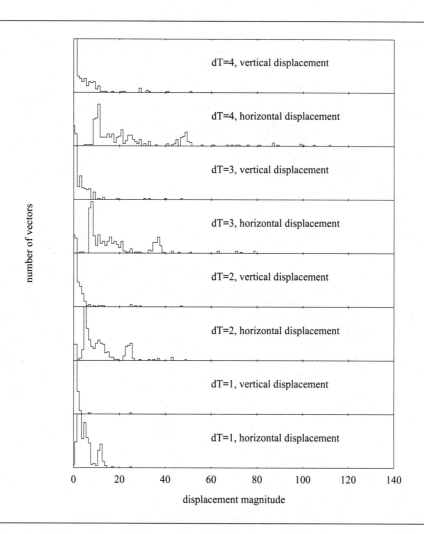

Figure 13.3: Histograms of motion displacement magnitudes for the flower garden sequence for four temporal distances, dT. The displacement unit is half-pel. Some counts near zero displacement are off-scale.

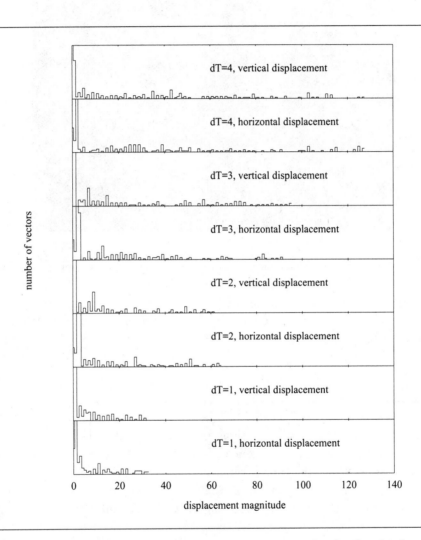

Figure 13.4: Histograms of motion displacement magnitudes for the football sequence for four temporal distances. The displacement unit is half-pel. Some counts near zero displacement are off-scale.

Figure 13.5: Mean average distortion over a 256x193 pel search range for a macroblock from a picture in the table tennis sequence. The example was generated for macroblock (10,10) in picture 6, with prediction from picture 2. A blurred hand holding a paddle is clearly distinguishable. Darker areas have lower MAD.

pel. This sequence has mostly horizontal motion, as is evident from the histograms.

The histogram for the football sequence in Figure 13.4 is quite different, exhibiting fairly similar distributions both horizontally and vertically. In a few areas of the football sequence the motion was large enough that the displacements exceed the search range.

The lesson to be extracted from these two examples is that a search range of ±8 pels per unit temporal distance is marginally adequate, and a search range of ±16 pels per unit temporal distance is probably sufficient for most sequences. No matter what search range is chosen, there will be times when the search fails.

13.4   Fractional pel searches ◖

MAD is usually smooth and continuous when evaluated at full pel resolution, as can be seen in Figure 13.5. However, this is no longer true when MAD is evaluated at half-pel accuracy. At half-pel accuracy, interpolation is required

Figure 13.6: Mean average distortion for a half-pel accuracy search over a 64x64 pel search range, for the same pictures and macroblock as in Figure 13.5. A histogram equalization was used to make the interpolation minima visible.

to produce half-pel position values. This is a noise-reduction process, and the distortion measure develops a lumpy character with local minima at the interpolated points in the search space. This can be seen in some parts of Figure 13.6, where the grayscale representation of MAD for a half-pel search is shown.

In a full search to half-pel accuracy, the noise averaging tends to favor selection of fractional pel displacements. This is not necessarily a problem, however, in that the selection is based on a minimization of MAD. If the proportionality between MAD and coded bits holds for this situation, coded bits are also minimized by the selection.

13.5   Correlation between motion vectors ◑

The graphical representation of motion vectors given in Figure 13.8 illustrates the high degree of correlation present in most motion vector arrays. Two source pictures from the sequence from which Figure 13.8 was derived are shown in Figure 13.7. Most of the motion was caused by horizontal movement of the camera, and the tree in the foreground therefore has larger motion displacements.

Judging from the correlation between the vectors in Figure 13.8, most motion vectors are cleanly identified by the search. There are occasions, however, where spurious motion displacement is evident, as can be seen in Figure 13.9. In this example the search was for the two pictures shown in Figure 13.7. These pictures are separated by three intervening pictures and the sequence has a relatively large amount of movement.

13.5.1   Adding chroma to the distortion measure ◑

Most search techniques use only luma MAD, and for the vast majority of macroblocks this works well. However, the coded bits for the macroblock includes bits for chroma as well as luma, and it is reasonable to ask whether chroma distortion might be important too.

As discussed in [Pen96e], a good luma match occasionally leaves very visible chroma distortion, suggesting that chroma should be included in the distortion criterion. Including chroma distortion is a reasonable idea, but it does introduce some complications because of the different sampling resolutions for luma and chroma.

Supposing for the moment that a full search at full-pel accuracy is performed, adding chroma to the criterion would require interpolation of chroma to half-pel positions. Interpolation is also a noise smoothing operation, and thus would favor selection of the interpolated points. Searching the luma at half-pel accuracy with chroma included in the distortion criterion

(a)

(b)

Figure 13.7: Picture 2 (a) and picture 6 (b) from the flower garden test sequence. The source pictures are used in the search for motion displacement values.

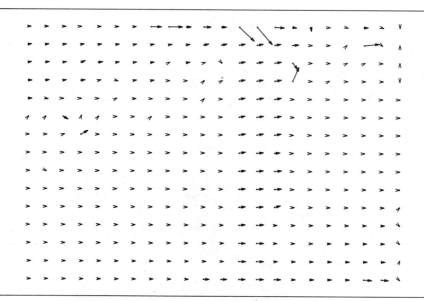

Figure 13.8: Motion vectors for flower garden picture 3 for prediction from 2.

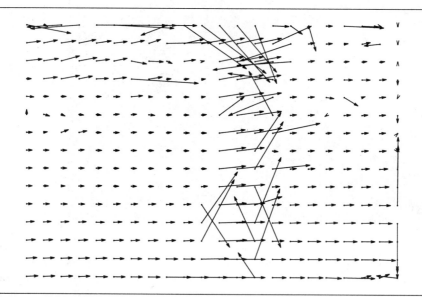

Figure 13.9: Motion vectors for flower garden picture 6 for prediction from 2, for a cost slope of 0.

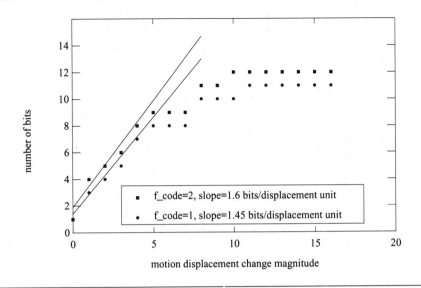

Figure 13.10: Code lengths for small motion displacement changes.

introduces still more complications.

A simple way of avoiding these complications is to simply replicate the chroma for the four Y samples rather than interpolating it. This effectively removes chroma from the mismatch estimate within each 2x2 luma block; it also greatly reduces the additional computational burden.

13.5.2   Tiebreaking ◑

One of the problems encountered with any search algorithm is what to do when two points have the same identical distortion. In MPEG, coding motion displacements costs more bits when the displacement changes are larger; therefore, all things being equal, smaller changes in displacements are better than larger changes.

The cost function developed in [Pen96e] attempts to quantify this coding cost. As can be seen in Figure 13.10, the cost of transmitting motion displacement changes is approximately linear for small changes, with a slope of about 1.5 bits/unit motion displacement. If this linear relationship is combined with the observed linear relationship between coded bits and MAD, a linear cost function results.

The cost function slope is the product of coded bits/unit motion displacement and MAD units/coded bit, giving a cost function slope of approximately 0.12 MAD units/unit motion displacement for a quantizer_scale of 20. This is in relatively good agreement with the empirically observed shallow minimum in bit rate at a cost function slope of 0.137.

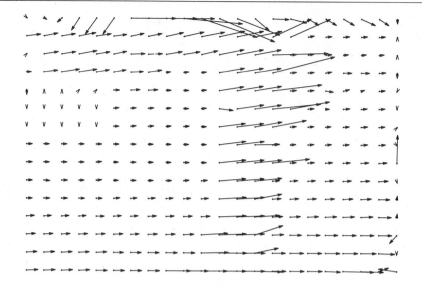

Figure 13.11: Motion vectors for flower garden picture 6 for prediction from 2, for a cost slope of 0.137. Figure 13.9 shows the same search without a cost function.

While a motion displacement cost function does not really avoid ties, it does mean that choosing between two displacements with equal distortions should not affect the coding efficiency. Figure 13.11 provides a graphical illustration of the improvement in motion vector continuity when a cost function is used.

13.5.3   Search failure ◐

While motion compensated prediction is extremely successful for most macroblocks, there will be times when the search fails. At the edges of the picture and in uncovered background areas there are often no usable predictions. If the motion is greater than the search range, the search will also fail.

Where the motion displacement search fails, the distortion usually is very large and the macroblocks should probably be coded with intra techniques. Although this sometimes requires more bits, the net result is usually an improvement in rate-distortion for the picture. Because of the coarser quantization typically used in P- and B-pictures, the distortions introduced by a bad prediction can be very visible.

In the experience of the authors, if only luma is used, search failure can be defined to occur for macroblock MAD values greater than a threshold set to the larger of 16 or the MAD value at the upper 1/8 of the MAD distribution in the picture. The fixed threshold of 16 is used to prevent needless intra coding at distortion levels that are handled well by nonintra techniques. The variable threshold is needed to avoid excessive use of intra coding in sequences with high activity.

If chroma and luma distortions are both used in the search (see Section 13.5.1), search failure is significantly reduced. The switch to intra coding of the macroblock can then be based solely on minimization of coded bits.

13.5.4   Search failure at scene changes ◑

Failure of motion-compensated prediction is virtually guaranteed at scene changes. Figures 13.12 shows histograms of MAD for picture 66, (a) when predicted from picture 63, and (b), when predicted from picture 67. As expected, the scene change between 66 and 67 causes a large increase in MAD. The motion vectors that result, shown in Figure 13.13, are badly scrambled.

In the experiments reported in [Pen96d], if the MAD at the start of the upper 1/8 of the distribution was above 8000, prediction from that picture was locked out. In that event, intra coding was forced for all macroblocks in P-pictures; for B-pictures, only the mode of prediction (forward or backward) that gave large MAD was locked out. One of the sequences, table tennis, had two scene changes, and this technique successfully detected them. Unnormalized MAD was always above 10000 for prediction across the scene change boundary, and below 7000 in all other cases.

13.6   Motion displacement search algorithms ◑

The use of motion compensation in the compression of video sequences was introduced in 1969 by Rocca[Roc72], and described more fully by Rocca and Zanoletti in 1972 [RZ72]. This work investigated the effect of applying motion compensation to improve the predictions made from a preceding picture in a sequence. The picture being coded was subdivided into zones using an algorithm based on contour identification, and a single translational motion vector was determined for each zone. Rocca and Zanoletti estimated that coding efficiency was improved by about a factor of two by motion compensation.

Most of the algorithms for motion estimation make the assumption that the motion is purely translational. Certainly, real motion also involves rotation and scale changes, but these enormously increase the complexity of the motion estimation process. It is widely held that if the regions associated with a vector are small enough, simple translation works quite well

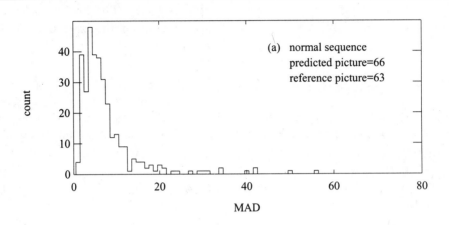

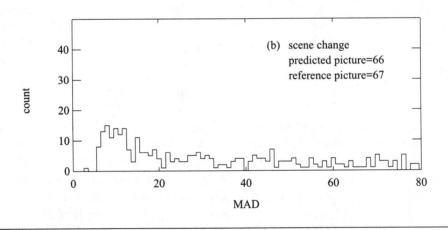

Figure 13.12: Histograms of MAD values for pictures from the table-tennis sequence. A scene change occurs between pictures 66 and 67.

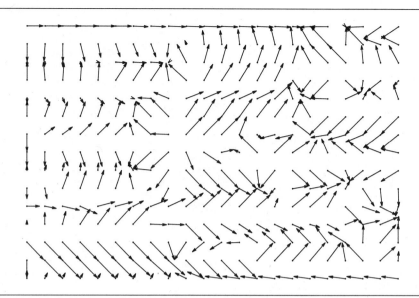

Figure 13.13: Motion vectors for picture 66 of the table tennis sequence for prediction from 67. A scene change occurs between pictures 66 and 67.

most of the time. It is possible, however, that significant coding gains might be realized with more complex types of motion compensation. However, all of the motion estimation concepts considered in this report involve only translational motions.

There are two classes of motion estimation techniques, pel-recursive and block matching [MPG85]. In pel-recursion the motion vectors are computed by an iterative process, using intensity gradient and frame difference information. In block matching the motion vectors are determined by computing some measure of distortion and selecting the vector that minimizes this distortion. An oversimplified view of the pel-recursion process is sketched in Figure 13.14, which is adapted from [MPG85].

In 1979, Netravali and Robbins [NR79] published the first pel-recursive algorithm for motion estimation. Also in 1979, Netravali and Stuller [NS79] published a paper applying motion compensation to DCT-based coding. Netravali and Stuller investigated two approaches. The first was an algorithm in which motion vectors were calculated recursively from transform coefficients in the preceding and present pictures. In this approach the motion vector information was predicted and did not have to be coded. The second approach used a technique of Limb and Murphy's [LM75] to calculate motion displacements that were explicitly transmitted to the decoder. Interestingly enough, the recursive scheme had slightly better performance (and a DPCM-based predictive scheme was slightly better yet). All of the coding

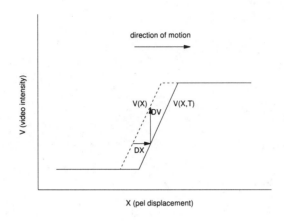

Figure 13.14: Illustration of pel-recursive calculation of motion displacement. The frame difference, DV, and the intensity gradient can be used to calculate the motion displacement, DX, provided that the displacement does not exceed the width of the object boundary.

schemes showed large improvements in coding efficiency (about 20% to 40%) when motion compensation was used.

The assumption that the distortion is smoothly varying and continuous is crucial to the pel-recursive algorithms, since recursive estimation depends on having well-behaved intensity gradients. The motion estimation computations use intensity variations at the boundaries of moving objects, and thus will fail when displacements exceed the widths of the boundaries. Indeed, even the most stable of the pel-recursive estimation processes become unstable when the motion displacement is so large that the intensity gradient changes sign [MPG85].

13.7 Fast algorithms for motion estimation ◑

That much of the literature about motion estimation involves fast search algorithms should come as no great surprise to the reader. While the improvement in coding efficiency when motion compensation is used is impressive, the computational burden of calculating motion vectors is perhaps even more impressive. Even with block matching and relatively coarse blocks, motion estimation remains by far the most computationally intensive aspect of the encoding process. If a full search is performed over a reasonably large area, say ±64 pels, 16,384 16×16 pel error calculations must be made for each macroblock for a full pel accuracy search (ignoring edge effects), and four times that number must be made for a half-pel accuracy search. A search over this range for a *common intermediate format* (CIF) picture of 360×240 pel resolution requires on the order of 10^{10} integer operations. With careful design, however, fast search techniques can reduce the computational burden by one to two orders of magnitude.

Many of the fast search block matching algorithms make the assumption that the distortion is a slowly varying continuous function. At least for the macroblock examined in Figure 13.5, this is a reasonable assumption, provided that the motion displacement sampling is not too coarse. The error image in Figure 13.5 can be regarded as a convolution of the reference picture with a 16×16 mask of samples from the current picture. When there is no overlap with the correct 16×16 matching point, the convolution produces a blurred representation of the reference picture that conveys little information about motion. Consequently, if the fast search sampling pattern is on a grid so sparse that the 16×16 mask never overlaps the correct area, the correct motion displacement will probably not be found.

13.8 A survey of fast search algorithms ◑

In 1981 Jain and Jain [JJ81] published a paper on block matching that is very relevant to MPEG. It describes a fast search based on direction of minimum distortion, in which the distortion metric, mean square error (MSE),[2] is computed only for a sparse sampling of the full search area. Figure 13.15 gives a sketch of the evolving search pattern for a relatively modest search range of ±6 pels. This approach, known as the 2-D logarithmic search, is a generalization of a well-known 1-D logarithmic search technique [Knu73].

Almost simultaneously with Jain and Jain, Koga, et al[KLH+81] published a very similar algorithm known as the three-step search procedure. The Koga, et al., approach differed from Jain and Jain's in two significant respects: the use of mean absolute distortion (MAD) instead of MSE, and a somewhat different search pattern. The three-step search is illustrated in Figure 13.16.

In 1983 Kappagantula and Rao [KR83] published the modified motion estimation algorithm (MMEA), an approach that is similar to the three-step algorithm, but computationally more efficient.

In 1984 Srinivasan and Rao published a search procedure called the conjugate direction search that is even more computationally efficient. Using groups of three displacements, this algorithm searches in the horizontal and

---

[2]According to this paper, MSE works significantly better than cross correlation when the area is small and the motion is not pure translation.

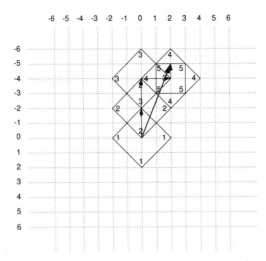

Figure 13.15: The 2-D logarithmic search for motion displacements. Each of the diagonal squares contains five search points at two-pel intervals. If the search selects one of the corners, the square is shifted and the search repeated. When the search selects a center point, a final search is run over a 3x3 pel array (the corners of which are labeled 5 in this example).

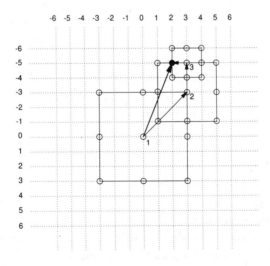

Figure 13.16: The three-step search for motion displacements. With each step the search range is decreased.

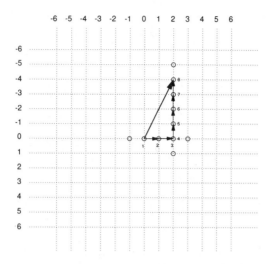

Figure 13.17: The simplified form of the conjugate direction search. For each direction, the search progresses in the direction of the smaller distortion until until a minimum is found.

then the vertical direction until a minimum in distortion is found. It is illustrated in Figure 13.17.

In 1985 Mussman, et al., [MPG85] published a very readable review of motion estimation techniques for both pel-recursion and block matching. It includes a very nice discussion of convergence and range of stability of the various pel-recursion techniques. It also recommends the use of MAD for the matching criterion, primarily because MAD is simple and has performance equivalent to other measures.[3]

Since the Mussman, et al. review, a number of refinements and variations of fast search algorithms have been published. All of these fast search techniques trade off quality of motion estimation for a smaller computational burden. Most of the authors use full search as the definition of best match; a few, however, claim results better than full search. This might indeed be the case if full search is limited to integer pel accuracy; otherwise, full search can only be improved upon if coded bits or visual distortions are not minimized when MAD is minimized. None of the authors provides evidence to that effect.

---

[3]Another useful review of fast search strategies can be found in Clause D of the MPEG-1 standard.

13.8.1   Fast search algorithms based on sparse sampling ◑

The orthogonal search algorithm by Puri, et al. [PHS87], is a form of logarithmic search. At each step, four search positions are analyzed — two horizontal and two vertical.

The cross search algorithm [Gha90] uses a crossed diagonal search pattern. The performance is comparable to the other fast search algorithms and the computational complexity is modestly better.

[CCJC91] describes a fast search algorithm suitable for implementation in a parallel processing environment. Similar to the Srinivasan and Rao approach, this algorithm searches the x and y axes independently.

[LWLS93] describes an extension of the three-step algorithm in which the window size is varied depending on the sharpness of the selection criterion. The search sequence starts with an x pattern and alternates with a + pattern. The selection criterion is MAD, and if the smallest and second smallest MAD values are very close, a larger search window is used for the next step.

Another variation of the three-step hierarchical search is described in [JCC94]. A subsampled search is used in which nine samples spaced at four-pel displacements around the center point are evaluated and the best is chosen. Then, the eight neighboring samples at two-pel displacements are evaluated. Again, the best is chosen and the eight neighboring samples at one-pel spacings are evaluated to determine the motion displacements.

13.8.2   Variable resolution search techniques ◑

A coarse to fine algorithm for motion estimation is described in [WC90]. Based on the Laplacian pyramid image coding technique [BA83], each level of the pyramid is constructed from a 4x4 average of the lower (higher resolution) level.

In a similar algorithm, called successive refinement [CR94], the search is first carried out with low-resolution block averages to get a coarse estimate of the motion vector.

13.8.3   Statistically sparse searches ◑

In [Gil88] an empirical search pattern is described that uses sparser sampling for motion displacements that are less probable. Vector quantization (VQ) concepts are applied to motion estimation in [Gil90], in order to reduce the number of search positions in a statistically optimal way. As might be expected, the reduced set of search positions tend to cluster around zero, and are much sparser at large displacements. Dynamic operation is provided

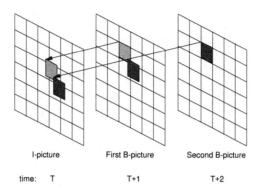

I-picture   First B-picture   Second B-picture

time: T   T+1   T+2

Figure 13.18: Illustration of a telescopic search for motion displacements.

by switching of vector-books, and a DC correction of the distortion is used
to adjust for changing illumination levels.

A search algorithm based on genetic concepts is described in [CL93].
In terms of the effect, the approach bears some resemblance to the VQ-
based scheme of [Gil90]. The major difference is that in the VQ scheme the
sampling points are developed a priori using a training session, whereas in
the genetic algorithm the sampling pattern is developed dynamically.

13.8.4   Searches dependent on spatial continuity ◑

The PPS (predictive pattern search)[ZZB91] scheme makes a full search for
one block in a superblock using the result to predict vectors for the remaining
blocks. A limited full search around the prediction values is then used to
refine the values.

[LZ93] uses motion field subsampling and pixel subsampling to reduce
the computations relative to full search. The algorithm is related to the
approach described in [ZZB91].

13.8.5   Telescopic search ◑

Temporal continuity of motion displacements from one picture to the next
is the basis for the telescopic search [ISO93a]. This technique uses motion
vectors from one picture to predict vectors for the next.

Telescopic search uses the motion displacements determined for the preceding picture to decrease the search range. Suppose, for example, that a video sequence consists of an I-picture followed (in display order) by three B-pictures and a P-picture. If the search is done independently for each picture, the P-picture — at a temporal distance of four picture times from the I-picture — requires a much larger search range. If instead, the motion vectors determined for the first B-picture could be used as the starting point in searching for the vectors for the next B-picture, the temporal change from one starting point to the next would be only one picture time.

Exactly how telescopic search should be implemented is not described in [ISO93a]. A motion vector is an offset to the predicting array, relative to a macroblock position in the picture being coded. Therefore, referring to Figure 13.18, there are at least two ways of obtaining the motion displacements for the second B-picture. One way is to estimate motion displacements independently for each B-picture, using the nearest neighbor (which needs the shortest search range) to provide motion displacements that are extrapolated back to the predictive picture (the I-picture in Figure 13.18). A second way is to estimate the motion relative to the nearest neighbor and use an appropriately weighted average of the motion displacements already determined for that picture to find an initial estimate of the motion displacements. In both cases the initial estimate provides only a starting point for a search in the I-picture. A backward telescopic search algorithm is described briefly in [LD94].

13.8.6   3-D (spatial/temporal) estimation ◑

In [dHB94] a recursive 3-D search (x,y,t) is defined in which motion displacements in three neighboring blocks that form a Y shaped pattern are used to estimate the current block. The two upper branches of the Y are motion displacements from the current picture, and the motion displacements for the bottom of the Y are from a block in a temporally adjacent picture. The number of computations is reduced by using a subsampled error sum.

13.8.7   Phase correlation search ◑

Girod [Gir93] uses a DFT-based "phase correlation" search to get a full-pel accuracy motion vector, following this with a localized full search to find the fractional pel values. Girod's paper is of interest for several reasons other than the novel motion estimation technique: It provides evidence that half-pel to quarter-pel accuracy is optimal and shows that more sophisticated filtering techniques provide only a slight gain in prediction efficiency relative to the simple bilinear interpolation defined in the MPEG standard.

13.8.8   Other search techniques ◑

Lee [Lee86] suggests fitting the image blocks with generalized discrete orthogonal polynomials, but provides no test data.

In [KP89] projection operators are used to estimate motion vectors. A significant reduction in computations is claimed, as well as better matching in noisy pictures.

In [XEO90, XEO92] pel-recursion algorithms are applied to block matching. Some computational reduction is achieved by using a threshold MAD for zero displacement to bypass the search altogether.

13.8.9   Improving the motion displacement estimate ◑

Since motion displacements are normally highly correlated, occasional unusual displacements may very well be caused by a failure of the search algorithm. One interesting method for removing these is the work by [KN90], in which median filtering is applied to the motion vector field. Median filtering suppresses impulses while preserving edge information, thereby removing vector values that are not correlated with neighboring values. This technique may be particularly appropriate for high-resolution picture sequences.

13.9   Comparison of fast search algorithms ◑

Most fast motion displacement search techniques belong to one or more of the following four classes: 1) subsampling the motion displacement space, 2) subsampling the array over which mean absolute distortion (MAD) is calculated, 3) using reduced spatial resolution for the first stage or stages of the search, and 4) using prediction to initialize the search and concentrating most of the search locally around that point. In this section an overview is presented of a quantitative analysis of these techniques by [Pen96b].

This quantitative analysis is based on the following assumptions:

1. The searches are to full-pel accuracy only, and the correct motion displacements are defined by full search at full pel accuracy using luma distortion.

2. Because of the proportionality between MAD and coded bits, a change in MAD due to search error can be used as a measure of the accuracy of the search.

3. Complexity of an algorithm is proportional to the number of operations. Some algorithms allow significant parallelism, but this is not included in the complexity estimates.

4. If half-pel accuracy is needed, the search procedures can be extended by a localized half-pel search. This represents a fixed overhead that will not affect the relative performance of the different techniques.

5. The motion displacement range for MPEG is significantly larger than algorithms such as the three-step approach are designed for, but subsampling of the search space represents a reasonable generalization of these algorithms.

For a `quantizer_scale` of 15 in Figure 13.1, the measured ratio between coded bits and MAD is approximately 17 bits/MAD unit per macroblock. Video in CIF has dimensions of 360x240 pels, and with left and right edges trimmed to integer macroblock dimensions, has 330 macroblocks per picture. Consequently, a search technique that increases MAD by only one unit per macroblock will increase the coded bits by approximately 5600 bits per picture, if the coding parameters are not changed.

This is quite significant compared to the average coding budget of about 40,000 bits per picture (1.2 Mbits/s). At normal MPEG rates, a 10% reduction in bit rate usually causes a very visible decrease in picture quality. Consequently, a fast search technique that increases MAD by only 0.7 units/macroblock or 230 MAD units/picture should cause a noticeable degradation in picture quality when coding at a fixed rate. Fortunately, most of the fast search techniques typically produce MAD increases of less than 100 units/picture.

Computational complexity is usually measured by counting operations. Ideally, one would like to estimate complexity theoretically, but some of the fast search algorithms are based on statistical optimizations that would be difficult to model theoretically. An alternative is to estimate relative complexity by measuring execution times for C language software implemented with coding techniques as similar as possible and compiled for the same machine at the same level of optimization. In effect, the software execution time provides a Monte Carlo estimate of the total number of operations. While there are some drawbacks to this approach, it is a relatively simple way to compare very different algorithms.

In Figure 13.19 the performance of a number of fast search techniques is compared graphically by plotting the increase in MAD as a function of relative complexity. For this plot, relative complexity is defined as the ratio of CPU time for a given fast search technique to CPU time for a full search. Empirically, the log-log plot straight line given by:

$$\Delta\text{MAD} = 1.5/\text{Complexity} \qquad (13.2)$$

appears to give an approximate lower bound on the increase in distortion for a given complexity. Within the range where a given algorithm works

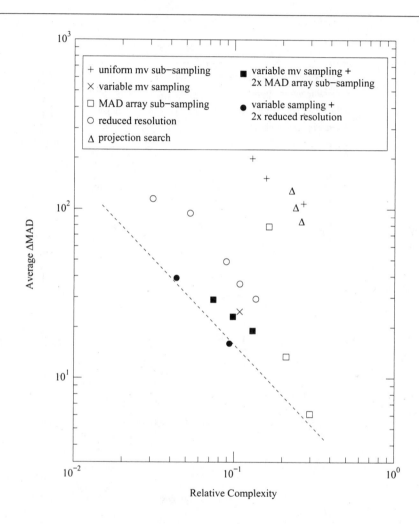

Figure 13.19: Average increase in MAD as a function of relative complexity, for the various motion displacement search techniques. The dashed line is obtained if the product of complexity and MAD is constant, equal to 1.5. By definition, full search has a relative complexity of 1.0.

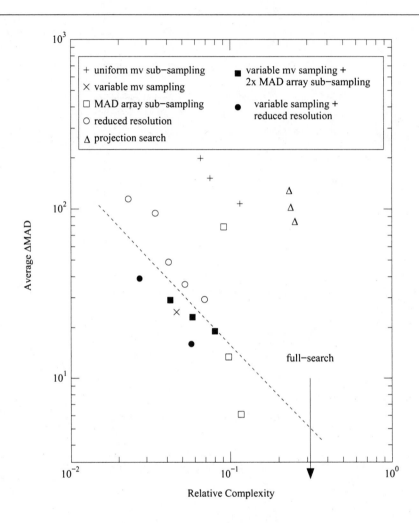

Figure 13.20: Average increase in MAD as a function of relative complexity, when the search is sped up by MAD summation truncation. The dashed line is carried over from Figure 13.19 to provide a reference.

well, varying search parameters such as the second stage search range tends to shift the operating point along this line. The constant in this equation therefore provides a useful figure of merit in judging the performance of these algorithms.

The complexity of these algorithms can be reduced significantly by truncating the distortion summation whenever the interim distortion sum exceeds the smallest distortion sum already computed for that macroblock.[4] Aborting the summation significantly improves the computational efficiency of all of the search techniques in Figure 13.19 except the projection based search, as can be seen in Figure 13.20.

If the search range is increased to ±64 (Figure 13.21), the relative complexity of the fast search techniques is modestly reduced, but the relative ranking is almost identical.

---

[4]M. Oepen and J. Steurer, unpublished work [EF95]. The technique is used in the ISO/IEC MPEG-2 software video codec [Gon96]; it was also suggested independently by J. Mitchell [Mit95].

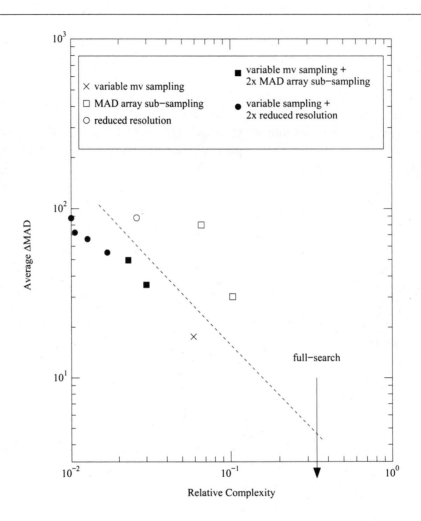

Figure 13.21: Average increase in MAD as a function of relative complexity for a search range of ±64 (with MAD summation truncation). The dashed line is again carried over from Figure 13.19 to provide a reference.

14

Variable Quantization

Variable quantization is one of the most difficult aspects of MPEG, and is an area where active research is still being carried out. It is the most important (and most effective) of the techniques for maintaining a fixed bitrate.

The encoder must choose the `quantizer_scale` (and if the default table is not used, a quantization table as well); therefore, this chapter is really about the encoder decision-making process. The decoder simply follows the decision sequence defined in the compressed data bitstream.

There are several aspects to variable quantization: Bitrate control is, of course, the primary rationale for varying the quantization at all. This can be done as a global process, maintaining a fixed `quantizer_scale` for an entire picture and varying that scale to achieve the desired rate for that picture type. However, the quantizer scale can be adjusted at the slice layer or at the macroblock layer, thereby achieving a finer granularity in the control and (hopefully) better picture quality as well.

Aspects of variable quantization that relate primarily to bitrate control will be discussed in Chapter 15. In this chapter we assume that the basic bitrate target for a given picture has been determined. Further, the primary focus here is on control of the `quantizer_scale` at the macroblock level.

What remains is a difficult problem: Namely, how to distribute the bits among the macroblocks within each picture so as to obtain the best picture quality. The problem is made more difficult by the following observation: When the distortion is increased in picture A, and picture A is the prediction for picture B, more bits will be needed to code picture B with a given `quantizer_scale`. However, this question is really more related to bit allocation among pictures. Given a target bit allocation for a given picture, the number of bits needed to code each macroblock of that picture should be a well-behaved function of macroblock complexity, assuming that this measure of macroblock complexity exists.

In fact, the basic premise of this chapter is that an activity metric or

313

masking function exists that can be used to calculate a `quantizer_scale` from intensity variations in a given macroblock, from intensity variations between the macroblock in the picture being coded and neighboring pictures, and possibly from other variables that might affect perception of quantization noise.

As we look at the various attempts to find such a metric, we note that there is no way of achieving a true optimum (or, at least, proving that the optimum has been obtained) — the problem is far too complex for that. It appears, however, that significant improvements in overall picture quality can be gained from variable quantization.

14.1   Literature review ◑

Variable quantization is based primarily on the concept of masking — the increase in threshold for a given stimulus in the presence of a masking signal. In real images, however, the number of intensity patterns is huge, and this makes the application of variable quantization quite difficult. The published work in this area is highly heuristic, as the review in this section should make clear. There are, however, some basic principles that recur in various forms in these papers.

Spatial masking: As a rule of thumb, a given stimulus will be harder to see in a very active block — one with very complex and strong spatially-varying intensities. The literature reviewed here describes the following methods for estimating activity: statistical techniques (i.e. variance), spatial edge detection and gradient measures, DCT coefficient energy, and coded bits.

Temporal masking: If the motion is large enough, temporal masking can be significant, and many schemes increase the `quantizer_scale` when the motion is relatively large. In most real video systems camera integration tends to blur the picture in areas of high motion, so the gains from coarser quantization of these regions will probably be modest. Later in this chapter some work will be presented showing that reduction of the `quantizer_scale` may be needed where relative motion between adjacent regions occurs.

Intensity dependence: Weber's law is cited in some papers as the justification for increasing the quantizer step size in very dark and very light regions. This is certainly true for true luminance, but is probably not appropriate for gamma-corrected luma. Perception of grayscale change in gamma-corrected pictures is quite uniform over the normal intensity range if the display is well-adjusted [Poy95b]. In the authors' experience, only in very light and very dark regions is there significant loss of grayscale intensity perception.

Pattern classification: Because of the huge number of possible intensity patterns in the 16×16 pel MPEG-1 macroblock, some researchers use classification techniques to separate the regions into categories with similar patterns. Activity measures such as the variance are then applied (and calibrated) to each category separately.

The division of a sequence into I-, P-, and B-pictures is arbitrary. Therefore, one should expect to find spatial masking effects in P- and B-pictures, and temporal masking effects in I-pictures. This literature review categorizes the papers according to the spatial masking measure used, but many of these papers also use other perceptual mechanisms for adjusting the `quantizer_ scale`.

In the reviews that follow in this section, only the basic ideas are presented. Many complex issues are necessarily treated too briefly, and the reader will need refer to the original literature for more details. The notation used here generally follows the notation used in the papers.

14.1.1 Still-picture coding ◑

Some of the earliest use of variable quantization is in still picture coding. Netravali and Prasada [NP77] review spatial masking effects and their application to picture coding. They also derive a masking function that is a weighted sum of slope magnitudes; this masking function is a more general form of the spatial derivative used in several papers reviewed here. Limb and Rubinstein [LR78] and Pirsch [Pir81] discuss the use of masking for DPCM coders.

Lohscheller [Loh84] used activity measures based on edge detection and variance to coarsen DCT quantization near strong edges; he also used a visually-weighted quantization table.

Pearlman [Pea90] developed an adaptive quantization scheme based on maintaining a constant block distortion; this is a somewhat different approach that is not consistent with scaling of the quantization to visibility criteria.

14.1.2 Masking functions based on spatial derivative ◑

In this section three papers, [PAC92], [PA94] and [CJOS94], are reviewed that use masking functions based on the spatial derivative. The first two describe their results in terms of a transmission factor, f, that has integer values in the range 0-175, with larger f values corresponding to larger quantizer step size. While these papers are actually more relevant to MPEG-2 than MPEG-1, the concepts are general.

14.1.2.1 Spatial derivative error ◐

In [PAC92] the bitrate control algorithm is based on a HVS-based distortion metric that invokes spatial masking, temporal masking, and decreased sensitivity in very dark and very light regions of the picture. The paper is oriented toward MPEG-2. It adapts the quantization for each 80×32 pel region of interest. Note that an 80×32 pel region is 20 16×8 pel macroblocks, a fairly large portion of the 720×288 pel field.

In this paper spatial activity is based on a spatial derivative error heuristic, where the spatial derivative error of the nth region, z_n, is defined as the sum over the region of interest of the difference in the spatial derivatives of the source and reconstructed pictures:

$$z_n = (1/2560) \sum_{i=0}^{79} \sum_{j=0}^{31} |\delta(I_{i,j}) - \delta(I'_{i,j})| \qquad (14.1)$$

where $I_{i,j}$ is the original picture intensity, and $I'_{i,j}$ is the reconstructed picture intensity. In this paper δ is the steepest slope magnitude of the pel with respect to any of the eight nearest neighbor pels.

A prediction of the spatial derivative error is calculated from a linear weighted sum of the luminance edge activity, average field luminance activity, a motion parameter, and average luminance. These are defined as follows:

The luminance edge activity is:

$$x_n = (1/2560) \sum_{i=0}^{79} \sum_{j=0}^{31} e_{i,j} \qquad (14.2)$$

where

$$\begin{aligned} e_{i,j} &= 1 \quad \text{if } \delta(I_{i,j}) > 15 \\ &= 0 \quad \text{if } \delta(I_{i,j}) \le 15 \end{aligned}$$

Average field luminance activity for the entire picture is defined by:

$$w = (x_{min} + x_{max})/2 \qquad (14.3)$$

where x_{min} and x_{max} are (essentially) the minimum and maximum luminance edge activities for the picture.

The motion parameter, m_n, is defined to be the normalized sum of the motion displacement vector magnitudes.

The average luminance, y_n, is defined by:

$$y_n = \sum_{i=0}^{79} \sum_{j=0}^{31} |I_{i,j} - 128| \qquad (14.4)$$

This is a measure of how far from neutral gray the pels are, on the average. The linear equation for the spatial derivative error, zp_n, then becomes:

$$zp_n = 15x_n + 0.02m_n + 3w + .002y_n + 1.7 \qquad (14.5)$$

The coefficients of this equation were obtained by measuring the values of the variables for several sequences coded with a quantizer step size that produced approximately uniform subjective quality.

A rate error, re_n, for region n is obtained by subtracting the predicted spatial derivative error from the actual spatial derivative error.

$$re_n = z_n - zp_n \qquad (14.6)$$

Then, f_{n+1}, the transmission factor for region $n+1$, is calculated from the rate error for region n and the transmission factor of region n.

$$f_{n+1} = f_n - re_n/5 \qquad (14.7)$$

where the constant 5 is numerical scaling factor. This last equation is one way of adjusting the average `quantizer_scale` for the entire picture. However, there must be other (unstated) constraints, since even a slight systematic deviation in re_n would cause the transmission factor to drift as n increases.

The connection to the masking effects measured in psychophysical experimentation is obscured by the sequence of differencing operations in these definitions. Perhaps the scheme works because differencing is a noise enhancing operation and noise is generally a masking stimulus. The algorithm also appears to reduce MSE, which is a puzzling result. Varying the quantization increases the distortion in one area in order to reduce it in another, and any change in MSE would have to be a second-order effect.

14.1.2.2   Vertical/horizontal spatial derivative ◑

For the system described in [PA94], the macroblock is defined to contain two luminance 8×8 blocks and two chrominance 8×8 blocks.

The value of the transmission factor, f, is calculated from

$$f = f_{pm} + f_{fld} \qquad (14.8)$$

where f_{pm} is a perceptual masking factor and f_{fld} is a field offset. The perceptual masking factor increases the `quantizer_scale` in macroblocks

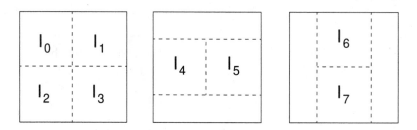

Figure 14.1: 4x4 sub-block patterns used for luminance activity factor in [PA94].

having a high number of luminance edges or containing significant motion. The field offset, calculated for each field, adjusts the mean quantizer_scale so that the bitrate remains within the bounds required to maintain the desired overall rate.

The perceptual masking factor, f_{pm} is assumed to be the sum of two parts:

$$f_{pm} = f_{act} + f_{mtn} \tag{14.9}$$

where f_{act} is an activity factor derived from the spatial activity (edges, etc.) in the macroblock and f_{mtn} is a motion factor.

The spatial activity is calculated from spatial derivatives in the macroblock. Two components are defined, one vertical and one horizontal:

$$\delta_v I = (1/(n(n-1))) \sum_{i=0}^{n-2} \sum_{j=0}^{n-1} |I_{i,j} - I_{i+1,j}| \tag{14.10}$$

and

$$\delta_h I = (1/(n(n-1))) \sum_{i=0}^{n-1} \sum_{j=0}^{n-2} |I_{i,j} - I_{i,j+1}| \tag{14.11}$$

These two spatial derivatives are used to compute the luminance activity factor using the following heuristic.

Each 8×8 block is divided into eight 4x4 subblocks, as shown in Figure 14.1. Eight parameters, $\lambda_0, \ldots, \lambda_7$, represent the luminance activity:

$$\begin{aligned} \lambda_i &= (1/2)(ln(\delta_h I_i) + ln(\delta_v I_i)) &&\text{for } i = 0 \ldots 3 \\ \lambda_i &= ln(\delta_v I_i) &&\text{for } i = 4, 5 \\ \lambda_i &= ln(\delta_h I_i) &&\text{for } i = 6, 7 \end{aligned} \tag{14.12}$$

The minimum luminance edge activity of the four corner 4x4 blocks, denoted by λ_{min}, is calculated. Then, the activity factor is given by:

$$f_{act} = 6\lambda_{min} - 0.7 \left[\left((1/8) \sum_{i=0}^{7})\lambda_i \right) - \lambda_{min} \right] \qquad (14.13)$$

The coefficients were chosen to maximize perceptual quality in several test sequences. The second term reduces the value of f_{act} in order to compensate for ringing effects because of coarse quantization at boundaries between high and low edge activity.

The motion factor is calculated as follows: Denote the motion of a macroblock by ρ, where the motion is derived from the luminance of a block in the previous field and the current field.

$$\rho = |N_n - N_{n-1}| + |S_n - S_{n-1}| \qquad (14.14)$$

where N_n, the average luminance edge activity of the first luminance block of the macroblock, is defined by:

$$N_n = (1/8) \sum_{i=0}^{7} \lambda_i \qquad (14.15)$$

and S_n, the average luminance value of the block, is given by:

$$S_n = (1/64) \sum_{i=0}^{7} \sum_{j=0}^{7} I_{i,j} \qquad (14.16)$$

The motion of the macroblock must reach a threshold before coding errors will be masked. The motion factor is then given by:

$$\begin{aligned} f_{mtn} &= 0 & \text{if } \rho \le 25 \\ &= (\rho - 25)/3 & \text{if } \rho > 25 \end{aligned} \qquad (14.17)$$

The threshold and numerical constants were derived from perceptual studies of several test sequences.

After obtaining values for f_{act} and f_{mtn} for the first luminance block, the process is repeated for the second luminance block. The minimum of these values is chosen as the value for f_{pm}.

The field offset is used to maintain the desired mean bitrate. This is obtained by predicting the number of bits the coder will produce for each 8×8 block of data. This prediction is based on a spatial frequency vector, V, that has six elements relating the DC and lower spatial frequency components of the block. These are used to develop a vector that is related to bit-rate by means of a codebook, following the techniques of vector quantization.

14.1.2.3 Spatial derivative classification ◑

In [CJOS94] macroblocks are classified using luminance intensity and gradient properties. First, a parameter, f_{det}, is developed from thresholded spatial derivatives.

For horizontal patterns define:

$$det_H = \sum_{i=2}^{16} \sum_{j=1}^{16} dH_{i,j} \qquad (14.18)$$

where

$$\begin{aligned} dH_{i,j} &= 1 \quad \text{if } |I_{i,j} - I_{i-1,j}| > TH1 \\ &= 0 \quad \text{otherwise} \end{aligned} \qquad (14.19)$$

and for the vertical patterns:

$$det_V = \sum_{i=1}^{16} \sum_{j=2}^{16} dV_{i,j} \qquad (14.20)$$

where

$$\begin{aligned} dV_{i,j} &= 1 \quad \text{if } |I_{i,j} - I_{i,j-1}| > TH1 \\ &= 0 \quad \text{otherwise} \end{aligned} \qquad (14.21)$$

Then,

$$f_{det} = max(det_H, det_V) \qquad (14.22)$$

f_{det} is used to calculate the complexity of the macroblock, and $TH1$ is a threshold value that is adapted to the average brightness of the macroblock, MB_{mean}, in the following way:

$$\begin{aligned} TH1 &= (MB_{mean}/4) + 8 \quad \text{if } TH2 < MB_{mean} \\ &= (MB_{mean}/4) + 4 \quad \text{if } TH2 \geq MB_{mean} > TH3 \\ &= (MB_{mean}/4) + 2 \quad \text{if } MB_{mean} \leq TH3 \end{aligned} \qquad (14.23)$$

The authors never define $TH2$ and $TH3$, but presumably these represent some reasonable segmentation of the grayscale range.

From f_{det} they identify three classes of macroblock:

$$\begin{aligned} \text{detail MB:} \quad & Q_{sub} = M \quad &\text{if } f_{det} > NO1 \\ \text{edge MB:} \quad & Q_{sub} = -M \quad &\text{if } NO2 > f_{det} > NO3 \\ \text{else:} \quad & Q_{sub} = 0 \quad &\text{otherwise} \end{aligned} \qquad (14.24)$$

$NO1$, $NO2$, and $NO3$ are presumably fixed thresholds and M is a parameter that determines the quantizer step size.

DCT activity is computed from the sums of the magnitudes of the coefficients. Defining the absolute sum for the kth 8×8 block in the macroblock, AS_k,

$$AS_k = (\sum_{i=0}^{7} \sum_{j=0}^{7} |16C_{i,j}/w_{i,j}|) - |C_{00}| \qquad (14.25)$$

where $w_{i,j}$ is the quantization matrix value, and $C_{i,j}$ is the corresponding unquantized DCT coefficient. Note that the DC coefficient is not included in the sum.

The activity, ACT, is the minimum of the luminance activities of the four 8×8 blocks in the macroblock.

$$ACT = \min(AS_k) \text{ for } k = 0 \ldots 3 \qquad (14.26)$$

A normalized activity is calculated from:

$$ACT_n = (2ACT + AVG)/(ACT + 2AVG) \qquad (14.27)$$

where AVG is the average activity of the previous picture of the same type (I-, P- or B-). Finally, the macroblock classification and activity are combined to give the quantizer step size, $MQUANT$:

$$MQUANT = Q \times ACT_n + Q_{sub} \qquad (14.28)$$

where Q_{sub} is defined by Equation 14.24 and Q is a reference step size.

14.1.3   Masking functions based on block variance ◑

[PA91] uses block variance both as a measure of activity in macroblocks and to classify macroblocks. A related approach is used in the MPEG-2 Test Model 5. Other aspects of [PA91] include:

14.1.3.1   Global variance ◑

Overall picture complexity is analyzed by averaging the variances of the individual 8×8 blocks to create a global variance. Each type of picture (I-, P-, B-) has a different relationship between subjective picture complexity and global variance.

14.1.3.2   Scene complexity ◑

In I-pictures scene complexity is estimated from spatial complexity; in P- and B-pictures complexity is estimated from how well the motion is compensated. Eight levels of complexity are recognized.

14.1.3.3   Bit allocation ◑

Bit allocation between I-, P- and B-pictures is a function of the scene complexity estimate. Scene complexity can be adjusted during a GOP and the bit allocation adjusted accordingly.

14.1.3.4   Coding artifact classification ◑

In [PA91] coding artifacts for low bitrate coding are categorized as follows: blockiness, blurriness, ringing and color bleeding. Blockiness occurs at the boundaries of the 8×8 blocks, owing to the coarse quantization of the low detail regions.[1] Blurriness is the loss of spatial detail in moderate and high-detail regions. Ringing and color bleeding occur at edges on flat backgrounds, where color bleeding is specific to strong chroma edges. In moving scenes these artifacts cause temporal busyness and a dirty appearance of uncovered backgrounds. If the artifacts are repetitive, they can give a flickering appearance.

14.1.3.5   Perceptual quality ◑

Perceptual quality of coded pictures is scored according to the type and degree of the artifacts. The visibility of each artifact is scored independently, and the individual scores are added to obtain a total score for the picture.

14.1.3.6   Statistical activity classes ◑

The variances of the subblocks are used to classify the macroblocks into two basics groups, VMAC and EMAC. VMAC (variance model activity class) is used to classify homogeneous macroblocks into 16 classes, using a nonlinear thresholding where the classes are more closely spaced at low variances. EMAC (edge model activity class) is used to classify inhomogeneous macroblocks into four classes, three for edges of various strengths and the fourth for relatively strong horizontal or vertical edges.

The use of statistical activity classes significantly extends the applicability of the variance as an activity measure, in that the choice of `quantizer_scale` is dependent on both the variance and the activity class. In fact, in [PA91] the actual assignment is integrated into the rate control system, with the bits needed for each macroblock class being calibrated by the data from the preceding picture of the same type. The actual string of quantizer values is chosen so as to achieve the best possible quality without exceeding the bit allocation for the picture.

---

[1]Blockiness is due almost entirely to errors in the five lowest frequency AC coefficients. It can be substantially reduced by predicting these coefficients from the nine DC coefficients of the block in question and its nearest neighbors.[Nis88, PM93]

14.1.4 DCT-based masking functions ◑

14.1.4.1 AC energy ◑

The earliest reference to variable quantization in transform coding that we have found is Reader's thesis[Rea73]. While not explicitly for DCT coding, many of the concepts explored in this thesis are directly relevant to MPEG. Of interest here is the use of AC energy to adapt the coding parameters, and the use of block classification to separate the blocks into different spatial patterns.

14.1.4.2 Minimax DCT coefficient ◑

In [GV91] information derived from the DCT coefficients is used to control adaptive quantization. This paper also incorporates rate feedback into the calculation. The rate control feedback establishes a reference value q_p^0 for the quantizer factor. A minimax calculation based on coefficient energies then establishes a second quantizer factor, q_p^{low}, that is used to calculate the actual quantizer_scale.

The quantization process is separated into two steps. First, the quantization is carried out with the quantizer_scale set to 8 (the default). If $C_{i,j}$ is the unquantized coefficient and $w_{i,j}$ is the quantization value for that coefficient, the first step gives a quantized coefficient, $Cw_{i,j}$, of:

$$Cw_{i,j} = C_{i,j}/w_{i,j} \tag{14.29}$$

The second step of the quantization process introduces the quantizer_scale, q_p:

$$Cq_{i,j} = int[8Cw_{i,j}/q_p + k/2] \tag{14.30}$$

where int signifies a rounding down to the nearest integer, and k is 0 or 1, depending on the type of macroblock. For simplicity, the coefficients are assumed to be positive. If integer calculations are used, appropriate scaling factors must be used to preserve precision.

The maximum energy for a block is obtained by finding the maximum for each 8×8 luma block, b, of the partially quantized DCT (quantized with quantizer_scale=8):

$$Cw^b_{max} = max_{i,j}|Cw^b_{i,j}| \tag{14.31}$$

The minimax energy, $Cw_{minimax}$ is the minimum of Cw^b_{max} over the four luma blocks, $b = 1, \ldots, 4$.

$$Cw_{minimax} = min_b Cw^b_{max} \tag{14.32}$$

The relationship between `quantizer_scale` and the minimax energy is postulated (see Equation 14.30) to be such that:

$$8Cw_{minimax}/q_p^{low} + k/2 = \text{const.} \qquad (14.33)$$

This is of the form:

$$q_p^{low} \propto Cw_{minimax} \qquad (14.34)$$

This equation tells us that the `quantizer_scale` should be proportional to the smallest value of the maximum coefficient amplitude for the four luma blocks in the macroblock.

14.1.4.3   Coefficient energy sum ◑

In [CJK+93] an activity index is developed from luminance DCT activity, an approach that is also found in [MN89]. In [CJK+93] macroblock activity is defined as follows:

$$\sigma_i = (128/(232 \times max(DC, DC_{min}))) \sum_{j=0}^{3} \sum_{k=6}^{63} C_{i,j}^2 \qquad (14.35)$$

where σ_i is the activity index for the ith macroblock, j is the index of the 8×8 luminance block in the macroblock, k is the index of the kth DCT coefficient, and the $max(DC, DCmin)$ term increases the activity index when the average luminance of the macroblock, DC, is small (DC_{min} is a constant, typically set to 512). The first five AC coefficients are not included in this activity measure, since they represent low frequencies "which are not strongly correlated with human perception activity."

The quantizer step size, q_i, is given by:

$$\begin{aligned} q_i \ &= 2^c\sigma_i && \text{for I-pictures} \\ &= (Q_P/Q_I)2^c\sigma_i && \text{for P-pictures} \ . \\ &= (Q_B/Q_I)2^c\sigma_i && \text{for B-pictures} \end{aligned} \qquad (14.36)$$

where Q_I, Q_P, and Q_B are the target quantizer step sizes for I-, P- and B-pictures, respectively, and c is a parameter that controls the overall bitrate.

As the authors of the paper note, one problem with this metric is the inability to detect boundaries between very high and very low activity. This can cause annoying artifacts associated with coarse quantization in the regions encompassing these edges. The problem is alleviated by selecting a threshold σ_e, where σ_i is changed to σ_e if any neighbor of the ith macroblock has a value less than this threshold. The choice of σ_e is based on a heuristic histogram analysis that incorporates a bitrate dependence.

The distribution of σ_i is modified such that very small and very large values are clamped at bounds set at the edges of the low and high tails of the distribution. Again, this involves some reasonable heuristics.

The quantizer tables used for this work are somewhat modified from the defaults. In addition, a thresholding of the unquantized DCT coefficients is performed in order to increase the number of zeros. This is done by a reverse scan of the unquantized coefficients, the loop being broken when a coefficient exceeds the threshold magnitude set for that coefficient.

14.1.4.4   Macroblock classification using the DCT ◑

In [CLCR93] macroblock classification is done in the DCT domain. A non-standard macroblock is used consisting of a 4x2 array of 8×8 luma blocks and two chroma blocks. The following energy measures are used:

$$
\begin{aligned}
E_h &= \text{horizontal energy} \\
E_d &= \text{diagonal energy} \\
E_v &= \text{vertical energy} \\
E_a &= \text{average high frequency energy} \\
E_m &= \text{minimum value of } E_h, E_d, \text{ and } E_v \\
E_M &= \text{maximum value of } E_h, E_d, \text{ and } E_v \\
E_{m/M} &= \text{ratio between } E_m \text{ and } E_M
\end{aligned}
$$

These are calculated from the DCT coefficients using the following template:

```
d   d   h   h   h   .   .   .
d   d   h   h   h   .   .   .
v   v   d   d   .   .   .   .
v   v   d   d   .   .   .   .
v   v   .   .   .   .   .   .
.   .   .   .   .   .   .   .
.   .   .   .   .   .   .   .
.   .   .   .   .   .   .   .
```

where d, h and v refer to diagonal, horizontal and vertical terms, respectively.

The energy terms are the following sums:

$$
\begin{aligned}
E_h &= \text{sum over h elements} \\
E_v &= \text{sum over v elements} \\
E_d &= \text{sum over d elements} \\
E_a &= \text{Avg}(E_h, E_v, E_d) \\
E_m &= \text{Min}(E_h, E_v, E_d) \\
E_M &= \text{Max}(E_h, E_v, E_d) \\
E_{m/M} &= E_m/E_M
\end{aligned}
$$

Since E_a represents the average high-frequency energy of a block, it is used to segment the blocks into low activity and high activity; high-activity blocks are further divided into texture blocks and edge blocks using E_m and $E_{m/M}$. The classification is as follows:

Smooth block: $E_a < T_1$

Texture block: $E_a \geq T_1$ and $E_m > T_2$ and $E_{m/M} > T_3$

Edge block: If a block satisfies neither smoothness nor texture criteria, it is an edge block

After the block classification, the activity of the macroblock is classified according to the distribution of blocks within it.

Let N_s be the number of smooth blocks, N_e be the number of edge blocks and N_t be the number of texture blocks. For the eight blocks within the macroblock used in this paper, there are 45 possible combinations of smoothness, texture, and edge blocks. Ordering these as (N_s, N_e, N_t), the combinations are:

(8,0,0)
(7,1,0)(7,0,1)
(6,2,0)(6,1,1)(6,0,2)
(5,3,0)(5,2,1)(5,1,2)(5,0,3)
(4,4,0)(4,3,1)(4,2,2)(4,1,3)(4,0,4)
(3,5,0)(3,4,1)(3,3,2)(3,2,3)(3,1,4)(3,0,5)
(2,6,0)(2,5,1)(2,4,2)(2,3,3)(2,2,4)(2,1,5)(2,0,6)
(1,7,0)(1,6,1)(1,5,2)(1,4,3)(1,3,4)(1,2,5)(1,1,6)(1,0,7)
(0,8,0)(0,7,1)(0,6,2)(0,5,3)(0,4,4)(0,3,5)(0,2,6)(0,1,7)(0,0,8)

In the subjective testing these were reduced to four classes:

Group 1: Smooth blocks dominate.

Group 2: Edge blocks dominate.

Group 3: Texture blocks or complex edge blocks dominate.

Group 4: All blocks are evenly distributed.

The testing showed that for a given quantizer_scale, the groups can be ordered in terms of decreasing perception of distortion, assigning an activity index as follows:

1								
2	4							
3	6	9						
5	8	12	13					
7	11	21	23	26				
10	15	22	25	27	33			
14	17	24	28	32	34	39		
16	19	29	31	35	38	40	43	
18	20	30	36	37	41	42	44	45

This index is then related to a "perceptual activity parameter" by means of a quadratic function, the details of which are not given in the paper. The activity is used, in combination with a buffer fullness parameter, to control the quantization.

Although this scheme is for a more complex macroblock structure, it can presumably be adapted to MPEG-1.

14.1.5 Overhead in switching `quantizer_scale` ◗

If the `quantizer_scale` is changed without regard to the cost of changing it, the coding may very well be quite suboptimal. In [OR94] the overhead of changing the `quantizer_scale` versus leaving it unchanged is analyzed. The casting of the problem in terms of a theoretical rate-distortion framework means that a relatively simple distortion metric such as MSE must be used.

14.1.6 MPEG-2 Test Model 5 ◗

An implementation of Test Model 5 (TM5) of MPEG-2[ISO93b] is described in [EF95]. The section of interest here is the model for variable quantization.

The basic idea in TM5 is to maintain a fixed ratio between the average quantization parameters, $Q_{avg,i}$, $Q_{avg,p}$, and $Q_{avg,b}$, where these are defined to be the arithmetic mean of the `quantizer_scale` over all macroblocks in I-, P-, and B-pictures, respectively. The ratios are controlled by two model parameters,

$$K_p = Q_{avg,p}/Q_{avg,i} \qquad (14.37)$$

and

$$K_b = Q_{avg,b}/Q_{avg,i} \qquad (14.38)$$

K_p and K_b are defaulted to 1 and 1.4, respectively.

The relationship between the coded bits, T, in a picture and the average quantization parameter is

$$T Q_{avg} = X = \text{constant} \qquad (14.39)$$

where X is a function of the picture type, and is computed from the previous picture of the same type. The bit allocation is updated after each picture is coded.

In TM5 the `quantizer_scale` is proportional to buffer fullness, d_j,

$$Q_j = A d_j \qquad (14.40)$$

where A is a constant determined by feedback loop control parameters.

Q_j is also modulated by a measure of the local spatial activity that is based on the minimum variance of the four luma blocks in the macroblock.

The spatial activity, ACT, is normalized by an average value, AVG, obtained for the previous picture and mapped to the range 0.5 to 2.0 by a nonlinear function:

$$\text{quantizer\_scale} = Q_j((2ACT/AVG) + 1)/((ACT/AVG) + 2) \quad (14.41)$$

To avoid needless expenditure of bits, the `quantizer_scale` is only changed when the difference exceeds a threshold.

14.1.7   Coded bits ◑

The classic Chen and Pratt paper on the scene adaptive coder [CP84] uses rate buffer fullness to control the scaling of the coefficients. More recently, [FAE+94] suggests the use of coded bits as an activity measure for MPEG. Coded bits is a straightforward measure of activity, but the computational overhead of recoding may be a deterrent to its use. Some data on using coded bits as a masking function will be presented later in this chapter.

14.2   Perceptual studies ◑

Many of the papers discussed in the literature review use perceptual quality studies to adjust their masking functions and classification schemes to better match to actual video sequences. Generally, these involve subjective evaluation of quality of video sequences. While few details about these perceptual studies are found in the literature reviewed in Section 14.1, the approach generally taken is to adjust the various parameters to achieve a uniform perceptual quality. In [PAC92] the adjustable constants are fitted by a regression analysis by measuring the different variables after coding several sequences to approximately equal quality. In [PA91] this is carried quite a bit further, the various artifacts being classified and the degree of subjective impairment determined for each class separately.

In this section we discuss results from a study by [Pen96c] of the visibility of distortion in individual macroblocks. The study is oriented toward spatial masking in I-pictures, and thus, addresses only part of the problem.[2]

The macroblocks for this study were chosen from a picture from the mobile sequence. Although this is a preliminary study and only 30 macroblocks were used, they were selected to have a reasonably wide range of spatial patterns. The macroblocks were coded using `quantizer_scale` values from 0 (the original source data) to 31, and observers were requested to rank the impairment of each block on a six-point impairment scale. Each

---

[2]The material from [Pen96c] is reproduced here with permission.

macroblock (luma only) was displayed in four rows of eight blocks in order
of increasing quantizer scale surrounded by a 48x48 pel surround of source
data. A viewing distance of six times the CIF picture width was used.

The goal of this study was to measure the effectiveness of some of the
simpler masking functions reviewed in the preceding section. The masking
functions tested were minimum block variance, coded bits, minimax DCT
coefficient energy, and spatial derivative convolutions.

The most effective of the simple masking functions tested was a variant
of the spatial derivative masking functions. This spatial derivative masking
function borrows heavily from the ideas reviewed in Section 14.1. It uses a
convolution based on the Laplacian operator:

$$\delta I = \sum_{i=1}^{6} \sum_{j=1}^{6} (L(I_{i,j}))^2 \qquad (14.42)$$

where L is a 3x3 Laplacian operator defined by:

$$L(I_{i,j}) = I_{i,j} - (1/8) \sum_{m=-1}^{1} \sum_{n=-1}^{1} I_{i+m,j+n} \qquad (14.43)$$

Note that the summation in Equation 14.42 does not use intensity values
outside of the 8x8 block.

As in the minimax calculation of [GV91], the minimum block activity
for the four blocks of the macroblock becomes the masking function value.

Figure 14.2 shows the measured thresholds as a function of this spatial
derivative masking function. The median and the first and fourth quartile
values define the circle and error bars respectively for the measured values. The correlation coefficient for this function is 0.76, comparable to that
measured for individual observers relative to the median.

The correlation found for some other simple masking functions is given
in Table 14.1. All of the masking functions tested were computed separately
for each of the four blocks, and the minimum value among them was used
as the masking function value.

While it appears that the spatial derivative masking function has the
best correlation, one should not infer too much from the data in Table 14.1.
First, the measurements are preliminary and incomplete; second, an important element — macroblock classification — is missing. All of the masking
functions tested are fairly comparable, and the slight edge achieved by the
spatial derivative might disappear when macroblock classification is added.

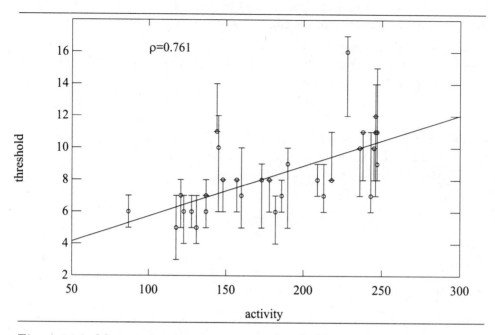

Figure 14.2: Measured `quantizer_scale` thresholds as a function of activity derived from a spatial derivative masking function.

Masking function	correlation coefficient
spatial derivative	0.76
DCT maximum energy (1-63)	0.66
DCT maximum magnitude	0.65
DCT maximum energy (6-63)	0.65
DCT sum of magnitudes	0.62
block variance	0.59
DCT sum of energies	0.58
coded bits (`quantizer_scale`=4)	0.53

Table 14.1: Linear correlation coefficients for simple masking functions. The DCT coefficient energy is defined here to be the square of the coefficient magnitude.

14.3   Variable quantization and motion ◑

14.3.1   Temporal masking ◑

A number of the models for variable quantization reviewed in Section 14.1 incorporate temporal masking effects [PA91, PAC92, LDD$^+$92, PA94]. These generally take the form of a term that increases the `quantizer_scale` when significant motion is detected relative to a preceding picture. This is certainly appropriate for large motions, and is an important effect. The effects can be particularly large at scene changes, where temporal masking makes even large distortions hard to see [LD94].[3] This is fortunate, because the failure of the motion displacement search at scene changes (see Chapter 13) usually decreases the quality of the prediction. If the bit allocation is fixed, the average `quantizer_scale` value must increase at a scene change. Note that leading B-pictures in the first GOP of a sequence have similar properties.

Although temporal masking might possibly be exploited in I-pictures, the I-picture is usually the foundation for the GOP. A reduction in quality of an I-picture reduces the accuracy of the predictions obtained from it, increasing the bitrate of dependent pictures.

14.3.2   Relative motion between adjacent macroblocks ◑

For modest motion, another effect can become important that requires a *reduction* in the `quantizer_scale` for P- and B-pictures. In P- and B-pictures, an object moving relative to a background will tend to drag the background with it. In the fixed and relatively large macroblocks of MPEG-1, the motion displacements for the macroblock are usually determined by the edges of the moving object in the macroblock. If the `quantizer_scale` is coarse (and it usually is), the background texture in the macroblock is taken unchanged from the prediction, and tends to move with the object in the decoded picture sequence. This gives rise to a halo of spurious motion of the background — the effect is sometimes called *mosquitos* because it gives the appearance of a cloud of mosquitos around the object.

Variable quantization can greatly reduce this effect [Pen96d]. The motion displacements derived for motion compensation can also be used to determine where there is relative motion between two macroblocks. If there is, the `quantizer_scale` should be decreased relative to the average for the picture, thereby allowing the false background motion to be corrected.[4] In

---

[3]In the author's experience chroma distortion at a scene change can be quite visible in the rare cases where the prediction is taken from the wrong scene.

[4]Another approach to this problem is suggested in the MPEG standard (see Clause D.6.4.3.1), where some some rules are given for switching to prediction without motion

the simplest variation of this scheme, relative motion is detected between a block and its four nearest neighbors, and a simple thresholding is used to detect when to decrease the `quantizer_scale`. Both a minimum and a maximum threshold are used, since large relative motions should exploit temporal masking effects.

14.4   Summary ○

Variable quantization is the most important parameter for rate control in MPEG. It also is a key to achieving higher picture quality at a given bitrate. The models for variable quantization range from the relatively simple to the quite complex and are characterized by a wide diversity of approaches. All use spatial and/or temporal masking effects and all are based on heuristic, albeit reasonable, assumptions regarding the structure of the masking function.

Impressive reductions in bitrate for a given visible distortion from the application of variable quantization are claimed — on the order of 30% or more — in the literature and in personal communications to the authors [NP77, Gon91]. Nonetheless, given the current incomplete state of research in this area, significant further improvements are quite possible.

---

compensation. Alternatively, the macroblocks can be coded using intra techniques, but generally that is more costly for a given picture quality.

15

Rate Control in MPEG

Chapter 14 discusses techniques for varying the quantization so as to minimize visible distortion for a given bit rate. Spatial and temporal masking effects tend to reduce the visibility of artifacts in more active areas — areas with strong temporal and/or spatial variations — and these effects can be used to improve the overall perceptual quality of a video sequence. Variable quantization is also, however, the most important mechanism for controlling MPEG bitrates. In this chapter the discussion turns to the methods by which variable quantization and other techniques are used to control the bitrate of MPEG data streams. In most cases the two uses of variable quantization — for improving quality and for controlling the bitrate — are complementary.[1]

A number of other aspects of encoder decision making will be considered in this chapter. These include GOP structure, bit-allocation between I-, P- and B-pictures, macroblock type selection, feedback to control bitrate, and buffer overflow and underflow.

This chapter starts with a review of literature relating to MPEG rate control. Then, buffer overflow and underflow, the idealized video buffering verifier (VBV) decoding model for regulating decoder buffer fullness, and proportional rate control feedback systems are discussed. Finally, some measures an implementer can take to obtain a robust rate control system that can handle unexpected extremes in picture complexity are described.

15.1 Literature review ◗

In this section we discuss the published literature on rate control. Where appropriate, the discussion of these papers in Chapter 14 is referenced.

---

[1]Some of the material in this chapter is from [Pen96d, Pen96c]. It is reproduced here with permission

15.1.1   The rate buffer ◑

Unless the instantaneous coding rate is constant, a buffer is needed to smooth the variable output rate and provide a constant rate output. Constraining the instantaneous coding rate to be constant requires, among other things, inefficient fixed-length codes and other constraints that very negatively impact picture quality. Consequently, buffering is essential if reasonable quality is to be obtained at the relatively low bitrates characteristic of MPEG-1.

The use of a buffer to smooth out variations in the instantaneous output rate and provide feedback for rate control can found in some of the earliest work on motion video coding [Has72, HMC72, Rea73], as well as in more recent work such as the scene adaptive coder [CP84].

The constraints on buffer size are described quite clearly in Reader's PhD thesis [Rea73]: "... the upper size of the buffer is constrained by consideration of the delay introduced by the buffer. In the picture telephone application, a buffer which is too large will introduce a delay long enough to impede the two way flow of conversational information. However, use of a small buffer renders the system liable to overload should fast movement occur."

These constraints remain valid today. However, for non real-time applications the concern with conversational delay is replaced by concern about the cost of buffering in hardware implementations. If the MPEG-1 sequence header `constrained_parameters_flag` is set, the buffer size needed by the decoder may not be larger than about 40 kbytes for the idealized decoder model described later in this chapter.

15.1.2   p×64 reference model and generalizations ◑

In [PA90] a description of the reference model developed for H.261 (CCITT p×64) and a generalization of this for MPEG are given. The paper also gives useful data on various GOP structures.

The p×64 codec uses only I-pictures and P-pictures. A FIFO buffer is assumed between coder and channel, and several times per picture is monitored for fullness. The quantization parameter, Q_p, (proportional to quantizer step size) is increased or decreased according to whether the buffer is relatively full or empty.

This basic approach must be extended for MPEG, because bit allocation depends on picture type. The use of a default allocation is suggested in [PA90], although the authors also indicate that the bit allocation can be adapted by unspecified techniques to match the needs of a particular scene.

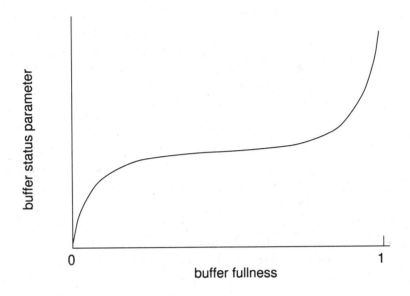

Figure 15.1: The s-shaped curve characteristic of rate control systems.

15.1.2.1   S-shaped control characteristic ◐

In a brief discussion of rate control in [CLCR93] rate control is based on a buffer status parameter that has a dependence on buffer fullness as in Figure 15.1. The general s-shape of this curve is characteristic of rate control systems with buffering constraints. The buffer status controls the quantization step size, and thereby, the rate. When the buffer is nearly empty, the rate must be aggressively increased in order to avoid underflow; when the buffer is nearly full, the rate must be aggressively decreased in order to avoid overflow. In the mid-regions, parameters can be modified less, thereby keeping picture quality more uniform.

15.1.3   MPEG-2 Test Model 5 ◐

In [EF95] a description is given of a software MPEG-2 video codec that is based on Test Model 5 (TM5) for MPEG-2 [ISO93b]. Aspects of this paper have already been discussed in Section 14.1, and the following description builds on that discussion.

The rate control system in TM5 has three layers operating at the GOP, the picture and the macroblock. At the highest layer, a bit allocation per picture is determined, adjusted for any excess or remaining bits from preceding GOPs. The relationship between average `quantizer_scale` and coded bits is assumed to be an inverse proportionality, and the proportionality constants for a given picture are predicted from the previous picture of the same type. The bits are divided equally among all macroblocks, and the `quantizer_scale`, Q, for the macroblock is defined to be proportional to buffer fullness. Thus:

$$Q = (31 \times B_f)/r \qquad (15.1)$$

where B_f is the encoder buffer fullness.[2] r is a reaction parameter, a constant that controls the rate at which the `quantizer_scale` follows variations in buffer fullness. The linear relationship is a limiting case of the postulated s-shaped characteristic in Figure 15.1.

Bit allocation in TM5 is controlled by keeping the average `quantizer_scale` ratios between I- and P-pictures and I- and B-pictures constant. These ratios are defaulted to 1 and 1.4 respectively.

15.1.3.1   Rate control with DCT minimax activity ◑

In [GV91] a heuristic is suggested that links the `quantizer_scale` calculated from minimax DCT activity (see Section 14.1) to buffer fullness.

The reference `quantizer_scale`, q_p^0, is chosen on the basis of buffer fullness, and thus is large when the buffer is nearly full. A second `quantizer_scale`, q_p^{low}, is computed from the minimax DCT activity, and the actual `quantizer_scale`, q_p, is then calculated from the following heuristic relationship:

$$q_p = min\left[q_p^0, max\left[q_p^{low}, int\left[1 + 30(q_p^0/31)^\alpha\right]\right]\right] \qquad (15.2)$$

A value for α of 2.5 is reported to work well.

15.1.4   Rate control based on recoding ◑

The rate control system in [DL95] is based on recoding. The basic unit for the control system is the slice, where in this work the slice is defined to be a macroblock row. A variable quantization scheme such as [GV91] is invoked for the macroblock level, but is not actually used in the studies. Similarly, the target bits for a picture is postulated to be available from an algorithm such as in [PA91].

---

[2]This should not be confused with the *decoder* buffer occupancy of the VBV model discussed in Section 15.3 This point will be discussed further in Section 15.2.

Global rates as a function of `quantizer_scale` and picture type are shown, and are modeled with an equation of the form:

$$R = \alpha + (\beta/Q^\gamma) \quad \text{where } 0 \leq \gamma \leq 2 \qquad (15.3)$$

where R is the rate, α and β are constants, and $\gamma = 0$ for a Gaussian random variable. Typically, I-pictures are bounded by $0.5 \leq \gamma \leq 1$ and P-pictures by $0.5 \leq \gamma \leq 1.5$, which is reasonably consistent with the approximately inverse proportionality used in MPEG-2 TM5 and other systems.

The rate control system is used to adjust a reference `quantizer_scale` at each slice. A trial encoding is performed using the reference scale and if the bits generated matches the target rate, that result is used; otherwise, the quantizer scale is adjusted up or down by an integer amount, depending on whether the rate is too high or too low. The amount of the adjustment is not computed from Equation 15.3; rather, it is an integer of value 1, 2 or 4, approximately proportional to the reference `quantizer_scale`. If necessary, the picture is recoded with the revised `quantizer_scale`.

15.1.5   VBR rate control ◗

In [PAC92] and [PA94] VBR (variable bitrate) rate control algorithms are considered. While the use of rate control in a VBR coder would seem to be an oxymoron, the intent of this work is to "maximize the perceived quality of the decoded image sequence" while "maintaining the output bitrate within permitted bounds". The goal is really bitrate limiting rather than precise rate control, since the intent is to multiplex many coded bitstreams simultaneously over a network. Further discussion of these papers can be found in Section 14.1.

15.1.6   Recursive prediction with recoding ◗

Some aspects of [CJK+93] have already been reviewed in Section 14.1. In this paper rate control is achieved by adjusting the parameter c in Equation 14.36. The authors note that according to rate-distortion theory, rate tends to be linearly related to the logarithm of the quantizer step size. Therefore, from the relationship in Equation 14.36, rate should be proportional to c. Define R_x as the bit count for coding mode x. Then, approximately,

$$R_x = a_x + b_x c \qquad (15.4)$$

where a_x and b_x are constants, but are dependent on the picture type and activity. (Note that b_x is negative.) These constants are recursively updated to track changes in scene activity.

The weighted sum of the bitrates for the pictures that make up the GOP are added together to create a similar equation for the entire GOP. Thus, for a GOP of N pictures with $M - 1$ B-pictures between I- and/or P-pictures (where M is a factor of N) the GOP bitrate, R_{GOP}, is given by:

$$R_{GOP} = R_I + (N/(M-1))R_P) + (N - (N/M))R_B \\ = a_{GOP} + b_{GOP} \times c \tag{15.5}$$

where a_{GOP} and b_{GOP} represent the rate control constants for the GOP. If Equation 15.5 is solved for c with R_{GOP} set to the target bitrate for the GOP, an optimal value of c is obtained that meets the target bitrate for the full GOP.

The bitrate for different coding modes is quite different, and the buffer occupancy is therefore strongly dependent on the position of the current picture in the GOP. However, for constant rate, the buffer occupancy at the end of the GOP should be approximately equal to the occupancy at the beginning. At any point within the GOP the buffer occupancy at the end of the GOP can be estimated from the current buffer occupancy. Consequently, the optimal value of c can be revised after coding each picture in the GOP.

After arriving at bit-allocations in this manner, the picture is coded using the quantizer step sizes calculated from Equation 14.36. Note that iterative recoding is used to achieve the desired bitrate.

15.1.7 Complexity prediction ◑

In [CJOS94] complexity of a segment of a picture is estimated from the coding of the same segment in the preceding picture of the same type. Bits are distributed in the current picture so as to be proportional to that complexity measure.

The complexity of the segment, α_n, is given by the product of two ratios:

$$\alpha_n = (A_n/(B/N)) \times (c_n/d) \tag{15.6}$$

where A_n is the bits required to code segment n, B/N is the average number of bits required to code the segments in the picture, c_n is the average quantizer_scale for the segment, and d is the average quantizer_scale for the picture. Note that if the inverse proportionality between bits and quantizer_scale used for MPEG-2 Test Model 5 is valid, α will be independent of the quantizer_scale, but will increase as the coding complexity of the segment increases.

A reference quantizer, Q, is calculated from encoder buffer fullness using a relationship that incorporates a multiplicative activity term to the relationship used in MPEG-2 Test Model 5 (see Equation 15.1):

$$Q = (k_n \times B_f \times 31)/r \tag{15.7}$$

where B_f is the fullness of the encoder buffer, r is twice the ratio of the bitrate to the picture rate, and k_n is empirically related to the activity by the following approximate relationships:

$$k_n = \sqrt{1 - (\alpha_n - \alpha_{avg})^2} \quad \text{if } (\alpha_n - \alpha_{avg}) \geq 0$$

$$k_n = \sqrt{1 + (\alpha_n - \alpha_{avg})^2} \quad \text{otherwise}$$

(15.8)

15.1.8   Adaptive GOP ◑

In [LD94] an adaptive GOP is developed that adjusts the GOP structure in accord with five distance measures between pictures. These measures include several different histogram differences, block variance difference and motion compensation error.

If the distance measure between two adjacent pictures, f_{n-1} and f_n is above a given threshold, a scene change is detected. Then, f_{n-1} is declared a P-picture and f_n an I-picture, but both are coarsely quantized to take advantage of temporal masking.

If the distance accumulated over the course of several pictures exceeds a threshold at picture f_n, a scene segmentation point is detected. In this case scene f_{n-1} is declared a P-picture and coded with normal P-picture quantization. In this case, adjacent B-pictures on either side are better predicted, and can be coded with a lower bit count.

15.1.9   Adaptive bit allocation ◑

The paper by Puri and Avarind [PA91] describes a comprehensive approach to the MPEG coding process, and aspects of this paper relating to variable quantization have already been reviewed in Section 14.1. Block variances are used to estimate activity in various regions of each picture, and averages of block variances are then related to four levels of picture complexity as shown in Table 15.1.

With four categories of picture complexity and three picture types, (some repeated many times within a GOP), scene categorization could be quite complex. However, it is reduced to eight levels by combining various picture complexity ranges. Different bit allocations are then used for each scene complexity level.

As discussed in [PA91], bit assignment is a function of the details of the GOP structure and the desired bitrate. For a bitrate of 1.15 Mbits/s and a GOP with IBBPBBPBBPBBP... structure the bit allocation ratios range from I:P=1.7 and P:B=5 for simple scenes with low motion to I:P=2.3 and P:B=10 for complex scenes. Given the large number of possible variations in MPEG coding techniques, these values are reasonably consistent with

Picture Complexity	Variance of I-picture	Variance of P-picture	Variance of B-picture
Low	0-499	0-99	0-49
Medium	500-999	100-249	50-124
High	1000-1499	250-499	125-199
Very high	≥ 1500	≥ 500	≥ 200

Table 15.1: Picture complexity categories (after [PA91]).

Annex D of the MPEG-1 standard, where an I:P ratio of 3 and P:B ratios of 2-5 are suggested.

The rate control system developed in [PA91] is for a causal system in which the `quantizer_scale` is determined without access to information about macroblocks not yet coded. In order to meet this constraint, information from a previous picture is used to predict the quantization in the current picture. Because this does not guarantee a rate, the buffer is monitored during the encoding of the picture. If too many bits are being generated, an additional increment is added to the quantizer step size, the size of which is determined by buffer fullness. Because of the causality requirement, the picture used for gathering statistics for a P-picture may be either a P-picture or an I-picture, whereas for I-pictures it is always a P-picture. Because of the relatively large coding cost of changing the `quantizer_scale`, the quantization of B-pictures is changed only at the slice header. The reader is referred to the original paper for more details.

15.2   Encoder and decoder buffering ◗

One source of confusion in MPEG rate control is the concept of buffer occupancy. In analyzing buffer occupancy, one must first define whether a buffer is an encoder buffer or a decoder buffer, as they have very different behavior. The literature cited in Section 15.1 generally assumes encoder buffers, whereas MPEG-1 mandates the use of the VBV, an idealized model of the decoder buffer.

An encoder buffer is filled in bursts as the pictures are coded, and emptied at a constant rate as the data are transmitted. A decoder buffer is filled at a constant rate as the transmitted data are received, and emptied in bursts as the pictures are decoded. The situation is sketched in Figure 15.2.

Referring to Figure 15.2, an encoder buffer overflows when too much coded data are produced and underflows when not enough data are produced. A decoder buffer overflows when not enough coded data are removed

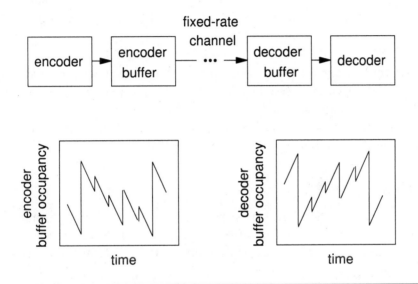

Figure 15.2: Encoder and decoder buffer occupancy. The encoder buffer is filled in bursts as the pictures are encoded and passed to the buffer, and emptied at a constant rate. The decoder buffer is filled at a constant rate and emptied in bursts as the pictures are removed and decoded.

and underflows when too much data are removed. Buffer occupancy has the intended meaning in both cases, but a full encoder buffer implies an empty decoder buffer.

In the rest of this chapter *decoder* buffer occupancy is used unless otherwise stated.

15.3   VBV decoding model ◑

For fixed-bitrate operation, MPEG defines an idealized model of the decoder called the *video buffering verifier*, or VBV. This model is attached (conceptually) directly to the output of an encoder when the encoder is generating a fixed-rate bitstream. If the fixed rate channel is omitted in Figure 15.2, the decoder sections become a VBV.

The VBV is used to constrain the instantaneous bitrate of an encoder such that the average bitrate target is met without overflowing or underflowing the decoder's compressed data buffer. Three parameters from the video sequence header and one from the picture header are used in this model. The video sequence header is described in Section 8.3, and the picture header is described in Section 8.5.

15.3.1   bit_rate ◑

When the MPEG bitstream is fixed-rate, the 18-bit **bit_rate** parameter in the video sequence header contains the nominal bitrate of the video data stream, in units of 400 bits/s. The nominal bitrate, R', in bits/s, is given by:

$$R' = \texttt{bit\_rate} \times 400 \qquad (15.9)$$

The **bit_rate** parameter is calculated from the true encoder bitrate, R, by:

$$\texttt{bit\_rate} = \lceil R/400 \rceil \qquad (15.10)$$

where $\lceil \ \rceil$ indicates a rounding up to an integer value.

Since encoder and decoder may have slightly different clocks and bit rates, the decoder must be able to periodically synchronize the decoding with the incoming bitstream. This can be done by means of the **vbv_delay** parameter discussed in the next section.[3] Variable bitrate data streams are signaled by setting the **bit_rate** parameter to 0x3FFFF. In this case, buffer status has no meaning and the VBV decoder model for buffer occupancy is not used.

---

[3]The MPEG standard recommends using system time stamps when streams are multiplexed (see Clause D.5.1.6 in ISO/IEC 11172-2).

nominal picture rate	exact values	P	picture_rate
forbidden	–	–	0000
23.976	24/1.001	23.976023976024	0001
24	24	24	0010
25	25	25	0011
29.97	30/1.001	29.97002997003	0100
30	30	30	0101
50	50	50	0110
59.94	60/1.001	59.9400599400599	0111
60	60	60	1000
reserved	–	–	1001
. . .	. . .	. . .	. . .
reserved	–	–	1111

Table 15.2: Interpretation of picture rate codes.

15.3.2 picture_rate ◑

Picture_rate is a four-bit parameter in the video sequence header that is interpreted according to Table 15.2. This table is similar to Table 8.3, but includes exact values for the picture rate. picture_rate determines the rate at which decoded pictures must be available for display, and thus, the rate at which compressed pictures must be removed from the compressed data buffer and decoded. For a constrained parameter environment, picture rates must be less than or equal to 30 pictures/s.

15.3.3 vbv_buffer_size ◑

vbv_buffer_size sets an upper bound on the size of the buffer needed by a decoder. This 10-bit parameter in the video sequence header gives the buffer size in units of 2048 bytes. Thus, the buffer size in bits is given by

$$B = 16 \times 1024 \times \texttt{vbv\_buffer\_size} \qquad (15.11)$$

For a constrained parameter environment, vbv_buffer_size must be 20 or less. Therefore, the hypothetical decoder buffer must be less than or equal to 327,680 bits (40,960 bytes). vbv_buffer_size sets a lower bound on the buffer size a decoder may have and still successfully decode the sequence. It therefore sets an upper bound on the number of bits that the encoder can buffer in the VBV.

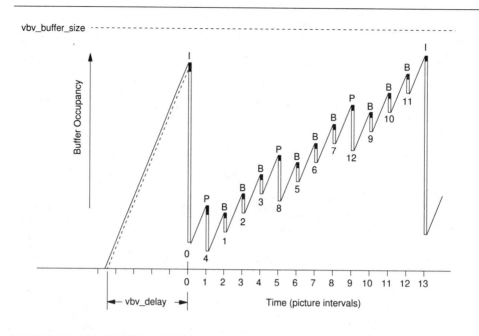

Figure 15.3: Buffer occupancy for a GOP at the start of a sequence.

15.3.4   vbv_delay ◗

The decoder compressed data buffer must be managed such that it neither overflows nor underflows. This is accomplished by means of a parameter that tells the decoder exactly when to remove a given picture from the buffer and decode it.

Given a constant *true* bit rate, R, picture rate, P, and buffer size, vbv_buffer_size, overflow and underflow are controlled by the picture header parameter, vbv_delay. vbv_delay sets the time between receipt of the first bit of coded data for the picture in the buffer and the start of decoding of that picture. It therefore sets the decoder buffer occupancy in Figure 15.2 at the start of decoding of each picture.

The 16-bit integer value of vbv_delay in each picture header is calculated from:

$$\text{vbv\_delay} = (90{,}000 \times B_n^*)/R \qquad (15.12)$$

where R is the full precision encoder bitrate, measured in bits/s.[4] B_n^* is the number of bits that should accumulate in the buffer before the compressed data for the picture is removed. (The bit count starts at the bit immediately following the picture_start_code.) Therefore, the vbv_delay is the number of periods of the 90kHz system clock the VBV model decoder waits

---

[4]Note that this is not R', the bitrate in the compressed bitstream.

before extracting the picture data from the buffer. The extraction of data from the buffer is assumed to be instantaneous.

Pictures are removed at fixed encoder clock intervals, as determined by the `vbv_delay`.[5] Since the true encoder rate and encoder clock are used in calculating the `vbv_delay`, the `vbv_delay` is really a measure of the amount of data in the buffer, i.e. the buffer occupancy. Note that in determining the total buffer occupancy, allowance must be made for the additional bits in the `picture_start_code` and in any higher layer headers and codes immediately preceding the picture. These are considered to be part of the picture in this idealized decoder model.

The `vbv_delay` can be used to correct any slight discrepancies between encoder and decoder timing at the start of each picture decoding. The means by which these small discrepancies are handled is left to the particular implementation, but they do create additional tolerances that must be considered in the design of a decoder. If the decoder picture rate cannot be adjusted, a means for repeating and discarding pictures must be provided.

Figure 15.3 provides a plot of decoder buffer occupancy as a function of time, beginning with an empty buffer at the start of receipt of a sequence. The ordinate is buffer occupancy and the abscissa is time, in units of encoder picture intervals. The narrow vertical bars represent the instantaneous removal of the picture data by the VBV, with compressed data in white and headers in black. The connecting constant slope ramps show the filling of the buffer at a constant rate.

The long delay between receipt of the first data and the start of decoding is needed in order to load most of the data for the first two pictures - in this case, an I-picture and a P-picture.[6] Without this delay, a buffer underflow would occur. After being nearly emptied by the removal of the first I-picture and P-picture, the buffer gradually refills. At the start of the I-picture of the next GOP it is back approximately to the occupancy it had at the start of decoding.

15.4   Rate control ◑

MPEG has a rich array of possible methods for controlling rate. A fundamental parameter is the `quantizer_scale`, as this controls the instantaneous bitrate in coding macroblocks. In Chapter 14 a number of techniques for using the `quantizer_scale` to improve overall picture quality for a given bitrate are reviewed. We noted in that chapter and we note here that variable

---

[5]Although there is no explicit statement that the removal is at fixed intervals, the equations in the MPEG standard allow for no other interpretation.

[6]The MPEG standard recommends filling the buffer almost completely (see Clause D.6.1.3).

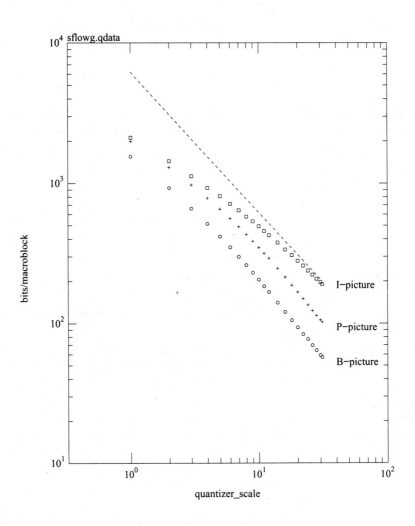

Figure 15.4: Relationship between bitrate and `quantizer_scale` for the flower garden sequence.

quantization is the most important and most effective of the techniques for maintaining a fixed bitrate. bitrate control can be done at many levels, and the choice may depend on bitrate, buffering requirements and complexity of implementation.

In much of the work reviewed earlier in this chapter, rate control is based on an equation of the form:

$$B = M/\texttt{quantizer\_scale} \qquad (15.13)$$

where B is the bits produced in coding the picture and M is a constant that is a function of picture complexity.

As can be seen from the experimental data in Figure 15.4, there is some deviation from strict inverse proportionality over the full `quantizer_scale` range of 1 to 31. Furthermore, different pictures have slightly different characteristics. However, the inverse proportionality is a good assumption for small changes in `quantizer_scale`. In causal systems the proportionality constant in Equation 15.13 is obtained from pictures already coded. If recoding is allowed, the constant is obtained from a trial encoding at some suitable `quantizer_scale`.

Feedback control and recoding can be done at the individual macroblock level, slice level, picture level, and even at the GOP level. There is a difference, however, between feedback at the macroblock level and at higher levels.

Feedback at higher levels involves the control of a reference `quantizer_scale` and the variation of that at the macroblock level by means of a masking function. Then, Equation 15.13 or some equivalent controls a reference `quantizer_scale` and the masking function controls the `quantizer_scale` relative to this reference at the macroblock level.[7]

Feedback at the macroblock level implies a highly localized rate control. Unless the masking function is derived from coded bits (not necessarily the best choice, as seen in Table 14.1), the masking and rate control procedures will sometimes work at cross purposes. Resolving this conflict requires a balancing of rate control and masking function requirements.

Note that feedback at higher levels than the macroblock may use a continuous value for the reference `quantizer_scale`. Only when the actual coding is done is the `quantizer_scale` restricted to a 5-bit integer.

15.5 Buffer occupancy and direct rate control ❶

In [EF95] and [CLCR93] (see Figure 15.1 and Equation 15.1) rate control is based on encoder buffer occupancy . The advantage in using buffer occu-

---

[7]Implicit in this is an assumption that the masking function can be scaled to apply to a range of visible distortion levels.

pancy is that the control of buffer occupancy is the end goal of rate control. As pointed out in [CJK+93], buffer occupancy depends on the position of the picture in the GOP. This will be discussed further in Section 15.6.

The rate can also be directly controlled with respect to a target rate, an approach that will be developed in this section. The advantage in controlling rate relative to a target rate is that a specific target rate can be used for each picture type, and even for each individual picture in a GOP. In this approach, however, any excess or deficiency in bits must be carried over from one control unit to the next. Otherwise, residual errors will accumulate and cause overflow or underflow.

15.5.1 Rate feedback ◑

In MPEG-2 TM5 [ISO93b, EF95] the product of rate and `quantizer_scale` is used to estimate complexity. This product can also be used as the basis for feedback control based on rate by calibrating the constant M in Equation 15.13. Thus, if B is the bits produced by coding at `quantizer_scale` Q,

$$B = \frac{(B_{trial} * Q_{trial})}{Q} \tag{15.14}$$

where B_{trial} is the bits produced by coding at `quantizer_scale` Q_{trial}. B_{trial} and Q_{trial} might be estimated from prior pictures as in [PA91], or might be obtained from a trial encoding. If trial encoding is used, the encoder is noncausal but the calibration is more precise.

If some elasticity is allowed in the rate, the equation for Q can be extended as follows:

$$Q = Q_{trial} \left(1 + \frac{K(B_{trial} - B_{target})}{B_{target}} \right) \tag{15.15}$$

where K is a constant with a function similar to the reaction parameter in [EF95]. $K = 0$ is the limiting case of no feedback; $K = 1$ gives:

$$Q = \frac{(B_{trial} * Q_{trial})}{B_{target}} \tag{15.16}$$

which will produce exactly the desired bitrate (within the accuracy of the assumed inverse proportionality between bitrate and `quantizer_scale`).

If the feedback is at a level higher than the macroblock, Q is a reference `quantizer_scale` for the `quantizer_scale` values determined by the masking function.

If the feedback is at the macroblock level, Q_{trial} is presumably derived from a masking function and a reference `quantizer_scale` and is then adjusted by the rate control equation. In this case, since K determines the strength of the rate control feedback, it is a way of balancing conflicting masking and rate control requirements. The tradeoffs here are complex and feedback at the slice level or higher appears to be simpler.

15.5.1.1 Error carry ◑

Although these equations are cast in terms of a target rate rather than buffer occupancy, the buffer occupancy must be included or the occupancy may slowly drift. As noted above, one way of incorporating buffer occupancy is to carry forward any residual rate errors, including them in an appropriate manner in either B_{trial} or B_{target}. Note that the accounting procedures for counting and apportioning the bits must be exact.

15.5.1.2 Interleaved rate control systems ◑

One of the problems with a simple error-carry for maintaining the rate is that I-pictures typically have a much larger residual rate error than B-pictures. If the rate error is simply carried forward from an I-picture to a P-picture, or from a P-picture to a B-picture, the rate control system of the picture receiving the bits can be so strongly perturbed that quality becomes very uneven.

One way of avoiding this is to interleave three independent rate control systems, one for each type of picture [Pen96b]. The perturbations from the rate error carry are then commensurate with the normal rate for each picture type and the picture quality is more uniform. At the end of each GOP, re-balancing is sometimes needed. Experimentally, summing the rate errors and re-apportioning the bits in proportion to the total coded bits for each picture type seems to work well.

15.5.1.3 Quantizer switching overhead ◑

The feedback equations above do not include the coding cost for switching the `quantizer_scale`. This overhead can be quite large for B-pictures.

One simple way of reducing the switching overhead is to require that the estimated change in rate be above a threshold determined by the cost of changing the `quantizer_scale`. More pragmatically, Puri and Avarind [PA91] advocate switching the `quantizer_scale` for B-pictures only at the slice header. Optimal switching based on a rate-distortion analysis of the problem is treated in [OR94].

15.5.2 Effects of quantization ◑

Because the `quantizer_scale` is an integer, the inverse relationship between rate and `quantizer_scale` in Equation 15.13 is not perfectly continuous and monotonic. This has some important implications for a control system based on proportional feedback and recoding. Two techniques can be used to calculate the reference `quantizer_scale`:

Estimation: An initial value is chosen for the `quantizer_scale` and the number of bits produced with that value calibrates the complexity of the macroblock. Equation 15.13 is then used to calculate the desired `quantizer_scale` from the desired rate.

Interpolation: If trial codings have been done for two `quantizer_scale` values and the bitrates straddle the desired value, interpolation can be used to estimate the complexity at the desired bitrate.

Deviations from linearity and the slightly nonmonotonic and discontinuous behavior make extrapolation dangerous.

Some care must be taken to suppress limit-cycle oscillations in the rate control system. A limit-cycle oscillation occurs when the coder can not find a single stable operating point and switches back and forth between two values of the `quantizer_scale`. Discontinuities in the rate-`quantizer_scale` relationship are one cause of such behavior. Some systems use a loop counter to break the recoding loop. Other techniques for reducing the likelihood of limit-cycle oscillation are detection of repeating values, setting a tolerance range for the rate, and terminating the feedback loop after one interpolation.

15.5.3 macroblock_type selection ◑

One of the issues the encoder must deal with as part of the decision making process is the selection of `macroblock_type`. Five bits must be set, as shown in Table 8.7 in Chapter 8.

`macroblock_quant` is generally determined from local picture activity, as discussed in Chapter 14. `macroblock_pattern` is determined by coding the blocks of the macroblock, once the basic macroblock type and `quantizer_scale` have been selected.

Intra coding or nonintra coding, and if nonintra, prediction with forward, backward, bidirectional, or no motion compensation can be selected by minimizing a distortion criterion such as MAD or variance. In general, since MAD is proportional to coded bits to a good approximation (see Figure 13.1 in Chapter 13), it should work reasonably well for this decision making process.[8] A more accurate method, albeit more costly in terms of

---

[8]The MPEG standard provides a useful discussion of this (see Clause D.6.4), providing some heuristic rules for when motion compensated prediction should not be used. Of particular interest is the selection of prediction without motion compensation for certain conditions where minimization of MAD would indicate the use of motion compensation (also see Section 14.3.2).

computation and hardware, is do a trial encoding for each possibility and choose the one that produces the fewest bits.

Note that not all combinations of bits in `macroblock_type` are defined. For example, if `cbp` is zero, the `macroblock_quant` bit must not be set. It is not uncommon for the encoder to increase the `quantizer_scale` only to find that the `coded_block_pattern` bit is then zero. One way of handling this situation is to trap the illegal combination, code `cbp`=0, and reset the encoder `quantizer_scale` back to the (unchanged) decoder value before moving to the next macroblock.

15.5.4   Interactions between pictures ◗

One problem with variable quantization is the interaction between the coding of a picture and the coding of pictures predicted from that picture. Figure 15.5 shows the effect on P- and B-picture rates of varying the `quantizer_scale` of an I-picture. This dependency is one reason why the quality of I-pictures and P-pictures should be maintained at a higher level than B-pictures. In MPEG-2 TM5 [ISO93b, EF95], for example, the average `quantizer_scale` values are maintained at ratios of 1 and 1.4 for I:P and I:B respectively.

15.6   Buffer constraints ◗

If the buffer is large compared to the bits needed to code the I-pictures, it provides enough elasticity that a large variability in bits/picture can be tolerated. Feedback for bitrate control can then be done at the GOP layer, or perhaps at even larger intervals. In this case, causal coders can be used.

If the buffer is small compared to the peak bits per picture — perhaps only slightly larger than the bits anticipated for a single I-picture — the rate must be tightly constrained. In this limit, picture recoding may be the only way to guarantee that buffer underflow and overflow do not occur.

At the highest bitrate allowed with constrained parameters, I-pictures may barely fit in the buffer. Under these conditions the range of `quantizer_scale` values that does not produce buffer underflow or overflow is quite restricted. It is important in this limit to use more than five bits of precision for the reference `quantizer_scale`. If the value is restricted to five-bit precision, a change from 2 to 1 approximately doubles the bitrate.

At very low bitrates, buffer underflow becomes an important consideration. A very low rate is usually not sustainable if variable quantization is the only mechanism for reducing the bitrate. Additional data reduction is also usually needed, starting with the discarding of DCT coefficient information. At the lowest rates, discarding of pictures (i.e. repeating of an adjacent

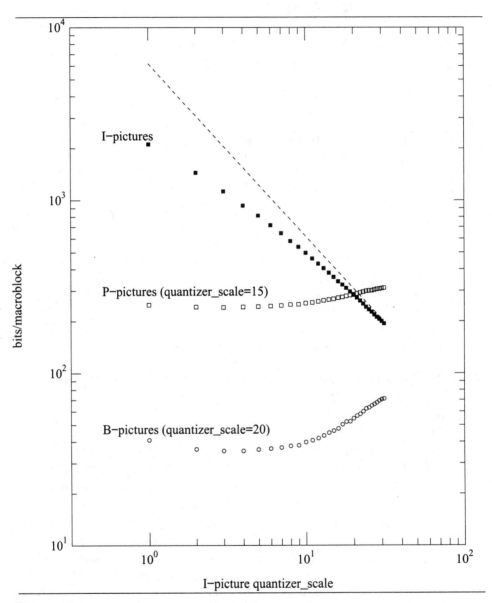

Figure 15.5: Effect of I-picture `quantizer_scale` on the coding of the I-picture and dependent P- and B-pictures.

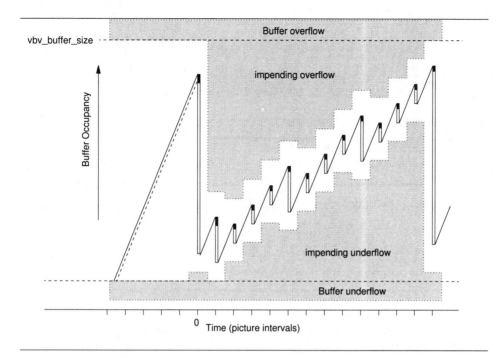

Figure 15.6: Rate control profiles for buffer management.

picture in the sequence) is also needed. Explicit repeating of pictures is not provided for in the MPEG-1 syntax, but an equivalent function is sometimes possible.

15.6.1 Rate control profiles ◑

If a simple buffer overflow and underflow test is used, the system is often not able to recover from problems that develop over the course of several pictures. For example, if B-pictures systematically exceed the desired rate, the buffer occupancy may be too low to accommodate a P- or I-picture, even when coded at the maximum `quantizer_scale`.

This situation can be mitigated significantly by setting up rate control profiles. Instead of fixed buffer bounds, upper and lower buffer occupancy profiles are established that provide early warning of developing problems. An example is plotted in Figure 15.6 for the same VBV data shown in Figure 15.3. The shaded areas labelled impending overflow and impending underflow represent conditions where overflow and underflow will occur if the bit allocation is not adjusted. The boundaries for these shaded regions are calculated from the GOP and the nominal target bitrates. When system operation falls into these shaded regions, steps should be taken immediately to correct the situation. If correction is delayed, recovery may be difficult or impossible.

Note that invoking specialized procedures when the buffer occupancy enters the shaded areas is one way of approximating the s-shaped control system characteristic of Figure 15.1.

15.6.2   Buffer overflow ◑

If the decoder buffer overflows, it is being filled faster than it is being emptied. Too many bits are being transmitted and not enough bits are being removed by the decoder. To correct this, the following steps should be taken (in order of preference):

Decrease the `quantizer_scale`.

Adjust bit-allocation.

Stuff bits.

Adjusting the bit-allocation usually reflects difficulties in maintaining the desired rates for each picture type. This happens automatically at the end of each GOP when the interleaved rate control system is used.

Bit stuffing is really a waste of bits, and should be done only as a last resort. Bit stuffing can be done as leading zeros in a start code, or with the explicit bit-stuffing code in the macroblock layer. Occasionally, the discontinuities in the rate-`quantizer_scale` relationship caused by quantization effects prevent the rate control system from achieving the desired rate. Then, bit stuffing is the only way to maintain buffer integrity.

15.6.3   Buffer underflow ◑

If the decoder buffer underflows, the buffer is being emptied faster than it is being filled. More bits are being removed by the decoder than are supplied by the transmission channel. Consequently, too many bits are being generated at some point in the coding sequence. The steps to be taken include (in order of preference):

Increase the `quantizer_scale`.

Adjust bit-allocation.

Discard high frequency DCT coefficients.

Repeat pictures.

Adjust the GOP structure.

Changing the `quantizer_scale` has already been extensively discussed and is included explicitly in the MPEG-1 syntax.

15.6.4 Adjusting the bit allocation ◑

Puri and Avarind [PA91] update the bit allocation between I-, P- and B-pictures a number of times during the coding of a GOP, a necessary step because of the causal nature of their encoder. They also consider the use of bit allocation in maximizing quality, decreasing the bit allocations in P- and B-pictures when temporal distances are small. Maintaining a constant ratio between the average `quantizer_scales` for different picture types as is done in MPEG-2 TM5 [ISO93b, EF95] has a similar effect.

Even when recoding is used, the rate control system will still need adjustment of bit allocation. For example, if an interleaved rate control system is used, the bit allocation is automatically adjusted at the end of each GOP so as to evenly distribute any rate errors. Under normal conditions this is a very minor reallocation that does not significantly perturb the operation of the rate control system.

However, for some of the steps needed to avoid buffer overflow, bit allocations sometimes are grossly violated. In this case it is important not to let the large perturbations in rate propagate only to the next picture of the same type. Instead, these bits should be reapportioned so as to increase the bit allocation to picture types where the rate cannot be maintained.

15.6.5 Discarding high frequency coefficients ◑

Discarding of DCT coefficients improves coding efficiency by triggering the EOB code earlier in the coding of the block. The effect on picture quality is similar to zonal or low-pass filtering. It can be made somewhat dependent on activity by selectively discarding coefficients only from more active blocks.[9]

15.6.6 Picture repeating ◑

Picture repeating is not specifically provided in the MPEG-1 syntax, but is possible for a restricted set of conditions. B-pictures can be repeated if they are adjacent to an I-picture, a P-picture, or another repeated B-picture. B-pictures can also be repeated without this constraint, but the displayed pictures will have the wrong temporal order.

If a B-picture is to be repeated, the motion vectors relative to the predicting picture are set to zero and all macroblocks (except the first and last) are simply skipped. The net result is a picture that is coded in about 30 bytes.

---

[9]It is also possible to truncate precision of the DCT coefficients in a manner similar to the JPEG successive approximation mode. However, this does not work well if the `quantizer_scale` is large. When the `quantizer_scale` is 31, a factor of two change in precision causes very visible artifacts.

15.6.7   Adjusting the GOP ◑

In [LD94] the GOP is adjusted when the temporal distance between pictures becomes too large. The GOP can also be modified when there are problems with rate control.

Typically, I-pictures are the most costly to code. If the end of the GOP is reached with a buffer occupancy such that buffer underflow is likely when an I-picture is coded, the GOP can be extended with additional P- and B-pictures. This decreases the average bitrate for the GOP; it also may cause problems with editing of the compressed video bitstream.

15.7   Stressing the rate control system ◑

If the `constrained_parameters_flag` is set and reasonable bit allocations are used, buffer management constraints are rarely invoked for bitrates of about 1.1 to 1.2 Mbits/s. However, most developers will want to test under situations where problems are routinely encountered — very low bitrates where even at the maximum `quantizer_scale` the rate cannot be maintained, and very high rates where the I-picture may not even fit into the buffer when a typical bit allocation is used. In these limiting cases the normal rate control feedback is sometimes completely overruled by buffer management constraints.

For example, the 30 pictures/second CIF resolution mobile test sequence cannot be coded at rates below about 0.9 Mbits/s without invoking buffer management procedures. With DCT truncation, the rate can be dropped to about 0.75 Mbits/s; with picture repeating, buffer overflow can be avoided down to about 0.3 Mbits/s. At this rate, however, picture repeating effectively reduces the picture rate to about 6 pictures/second.

For the maximum 1.856 Mbits/s rate and the maximum buffer size allowed with constrained parameters, buffer overflow or underflow may be encountered. Normally, this is handled by adjusting the `quantizer_scale`. When the desired rate cannot be achieved by adjusting the `quantizer_scale`, bit stuffing may be the only way to avoid buffer overflow.

16

MPEG Patents

Contributed by   Cliff Reader, Ph.D.
Director Strategic Marketing
Samsung Semiconductor Inc.
San Jose, CA 95134

This chapter explores the MPEG video patent situation. The MPEG patent problem is described, a brief history is given of the effort to create a collaborative patent pool, and some statistics about MPEG video patents are summarized. A contact point is provided for more information.

16.1   MPEG patent analysis effort ○

The establishment of a standard and all the associated elements of technology does not preempt the intellectual property rights (IPR) of the inventors of those elements. As a result, implementers of the standard need a license for each piece of IPR. The purpose of the MPEG patent analysis effort has been to provide a simple, reasonably priced mechanism for implementers to license MPEG technology. Its goal is the establishment of a pool of patent holders which ideally will be able to grant a single license covering all applicable intellectual property needed to implement MPEG decoders for the mass market. The scope covers video and systems technology for the MPEG-1 standard and for the MPEG-2 standard main profile at main level (MP@ML).

16.2   ISO position ○

The ISO policy on intellectual property in standards states that all owners of intellectual property rights to technology necessary to implement a standard shall provide a statement to ISO guaranteeing that licenses to use the IPR will be available on an equal basis to everyone at fair and reasonable rates. Should an owner not provide such a statement, ISO requires the standard to be redrafted excluding the pertinent technology, or failing that the standard is withdrawn.

Beyond this, ISO has no jurisdiction over this issue, and IPR matters are not a part of the ISO agenda.

16.3   MPEG patent problem ○

Traditionally, IPR for consumer products have been licensed by bilateral or perhaps trilateral agreements between major industrial corporations, often characterized by cross-licensing arrangements. In the case of MPEG however, the technology is too diverse. Many disciplines from different fields of engineering, different applications, different markets, and different segments of markets are involved. Owners of IPR include communication companies, semiconductor companies, consumer electronics companies, hardware systems companies, computer companies, and universities. It is very unlikely that such a group of organizations could negotiate a multilateral agreement, and even if they did, the time involved likely would stretch into years.

From an implementer's perspective, there are several problems. First, just being able to identify the holders of IPR.[1] Second, the time and expense of negotiating individual licenses with all the holders. Third, the aggregate cost of all the licenses.

The danger foreseen was that the market for MPEG-based systems could be stalled by uncertainty about the access to and cost of required IPR, and further that such a situation could be exploited by a provider of a proprietary solution.

16.4   Sponsorship of the patent effort ○

CableLabs has sponsored a patent effort as an interested user of MPEG technology [Tan92]. Baryn Futa, COO CableLabs, architected and led the program of work. A series of open-invitation meetings were initiated to confirm interest in both the licenser and licensee communities. These meetings were colocated with MPEG meetings for convenience, but were not part of the official MPEG meeting agenda. In parallel, CableLabs established a research and analysis effort to determine the exact position of IPR both legally

---

[1]The MPEG standards do contain annexes listing companies who have provided statements to ISO stating they may have technology relevant to implementation of the standard, and agreeing to the ISO terms and conditions. It should be understood however, that due to the limited jurisdiction of ISO, these statements have not been checked to validate relevance. The lists of these companies are provided near the end of this chapter.

and technically. CableLabs set up a team comprised of Dr. Ken Rubenstein, Esq. of the Law firm Meltzer, Lippe, Goldstein, Wolf, Schlissel & Sazer, P.C. and the author, then a private consultant. Following confirmation of interest by at least a considerable number of parties, and successful compilation of an initial list of relevant IPR, an invitation was issued by CableLabs for organizations owning IPR that they believed relevant to submit their lists of patents and join a process in good faith to establish a patent pool. This has led to substantive negotiations, with the formation of the pool expected for Fall 1996 and having a charter consistent with the ISO rules.

16.5   Historical context ○

The development of original technology had a significant influence on the process, and varied among the three parts of the standard. The critical part is video, which will be discussed here.

The original work in block transform coding started at the very end of the 1960s. The fundamental work was done up to around 1975, and then for the videoconferencing market, the work to produce practical implementations was done in the few years before and after 1980. The important point here is that the patents issued for this work have expired.

The market for image/video compression was disappointingly small, and there was a lull in the research. A few key researchers made important developments in the first half of the 1980s, and the lull was broken in the mid 1980s, building to a flood of developments in the few years around 1990, as the emergence of the H.261 standard, plus the commercial success of compact disc and then CD-ROM, indicated that for the first time, there would be a mass market for compression products. As a side note, the arrival of many newcomers in the 1980s meant that they duplicated much of the work done in the early 1970s.

16.6   Patent research and analysis program ○

The goals of the patent analysis were threefold. The first goal was to confirm that the set of patents needed to practice the standard was relatively large (greater than 20 for example). The second goal was to confirm that the patents were held by a diverse set of assignees who did not historically have a business relationship (and therefore did not have pre-existing bilateral licensing agreements). The third goal was to confirm that the assignees were themselves likely to practice the standard, and therefore were interested in enabling the market at the earliest possible date at the lowest cost, as opposed to being interested in developing a major revenue stream from royalties on the patents.

A definition was made for the purposes of organizing the search. Note that this definition was independent of any definition subsequently made by the members of the patent pool, it merely served to direct the background investigation. The basic definition was "those patents essential to practice the art". This was interpreted to mean several things in the context of the standard. It covers elements of technology that are:

- Normative for implementing decoders

- Normative for constructing bitstreams

- Essential within encoders for making bitstreams

In addition, other categories were defined for relevant, but not essential IPR:

- Encoder technology, especially for smart encoders

- Pre- and postprocessing technology

- Specific hardware implementation techniques

No attempt was made to evaluate the validity of the claimed patents. The patents were taken at face value.

For the searches, the net was cast very wide. Technologies were searched that were not a part of the standard, but which used common elements of technology and therefore potentially would contain relevant patents. For example, TV standards converters started using motion compensation some years before the MPEG effort was started.

16.7   Results of the video patent analysis ◑

Table 16.1 summarizes the results of the video patent analysis. The number of abstracts reviewed was more than 6,000 of which somewhat more than 10% were judged relevant to MPEG. These patents were reviewed in full. Of those, about 30 were judged essential, about 200 were related to MPEG, but not essential, and about 100 concerned hardware implementation of MPEG technology.

16.8   Patent holders in MPEG video standards ○

MPEG-1 Video (ISO/IEC 11172-2) Annex F (List of Patent Holders) states that information about patents can be obtained from the following companies (and provides their addresses): AT&T (USA), Aware (USA), Bellcore

Number of abstracts	>6,000
Number of patents	>750
European	≈ 25%
Asian	≈ 45%
American	≈ 30%
Number of assignees	115
Number essential	≈ 30
Number nonessential (total)	≈ 200
Pre/postprocessing	≈ 20
Smart encoder (subtotal)	≈ 150
Rate-buffer/quantizer	≈ 40
Motion estimation	≈ 60
Number hardware designs (total)	≈ 100
DCT	≈ 30
VLC decoders	≈ 30
System architecture	≈ 10

Table 16.1: Results of video patent analysis.

(USA), The British Broadcasting Corporation (UK), British Telecommunications (UK), CCETT (France), CNET (France), Compression Labs, Inc (USA), CSELT (Italy), CompuSonics Corp. (USA), Daimler Benz AG (Germany), Dornier GmbH (Germany), Fraunhofer Gesselschaft zur Foerderung der Angerwandten Forschung e.V. (Germany), Hitachi Ltd (Japan), Institute fuer Rundfunktechnik Gmbh (Germany), International Business Machines Corp. (USA), KDD Corp. (Japan), Licentia Patent-Verwaltungs-Gmbh (Germany), Massachusetts Institute of Technology (USA), Matsushita Electric Industrial Co. Ltd (Japan), Mitsusbishi Electric Corp. (Japan), NEC Corp (Japan), Nippon Hoso Kyokai (Japan), Philips Electronics NV (The Netherlands), Pioneer Electronic Corp. (Japan), Ricoh Co., Ltd. (Japan), Schawartz Engineering & Design (USA), Sony Corp. (Japan), Symbionics (UK), Telefunken Fernseh un Rundfunk GmbH (Germany), Thomson Consumer Electronis (France), Toppan Printing Co., Ltd. (Japan), Toshiba Corp. (Japan), and Victor Company of Japan Ltd. (Japan). Seven countries with six native languages are represented in the MPEG-1 video list. The worldwide distribution of patent holders is not surprising since technical experts from many companies create ISO international standards.

MPEG-2 Video (ISO/IEC 13818-2) Annex F (Patent Statements) also states that information regarding patents can be obtained from a list of companies who have submitted at least informal statements. In addition to many of the MPEG-1 listed companies, the list adds Belgian Science Policy Office, BOSCH, Columbia University, David Sarnoff Research Center, Deutsche Thomson-Brandt GmbH, Fujitsu Ltd, GC Technology Corp., General Instruments, Goldstar Co. Ltd., IRT, Nippon Telegraph and Telephone Corp., Nokia Corp, Norwegian Telecom, OKI Electric Industry Co., Ltd., QUALCOM Inc., Royal PTT Nederland N.V., PTT Research NL, Samsung Electronics Co. Ltd., Scientific-Atlanta Inc., SHARP Corp.,Siemens AG, Texas Instruments Inc., and TV/COM International. At least two more countries are now represented in this patent holder list.

16.9   For more information ○

For more information about joining the patent pool or obtaining an MPEG license contact:

> Baryn Futa, Esq., Chief Operating Officer,
>   Cable Television Laboratories
> 400 Centennial Parkway
> Louisville, CO U.S.A. 80027

16.10   Acknowledgments ○

Considerable work on the analysis has been performed by Evan Kahn Esq. of Meltzer, Lippe, et al. His thoughtful, independent thinking has proven invaluable.

17

MPEG Vendors and Products

This chapter provides a sampling of vendors and MPEG products. As the reader is undoubtedly aware, there is tremendous commercial interest in MPEG, and this list was obsolete even before the data collection was complete.

17.1   Vendor Participation ○

The information presented here was solicited from vendors attending the international MPEG meetings. A standard form was supplied for the information we needed for this chapter, but the actual information was supplied by the vendors themselves. Since we necessarily had to massage this information, final versions were returned to the vendors for approval. We believe, for this reason, that the information presented here is accurate, but we cannot guarantee this. In addition, although we made an effort to keep the information in this chapter complete and up-to-date, inclusion in this chapter required an active effort on the part of the vendors themselves. If a vendor failed to respond to our requests for product and company information, no information about those products will be found here.

The worksheet provided to the vendors is reproduced here in Figure 17.1, in part to give the reader an idea of exactly what information might be found in this chapter. The absence of a given product — say an MPEG encoder — under a vendor would indicate that the vendor had no encoder offering at the time the worksheet was filled out. However, in such a dynamic marketplace, product offerings will be changing almost continuously and the reader should contact the vendors directly to get the latest information.

1. Section heading (what version of your company name would you prefer as the section name?) (e.g., IBM):

2.* Company legal name (e.g., International Business Machines Corporation):

3.* Company short name (e.g., IBM):

4. Company mission statement (one sentence!):

5. Brief history (suggestions below-none required):
    a. Date founded:
    b. Type of company founded (consulting, SW, HW, chips, boards, an application):
    c. Original primary product:
    d. Current primary product:

6. Preferred corporate-level contact if different from 8 (probably will be included as a footnote):
    Name:
    Title:
    Company name:
    Street:
    City, state, ZIP:
    Country:
    Telephone:
    Fax:
    E-mail:

7. A paragraph describing your application and the person quoted, if desired (likely to appear as an edited quote):

Figure 17.1: (a) Vendor Worksheet for MPEG product information.

8.* Repeat 8a through 8e for each MPEG video or MPEG video-related product that has a different contact person (use as much room as needed):

    8a.* Complete name of MPEG video product(s)/MPEG video application(s)/etc.:

    8b.* Type (consulting, SW, HW, chips, boards, OEM, applications):

    8c.* Description of the MPEG video function supported and environment (MPEG-1, MPEG-2, encoder, decoder, I-frames only, I+P, I+P+B, etc. on which machines, platforms etc.):

    8d.   Additional information (e.g. audio and systems supported?)

    8e.   What resolutions, bit rates, frame rates are supported?

    8f.   What other standards do you sell? (JPEG, H.261, JBIG, MHEG,..)?

    8g.   If you support MPEG-2, what profiles and levels?

    8h.* Preferred contact:

    Name:

    Title:

    Company name:

    Street:

    City, state, ZIP:

    Country:

    Telephone: Fax: E-mail:

9. If you have any trademarks included in this material, please list them here with the appropriate words (e.g., XXX is a registered trademark of YYY Corp.):

10. Any additional comments/special instructions:

11. Is this complete information? (Yes/No)

12. Contact person to receive draft copy for review, revised versions, and/or for questions (not for publication):

    a.* Name:

    b.* Telephone:

    c.* Fax:

    d.   e-mail:

Note: There is no guarantee that any of your responses will be used. Include no proprietary or confidential material, please.

Figure 17.1: (b) Vendor Worksheet for MPEG product information.

Vendor	Video		Audio	Product	Page
	Profile	Enc			
Array	1+	y	y	H,S,B	367
AuraVision					368
C-Cube	1,2	y	y	H,C,B,S	369
CRS4					370
CompCore	1,ML+		y	D,S	369
Digital	1	y		S	371
DiviCom	1,MP,ML	y	y	B,S,T	371
ESC				T	372
FutureTel	1	y	y	B+	372
GI	MP,SP,ML,LL,+	y	y	C,B,E	373
HEURIS	1,2	y	y	T	374
HMS	1		y	S,P,T	374
Hughes	1			B	375
IBM	1,MP	y		C,B	375
Imedia	MP,ML	y	y+	H,S	376
Logician	1,2		y+	D,T	377
LSI	1,MP,ML+		y+	C,B	377
Siemens	MP,SNR	y		E	379
StellaCom	1,2	y	y	T	379
Sun	1+	y	y	S,B	380
TI	1+		y	S,B	380
Zoran					381

Table 17.1: Table of MPEG vendors in this chapter. Product codes are: H = hardware, S = software, C=chips, B=boards, D=chip design, E=end products, P=programming material, T=technical services (consulting, testing, design, etc.). If a "+" is shown, additional capabilities are offered.

17.2   Vendor overview ○

Table 17.1 summarizes the information provided in the individual sections describing each vendor. This summary is necessarily brief and incomplete (and in some cases may be out of date), but the reader can refer to the individual sections for more details. Most of the vendors supplied the name of a contact where more information can be obtained.

17.3   Array Microsystems ○

Array Microsystems, Inc. (Array) was founded in 1990 to "design, market, and manufacture chips and boards for high-performance DSP applications". Its original primary product was the a66xxx family of vector processing chips and VME/ISA bus boards for high-performance FFT-based spectral processing applications. Its current primary product is a a77xxx VIDEO*FLOW* family of multi-standard, programmable compression chips, ISA/VL/PCI bus boards, and software development tools for PC multimedia, video authoring, and video conferencing applications. "Array's corporate mission is to become the leading supplier of low cost digital motion video chip set solutions for professional and consumer multimedia markets [Kop94a]."

Array's a77100 Image Compression Coprocessor (ICC) and a77300 Motion Estimation Coprocessor (MEC) chips, a77500 VIDEO*FLOW* Hardware Evaluation Kit board, and a77800 VIDEO*FLOW* Software Development Tool Kit software support MPEG-1, H.261, and JPEG encoding and decoding. The software runs on Unix workstations. The Hardware Evaluation Kit's ICC/MEC-based circuit board runs on IBM-compatible PCs and plugs into either the VL or ISA bus. It encodes and decodes MPEG-1 video and systems in real time. With optional additional software, it also performs real-time MPEG-1 layer 1 or layer 2 audio encoding and decoding when used in conjunction with off-the-shelf sound cards based on the Analog Devices Personal Sound Architecture (TM). The ICC and MEC chips embody a unique vector dataflow computing architecture (patents applied for) which Array is using as the foundation for creating future programmable chip sets serving applications such as real-time MPEG-2 video encoding and very low cost video conferencing.

Dr. Bang W. Lee, DSP Manager, Micro Devices Business, Samsung Electronics Co., South Korea writes: "Due to its programmability, flexibility, and high performance, the initial VIDEO*FLOW* ICC/MEC chip set serves a range of high-end applications such as video authoring, video conferencing, and digital video broadcast. The VIDEO*FLOW* technology is also well suited for the quick development of a range of derivative products for many emerging applications. Samsung participated in the VIDEO*FLOW* development due to Samsung's need to serve a variety of applications such as HDTV, digital CATV, video CD players, videophones, digital VCRs, and camcorders [Kop94b]."

Contact: Paul Vroomen, Vice President, Marketing and Sales
          Array Microsystems, Inc.
          987 University Ave., Suite 6
          Los Gatos, CA, U.S.A. 95030
          V:+1 408-399-1505; F:+1 408-399-1506
          E:vroomen@array.com

17.4   AuraVision Corp. ○

The AuraVision Corp. (AuraVision) was founded in July, 1992 to develop large-scale integrated circuits that bring full motion video and audio to life on the PC. AuraVision's first video process, the VxP500 provided video capture with built in compression and full screen playback. AuraVision's mission is "to provide the enabling technology that will launch the next wave of the computer revolution by integrating full-motion, full-screen video with the processing power of today's computers [Sey94]."

The VxP201, VxP501 and VxP202 make up AuraVision's new generation of state-of-the-art multimedia processors. All three processors are pin compatible. The VxP201 MPEG Playback Processor & VxP501 MPEG Video Capture / Playback Processor is a video processor for a multiformat video playback and acceleration subsystem. Designed specifically for AVI and MPEG playback environments, the VxP201 incorporates full vertical and horizontal bilinear filtering with scaling, zooming, antialiasing and interpolation without the use of external line buffers or memory. The VxP201 has an optimized glueless interface to the most popular MPEG and JPEG audio and video decoders and contains on chip audio/video synchronization support to reduce host overhead. Offering full AVI compatibility, the VxP201 plays Indeo, Cinepak, Video 1, and all other AVI files without the need for additional hardware. For additional videoprocessing functions, the VxP501 can be a direct upgrade path for the VxP201.

AuraVision has developed reference designs for each of their MPEG video processors. These are designed to help computer hardware vendors deliver multimedial products to market faster and more cost effectively. A developers kit includes a board, schematics, or CAD files as well as a royalty-free software license for developing custom applications.

17.5 C-Cube Microsystems ○

C-Cube microsystems, Inc., a leader in digital video compression, designs and markets integrated circuits that implement international standards for the compression of digital video for consumer electronics, communications and computer applications such as Video CD, Direct Broadcast Satellite and multimedia computing [LG96].

C-Cube was founded in 1988 and is headquartered in Milipitas, California with offices in the United stares, Europe and Asia. From its inception C-Cube has been entirely focused on complete solutions for digital video and has delivered many leadership products.

C-Cube's CL480VCD is the leading solution for consumer electronics application of MPEG-1 (Video CD). C-Cube's CL9100/9110 chipset for MPEG-2 decoding and transport demultiplexing is used in many broadcast applications. C-Cube is the leading supplier of circuits and software encoding MPEG, including the CLM4100 MPEG Multimedia accelerator, the CLM 4500 MPEG-1 encoder, and the CLM 4700 Broadcast encoder solution for MPEG-2. Further information about C-Cube products can be found on the World-Wide-Web at http://www.c-cube.com.

> Contact: Mary Giani
> C-Cube Microsystems
> 11778 McCarthy Boulevard
> Milpitas, CA 95035
> V:+1 408-944-8628

17.6 CompCore Multimedia ○

CompCore Multimedia, Inc. was founded in 1993 as a Silicon Valley hardware and software solutions company. Its original primary product was an MPEG-1 video decoder that required only 10K gates plus RAM. Its current primary product is a single-chip MPEG-1 system (video and audio) that requires only 20K gates plus RAM. Its current mission is "to develop and license the most optimal and fine tuned core technology for multimedia [Hab94]." "MPEG Are Us" ™ is its trademark.

CompCore Multimedia, Inc. a "Chip-less Chip Company", provides technology licensing and design expertise for video compression based on the MPEG standard. Compcore's hardware products include both MPEG-1 and -2 gate level netlists for the complete (video, audio, and systems) solution. Its software products include MPEG-1 Real Time Software for the Pentium, MIPS, and PowerPC platforms.

Contact: George T. Haber, President
CompCore Multimedia, Inc.
1270 Oakmead Parkway, Suite 214
Sunnyvale, CA, U.S.A. 94086
V:+1 408-773-8310; F:+1 408-773-0432
E:mpeg@compcore.com

17.7 CRS4 ○

The Center for Advanced Studies, Research, and Development in Sardinia (CRS4) was established in November 1990 as a research center and non-profit consortium of public and private companies for the purpose of consulting, software and applications. The original emphasis was on know-how in mathematical modeling, high performance computing and networks, and scientific visualization. The emphasis has expanded to include know-how in multimedia and hypermedia authoring and applications, animation and picture/movie compression techniques, computational genetics, combustion, environmental modeling, parallel computer architectures, and computational chemistry and biology. "CRS4 is an interdisciplinary research centre developing advanced simulation techniques and applying them, by means of high-performance computing and networks, as well as scientific visualization and animation techniques, to the solution of large-scale computational problems [Fil94]".

The lack of documentation and prototype software tools greatly influences the acceptance of a new standard, but typically such information exists only in the form of informal Usenet News postings, press releases and miscellaneous papers and articles. This led to the birth, in 1992, of the MPEG repository, an archive freely accessible through the Internet, where most of the public-domain software, such as coders and decoders for many different architectures, is archived. When more written documentation about MPEG started to appear it became more critical to guarantee its access to a larger audience and in a more structured way. It was then decided to collect the list of frequently asked question (FAQ) related to the MPEG standard and to organize it as a hypertext document accessible through a World Wide Web browser. The collection, authored by more than 100 contributors and continuously updated, is maintained by Luigi Filippini, a researcher in the scientific visualization group of CRS4. This WWW site,

    http://www.crs4.it/ luigi/MPEG/mpegfaq.html

currently has more than 5000 world wide accesses per month.

>   Contact: Luigi Filippini
>   CRS4
>   Via Nazario Sauro, 10
>   Cagliari 09123, ITALY
>   V:+39-70-2796-253; F:+39-70-2796-283
>   E: Luigi.Filippini@crs4.it

17.8   Digital Equipment Corporation ○

Digital Equipment Corp. (Digital) has a mission to be "the world's leader in open client/server solutions from personal computing to integrated world-wide information systems [Bec95]." Digital Semiconductor's mission is "to provide industry leading Alpha CPU and PCI-based peripheral components and technology to the marketplace."

Digital has a full software-only MPEG-1 playback with compliant systems layer, 30 frames/s video and audio. High-quality, software-only, fully compliant MPEG-1 encode with throughput of 2 frames/s has also been developed.

Digital Semiconductor is investigating powerful silicon and software technology which will provide fully compliant support for MPEG-1, JPEG and H.261 codecs and a wide range of audio options on a single low cost PCI option card. The solutions will integrate into a wide range of desk-top platforms and support industry standard software application program interfaces.

>   Contact: Susan Yost, Marketing Manager
>   Digital Equipment Corp.
>   77 Reed Road
>   Hudson, MA, U.S.A. 01749
>   V:+1 508-568-6429; F:+1 508-568-4681
>   E:yost@pasta.enet.dec.com

17.9   DiviCom ○

DiviCom, Inc. (DiviCom) was founded in April, 1993 to make encoding and decoding systems. It provides software tools and consulting. DiviCom has as its company mission "the creation of exceptional products that merge video, audio, and data with digital compression and communication technologies [Loo94]."

Among the DiviCom's software tools are MPEG-1,2 encoding, decoding, multiplexing, and remultiplexing tools including editing, assembly, and format conversion. DiviCom is prepared to provide consulting in compression and communication system integration, MPEG, ATM, control, format conversion, and quality.

> Contact: Bob Natwick, Vice President, Sales and Marketing
> DiviCom, Inc.
> 1708 McCarthy Blvd.
> Milpitas, CA, U.S.A. 95035
> V:+1 800-286-1600; F:+1 408-944-6705
> E:bnatwick@divi.com

17.10   ESC Inc. ○

Encoding Science Concepts, Inc. (ESC) was established in 1994 by W. Pennebaker, one of the authors of this book. ESC's mission is to "provide consulting services and contract research and development in the area of image compression [Pen96a]."

Several of the advanced chapters of this book are based on technical reports issued by ESC. These reports may be purchased from ESC; send a note to the e-mail address below for a current list of reports.

> Contact: William Pennebaker
> Encoding Science Concepts, Inc.
> RR #2 Box 84A
> Shaftsbury, VT 05262
> V:+1 802-375-1186; F:+1 802-376-2913
> E:wmpenn@sover.net

17.11   FutureTel, Inc. ○

FutureTel, Inc. (FutureTel ™) was founded in 1993 as a manufacturer of digital audio/video and digital communications board products. FutureTel's original primary products were its PrimeView ™ real-time MPEG encoding board and its TeleMux ™ ISDN/T1 network interface card. Since then it has added the MPEGWORKS ™ comprehensive encode control applications and the Digital AV*C ™ Software Developer's Kit (SDK) to its primary products. FutureTel's mission is "to promote standards-based MPEG and network communications products to ensure compatibility across platforms and with other vendors [Mak94]."

FutureTel offers PC-based digital audio-video MPEG compression and communications products. PrimeView ™ provides a single-slot solution for real-time MPEG digital encoding from NTSC or PAL analog sources, as well as from digital sources. The TeleMux ™ network interface card supports T1/ISDN and other network standards. Together with the company's software, the two products can deliver full-motion, TV-quality MPEG files over local- and wide-area networks. Applications include video databases, video distribution, remote viewing, distance learning, interactive multimedia, VIDEO CD, and video-enhanced enterprise networks. OEM plans are available. FutureTel also offers a Digital AV*C Software Developer's Kit (SDK) and MPEGWORKS ™ , a comprehensive encode control application.

> Contact: Jeanne Wun, Senior Marketing Manager
> FutureTel, Inc.
> 1092 E. Arques Avenue
> Sunnyvale, CA, U.S.A. 94086
> V:+1 408-522-1400; F:+1 408-522-1404
> E:jeannew@futuretel.com

17.12 General Instrument Corporation ○

General Instrument Corp. (GI) was founded in 1923 as a components manufacturer for the emerging radio electronics industry. It is a "leading supplier of systems and associated equipment for broadband communications networks. [Tay94]" Its current primary products are Jerrold addressable cable systems (cable and wireless), VideoCipher® addressable satellite systems, DigiCipher® digital television communications systems and associated equipment (satellite receiver/decoders, cable/wireless set-top decoder), broadband amplifiers and cable-head end equipment, and coaxial and fiber optic cable.

GI's DigiCipher II sytem incorporates MPEG-2 video and MPEG-2 transport. The primary application for these products is for distribution of digital television signals to cable and satellite television subscribers [Eng94]. These chips, boards and end-products encode and decode MPEG-2 with dual-prime motion estimation, as well as some DigiCipher II extensions which support additional motion estimation techniques.

> Contact: Marc Tayer, Director of Business Development

General Instrument Corporation
2200 Byberry Road
Hatboro, PA, U.S.A. 19040
V:+1 215-957-8354; F:+1 215-956-6638
E:mtayer@gi.com

17.13   HEURIS Logic ○

HEURIS Logic Inc. (HEURIS Logic) was founded in December, 1991 as an
MPEG services company with expertise in software, hardware, and consulting. Its mission is "to provide digital video encoding service of the highest
quality, as well as creative and technically sound solutions for all of your
MPEG encoding needs [Qua94]".

HEURIS Logic's Digital Audio/Video MPEG Encoding Services remains
its primary product. It is a proprietary in-house system for MPEG-1 encoding with multiplexed audio and video from virtually any format. The compressed bitsteams are returned in virtually any format. Its proprietary technology includes optimization of the MPEG algorithm. In the future MPEG-
2 encoding will be offered. In addition, its staff provides hardware/software
development or consulting for multimedia or interactive video.

Contact: Jeanette L. DePatie, Director of Marketing and P.R.
HEURIS Logic
1758 Muegge Rd.
St. Charles, MO, U.S.A. 63303
V:+1 314-949-0415; F:+1 314-949-0415
E:quandt@umrgec.eec.umr.edu

17.14   Hartmann Multimedia Service ○

Hartmann Multimedia Service (HMS) was established in 1994 to provide
consulting services and MPEG hardware, and software. Its original primary
products were the Digital Video Mixer Box MFB-901, the TBC MFB-1000,
the Picture in Picture processor MFB-802, and the distribution of XingCD ™
software. Currently it has added VideoCD mastering on CD-ROM compliant with the White book for MPEG movies. HMS's mission is to "get the
best Multimedia Service to the customer [Har94]".

HMS has produced an MPEG INTERNET CD-ROM which features all
of the available MPEG movies and MPEG audio songs (Internet Under-
ground Music Archive; IUMA) in an HTML Hypertext document, ready to

browse directly via CELLO-browser from the CD-ROM drive. The effect
is as if the reader were browsing on Internet via WWW. PCs are used to
encode MPEG-1 system streams via XingCD ™ MPEG-1 encoder from
VCR tapes and other video sources. Its interactive CD-ROM discs use the
CELLO shell for Hypertext application and information browsing and are
compatible with most computer platforms.

> Contact: Stefan Hartmann, Diplom Ingenieur
>   Hartmann Multimedia Service
>   Keplerstr. 11 B
>   10589 Berlin, Germany
>   V:+49 30 344 2366; F:+49 30 344 9279
>   E:harti@contrib.de or leo@zelator.de

17.15   Hughes Aircraft Co. ○

The Visualization and Information Systems Laboratory of Hughes Aircraft
Company (Hughes) developed several MPEG-related products. The MPEG
Video Codec Card is a VME circuit board with software drivers and subsystem level applications. This circuit card is a real-time MPEG-1 codec which
is capable of simultaneous encoding and decoding. The encoding process
captures NTSC video and after MPEG-1 encoding, writes the bitstream to
the VMEbus. The decoder accepts the MPEG-1 bitstream from the VME-
bus and converts it to an NTSC signal. Hughes also developed software
tools to facilitate the use of this circuit card with different communications
network and mass data storage environments.

> Contact: Jim C. Williams, Multimedia Projects Leader
>   Visualization and Information Systems Lab.
>   Hughes Aircraft Company
>   Mail Stop: 617/T206
>   P.O. Box 3310
>   Fullerton, CA, U.S.A. 92634
>   V:+1 714-732-8713; F:+1 714-732-0810
>   E:0061753@ccmail.emis.hac.com

17.16   IBM Microelectronics ○

The Digitial Video Products (DVP) Group in the International Business Machines Corp. (IBM®) Microelectronics ™ Division in Endicott, N.Y. was

established in October 1992 with a mission "to develop the core competencies required to compete in the digital video arena and develop applications applying these competencies to the areas of cable TV, computers, telecommunications and regional Bell operating companies (RBOC) and broadcast [Kor94]."

In partnership with the IBM T.J.Watson Research Center in Hawthorne, N.Y. DVP produced its first prototype MPEG-2 decoder chip in November 1993 and first DVP product, an MPEG-2 full function decoder chip, in March 1994. DVP is focused on the MPEG-2 marketplace and intends to introduce a full family of decoder and encoder products as well as application cards and systems so as to service the whole of the MPEG-2 market place. "An increasing number of applications today demand digital video techniques to display high-quality full-motion pictures. With the support of IBM's T.J. Watson Research Center, we intend to deliver innovative products to this market early and often, serving the needs of a growing customer base seeking high-quality, low-cost solutions from a reliable source [Zur94]".

The decoder chip (IBM39-MPEGSD1-CF-A-40-C in the sales manual) decompresses MPEG-1 and MPEG-2 encoded video. The group also has an applications arm that develops card/board assemblies to specific customer requirements.

    Contact: Frank Zurla, Product Manager
              IBM Microelectronics
              1701 North Street
              Endicott, NY, U.S.A. 13760
              V:+1 607-755-9070; F:+1 607-755-9056
              E: zurla@gdlvm6.vnet.ibm.com

17.17   Imedia Corp. ○

The Imedia Corp. (Imedia ™) was founded in March, 1994, to develop and manufacture hardware and software solutions for the growing video-on-demand market. Imedia "develops and manufactures video compression, storage, and distribution systems using proprietary video coding, data formatting, and signal distribution technologies for video service providers to cost-effectively serve video to their viewers in an interactive manner [Tom94]." Imedia's current product is a video server system and the provision of encoded video, for storage on disk, using Imedia's concatenated interleaved format and variable bitrate allocation algorithm.

Imedia gives service providers the ability to offer the most popular, current hit movies in an enhanced, near video-on-demand manner. The Imedia

1000 Video Server and the Concatenated Interleaving Video Storage Format are hardware and software systems. They provide for maximum five-minute start-time intervals and VCR-functionality over a single 6-MHz channel. By combining its concatenated interleaving method with the encoding and compression process during data layout onto disk, Imedia is able to maximize picture quality while keeping the time from purchase to viewing to a minimum.

Contact: Dr. Adam S. Tom, Vice President,
            New Technologies and Business Development
            Imedia Corp.
            5854 Ithaca Place
            San Diego, CA, U.S.A. 92122
            V:+1 619-453-2510; F:+1 619-625-7959
            E:atom@imedia.com

17.18   Logician, Inc. ○

Logician, Inc. (Logician) was founded in 1991 for chip design consulting. Logician's mission is the "development of portable and customer specific hardware designs for MPEG and multimedia chips and systems, specializing in high-volume low-cost applications [Pin94]."

In addition to continuing the chip design consulting, Logician has portable customer specific chip subsystem designs for MPEG audio, video, and systems using Verilog/VHDL/Synopsys design methodology. Portable designs are tailored to customer requirements to produce reliable and sophisticated MPEG implementations on very short schedules.

Contact: Juan Pineda, Multimedia Hardware Architect
            Logician, Inc.
            466 Hill St.
            San Francisco, CA, U.S.A. 94114
            V:+1 415-641-9192; F:+1 415-641-9193
            E:juan@logician.com

17.19   LSI Logic ○

LSI Logic Corporation (LSI Logic) was founded in 1981 as a semiconductor design and manufacturing company. Its first primary product was an application-specific integrated circuit (ASIC). Since then RISC Microprocessors and Digital Video Integrated Circuits for interactive compressed digital television source decoding and channel encoding and decoding have been added to their product portfolio. LSI Logic has developed a CoreWare ™ methodology that enables customers to create Customer Specific Integrated Circuits (CSICs) that can include functional units such as Reed-Solomon Error Correction and/or MPEG-2 video decoding along with the customer's proprietary logic. "LSI Logic's goal is to provide complete I/C solutions for interactive compressed digital television systems [Man94]."

LSI Logic is shipping the L64002, a single-chip integrated MPEG-2 audio/video decoder including enhanced on-screen display (OSD). The chip allows full-screen 2 or 4 bits per pixel overlay while the chip is decoding. Graphic overlay colors can be selected from a palette of over 1600 colors. The chip supports pan and scan of both graphics and video. A complete PC system with LSI Logic L64002 evaluation boards and MPEG bitstreams installed is available. Also a credit-card-sized module can be purchased. This module includes the L64002 plus 16 Mbit DRAM with a surface mount connector that can be integrated into the user's system.

LSI Logic is also a Dolby ™ AC3 licensee and can incorporate this functionality into customer specific I/Cs. The transport stream is demultiplexed by a separate I/C, the L64007 Transport Stream Demultiplexer. The L64007 will also provide clock recovery. This chip could be customized to meet individual system requirements or integrated with other logic.

> Contact: Mary Mansson, Product Marketing Engineer
> LSI Logic
> 1525 McCarthy Blvd.
> Milpitas, CA, U.S.A. 95035
> V:+1 408-433-6779; F:+1 408-954-4939
> E:marym@lsil.com

For more information about MPEG Source Encoding contact:

> Bob Saffari, Product Line Manager
> LSI Logic
> 1525 McCarthy Blvd.
> Milpitas, CA, U.S.A. 95035
> V:+1 408-954-4655; F:+1 408-954-4939
> E:saffari@lsil.com

17.20   Siemens ○

The Eikona® Encoding Services of Siemens Ltd. (Siemens) in Australia
has a mission "to provide 'worlds best quality' MPEG standards based
video and audio encoding services and products to the international market
[Ste94]". Siemens Ltd. is a wholly owned subsidiary of its German parent,
Siemens AG.

The Siemens Eikona System does off-line MPEG-1 encoding of video and
audio to any chosen bit rate on an SBus platform. The Block Motion Estimator (BME) is a single-slot SBus card specifically designed to calculate
block motion vectors. Two SGS-Thomson block motion estimation VLSI
chips produce a total of 124,000 block matches per second. Software associated with the BME implements both extended window and telescopic block
motion vector searches up to +/-128 pixels (or greater in some cases). Its
Serial Digital Interface (SDI) is capable of sustaining real-time video rates
and controlling Sony (and other) Digital Video Recorders. Production companies use SDI to connect their existing serial digital equipment and produce
digital compressed audio and video in either CD-I, MPEG-2, or other video
compression standards.

> Contact: Phillip Stevens, Project Manager, Video Communications
> Siemens Ltd
> 885 Mountain Hwy.
> Bayswater 3153, Australia
> V:+61 3 9905 5706; F:+61 3 9905 3707
> E:phillip_stevens@siemens.com.au

17.21   StellaCom ○

StellaCom, Inc. (StellaCom) has a mission "to provide products and services in digital video, advanced television and associated computer and
telecommunications technologies [Bea94]." It was founded in 1985 as a
research/development and consulting company. It originally offered primarily consulting, but now its primary product is software for measuring
compressed digital video quality.

StellaCom offers digital video codec testing, consulting, and software. It
supports all of MPEG and includes audio and system testing.

> Contact: Dr. Guy W. Beakley, Vice President
> StellaCom, Inc.
> 1901 N. Moore Street, Suite 601
> Arlington, VA, U.S.A. 22209-1706

V:+1 703-243-1695; F:+1 703-243-1698
E:gb@StellaCom.com

17.22   Sun Microsystems, Inc. ○

Sun Microsystems, Inc. (Sun) was founded in 1982 to make Sun-1 Workstations hardware and software. Today its primary products are the SPARC-
stations ™ , servers, and Solaris. Sun's mission is "to be the world's premier
supplier of open network computing solutions, ranging from desktops to su-
per servers [Haa94]."

SunVideo ™ is an SBus card for use in SBus-based computers such as
the Sun SPARCstation ™ family of products. The board utilizes the C-
Cube CL4000 VideoRisc ™ processor. It encodes MPEG-1 video. Software
is available to create systems level streams in non-real-time using a real-time
capture of both audio and video into intermediate formats. A Sun-developed
algorithm CELL is also supported by the board.

> Contact: Jon C. Haass, Product Line Manager,
>   Multimedia Products, Product Marketing
>   Sun Microsystems, Inc.
>   2550 Garcia Avenue, MailStop MTV16-10
>   Mountain View, CA, U.S.A. 94043
>   V:+1 415-336-3877; F:+1 415-336-6042
>   E:jon.haass@sun.com

17.23   Texas Instruments ○

"Texas Instruments Inc. (TI), headquartered in Dallas, Texas, is a high-
technology company with sales and manufacturing operations in more than
30 countries. TI products and services include semiconductors; defense electronic systems; software productivity tools; printers, notebook computers
and consumer electronics products; custom engineering and manufacturing
services; electrical controls; and metallurgical materials [Kro94]."

TI products include a variety of highly integrated compression/decom-
pression devices optimized for high-volume applications in consumer electronics, computer, and telecommunications markets. In addition, TI provides semicustom solutions, standard dedicated-function digital signal processors (DSP) and programmable DSP devices. The TMS320AV220 Video
CD MPEG Decoder is a dedicated hardware MPEG-1 video decoder. It supports Video CD and Karaoke standards, provides direct interface to Sony

and Sanyo CD-ROM decoders, direct interface to TMS320AV120 MPEG audio decoder and TMS320AV420 NTSC Encoder. It includes an integrated
system layer decoder, audio input buffer and audio synchronization logic,
and real-time horizontal pixel interpolation and frame duplication.

    Contact: John Krog, Digital Compression Products,
              Product Marketing
              Texas Instruments
              8330 LBJ Freeway M/S 8316
              Dallas, TX, U.S.A. 75248
              V:+1 214-997-5495; F:+1 214-997-5763
              E:krog@msg.ti.com

17.24   Zoran Corp. ○

Zoran Corp. (Zoran) has a mission to "create value by developing, manufacturing and marketing high-quality, cost-effective VLSI solutions for digital
image enhancement, image compression, and audio processing applications
[Raz94]."

    The ZR36100 MPEG I System and Video Decoder chip is an integrated
MPEG-1 full motion video decoder. It also performs the function of a
system-layer decoder and audio-video synchronizer. In this role, it demul-
tiplexes the MPEG system bitstream, extracts the time stamps from the
individual video and audio bitstreams, buffers both the compressed audio
and video data, and provides the audio data to the input of an external audio
decoder. The above task is performed while maintaining full synchronization of audio and video, and compensating for any differences in processing
delay of the audio and video decoders.

    Contact: Abbas Razavi, Product Marketing Manager
              Zoran Corporation
              1705 Wyatt Drive
              Santa Clara, CA, U.S.A. 95054
              V:+1 408-986-1314, ext. 356; F:+1 408-986-1240
              E:abbas@zoran.com

18

MPEG History

The ISO moving picture standardization process started in 1988 with a strong emphasis on real-time decoding of compressed data stored on digital storage devices (DSM) such as CD-ROMs. That allowed for the possibility of a much more complex encoder which did not need to run in real time. The technical work for that MPEG effort was nearly complete (although the official standardization procedures required much more time) when a new project was started to target higher bits rates and better quality for applications such as broadcast TV. The two projects then became known as MPEG-1 and MPEG-2. An MPEG-3 project was anticipated to be aimed at HDTV, but MPEG-2 was shown to be capable of filling that need and MPEG-3 never occurred. For very low bitrates, a fourth project, MPEG-4, was started. However, MPEG-4 has now developed into a generic coding technique that is not limited to low bitrates.

This chapter points out some of the milestones in the development of these MPEG standards, especially the parts of MPEG dealing with video. The audio and systems subgroups usually met concurrently with the video subgroup, but they both had extra meetings that are not detailed in this chapter.

18.1 Organizational meetings ○

At the January 1988 meeting of ISO/IEC JTC1/SC2/WG8 in Copenhagen, Denmark, Working Group 8 was reorganized. New subgroups were formed for audio coding, CDI motion picture coding, and binary coding. The two original subgroups, graphics coding and photographic image coding (JPEG), also continued as official subgroups. Little time was devoted to the new subgroups since the committee's attention was focused on selecting a coding technique for JPEG. However, in the numbering of MPEG meetings, this meeting is listed as the first.

The new Moving Pictures Experts Group (MPEG) met for the first time at the May 1988 meeting of ISO/IEC JTC1/SC2/WG8 in Ottawa, Canada. About 15 delegates attended. This subgroup was concerned primarily with motion video in relationship to digital storage devices. A consensus was reached to target real-time decoding of video at about 1.5 Mbits/s bit-streams. The encoder did not have to be real-time.

The subgroups of WG8 met in Torino, Italy and then moved to London, UK, for more meetings as well as a full WG8 meeting in September, 1988. The MPEG committee hammered out goals and objectives. Only two delegates attended sessions in both locations.

At the next MPEG meeting, held in December 1988 in Hannover, Germany, the audio activity was started. The video group selected six video test sequences, table tennis, flower garden, mobile and calendar, football, cheerleaders, and bicycle provided by CCIR.[1] [2]

The February 1989 MPEG meeting was held in Livingston, New Jersey, USA. A matrix testing methodology was defined. The MPEG audio subgroup set a June 1, 1989 deadline for participation in the MPEG audio standard.

In May 1989 the Rennes, France, MPEG meeting was devoted to work on the proposal package descriptions. A few months later in July 1989 in Stockholm, Sweden the video package was finalized and the MPEG system activity was started.

18.2 Proposals and convergence ○

The proposals that included decompressed digital image sequences were available at the October 1989 MPEG meetings in Kurihama and Osaka, Japan, for review and the process to reach convergence was started. There was a press release for MPEG that offered a U-matic video tape demonstrating the current state of moving image coding within MPEG at 1.15 Mbit/s.

In January 1990, at the Eindhoven, Netherlands, MPEG meetings, the difficulty in achieving convergence became clear. Core experiments were defined to allow semi-independent testing and selection of the best elements from the proposals. In order to allow for verification of proposals, proposers were required to include a syntax and documentation of specific code-bit assignments.

---

[1]CCIR is now known as ITU-R.

[2]By late 1994 the number of video test sequences for MPEG-2 had increased to about fifteen.

WG7:	Computer Graphics Experts Group, formerly CGEG
WG8:	Technical Advisory Group (TAG)
WG9:	Bi-level Image Coding Group (parent of JBIG)
WG10:	Photographic Image Coding Group (parent of JPEG)
WG11:	Coding of Moving Pictures and Associated Audio (MPEG)
	subgroup under WG11 - MPEG video subgroup
	subgroup under WG11 - MPEG audio subgroup
WG12:	Multimedia and Hypermedia Experts Group (MHEG)

Table 18.1: Working Groups in JTC1/SC2 and later, JTC1/SC29.

At the following meeting in March, 1990 in Tampa, Florida, USA, results from the core experiments were discussed. Some of the core experiments in process were full-pel, half-pel, and quarter-pel interpolation in motion compensation (MC) and a MC size of 16×16. JPEG (visually-weighted quantization matrices) and H.261 (flat) quantization were compared on MPEG-intra pictures. The JPEG downloadable run-level VLC tables and the H.261 fixed VLC tables were considered. A proposal was made to define the syntax in pseudocode. Discussions were continued on a high level syntax to multiplex the video and audio elementary bitstreams. Attendance was up to 104 delegates.

The first video simulation model (SM1) was produced at the Tampa meeting and published.[3] Proposers were to document their improvements as incremental changes to this model, thereby making testing and verification easier. Some of the ways that SM 1 differed from the previous reference model 0 (RM0) were that all motion vectors were made half-pel, the H.261 loop filter was removed, visually-weighted quantization matrices were adopted for intra pictures (and flat quantization was retained for nonintra pictures), DC coding was done by a JPEG-style DC prediction, the fixed run-level VLC H.262 AC coding method was used, and MC was at the 8×8 block level for B-pictures. Both 176×144 (4:2:0) and 120×88 (4:1:0) chroma formats were supported.

At its plenary meeting in Washington DC, USA, in April 1990, ISO/IEC JTC1/SC2 restructured WG8 and its subgroups as shown in Table 18.1. In this table WG11 is MPEG, and its subgroups are MPEG video and MPEG audio.

In May 1990, at Torino, Italy, the MPEG video subgroup had significant discussions over whether motion compensation should be at the 8×8 block

---

[3]SM1 evolved from reference model 0 (RM0). The starting point for RM 0 was Reference Model 8 from H.261, as adapted for MPEG.

level or the macroblock level. The question to be decided was "Were better coding results worth extra complexity?" In July 1990, at Porto, Portugal, the "8×8 prediction war" (as some people called it) continued. For MPEG-2, support for interlaced video coding at higher resolutions (i.e., standard definition TV) was proposed. Finally, at the September, 1990 meeting in Santa Clara, California, USA, the 8×8 prediction controversy was resolved by only allowing motion compensation at the macroblock level. The video specification was frozen technically and a Working Draft (WD) created. The meaning of the bits could no longer change, but editorial improvements were still allowed to the documentation. The definition of requirements for MPEG-2 was begun at this meeting.

18.3   Standardization process ○

The December 1990 MPEG meeting held in Berlin, Germany, was devoted to editing WD 11172 (MPEG-1). Companies had informally interchanged bitstreams before this meeting without finding any problems in the MPEG video description. The MPEG video subgroup decided that a tutorial (which evolved into Annex D) would be written to explain some of the nonnormative aspects of MPEG-1. The buffer fullness measure was debated and led to the introduction of vbv_delay. The initial vbv_delay was defined to "insure the VBV will not overflow or underflow at a later time." Skipped macroblocks at the beginning of a slice were still permitted. MPEG-2 requirements were set. MPEG-2 proposal package description was provided and the proposal testing was scheduled for late 1991. Alternate_scan was proposed in anticipation of interlaced video. The relationship between the MPEG and the MHEG (Multimedia and Hypermedia Experts Group) charters were discussed. About 112 attendees participated in this Berlin meeting.

The Gulf War crisis caused the MPEG subgroups to cancel their scheduled March 1991 San Jose, California, USA, meetings. The May, 1991 MPEG meeting in Paris France was devoted to editing and cleaning up the WD. It had three parts: Systems, Video, and Audio.

The August, 1991, MPEG meeting in Santa Clara, California, USA, had an attendance list with 160 delegates representing 89 companies, universities, or government agencies from sixteen countries [Chi91]. Multiple attendees from each participating organization were almost a necessity, since the subgroups usually met concurrently. Editorial changes and corrections were made to the MPEG-1 WD so that it could be promoted to Committee Draft (CD) and the first international ballot started. In addition, the MPEG video subgroup focused on determining the MPEG-2 testing methodology.

Finally in November 1991, at the Kurihama, Japan, meeting, the working draft was accepted as the committee draft and the group recommended to its parent committee, ISO/IEC JTC1/SC2, that the three-month international ballot commence for CD 11172. At the same meeting the MPEG-2 proposals went through subjective testing. At the next meeting in January 1992, at Singapore the convergence work for MPEG-2 resulted in a Test Model 1. The MPEG-1 CD international ballot was still in progress.

The March 1992, meeting in Haifa, Israel, was the first meeting of the MPEG committee as an official working group under ISO/IEC JTC1/SC29. It kept its Working Group 11 designation. At this meeting the MPEG appropriate subgroups revised the CD 11172 in response to national standards bodies comments and created the Draft International Standard (DIS) 11172 document. A test model ad hoc group was formed.

In July 1992, at the Angra dos Reis, Brazil, MPEG meeting, basic agreement was reached on the nonscalable aspects of MPEG-2. Core experiments were designed to test the scalable extensions. The test model ad hoc group provided test model 1 (TM1). It was revised into test model 2 (TM2) at the meeting. An ad hoc group was formed for 10 kbit/sec audio/visual coding; this became the precursor to MPEG-4. The attendance list had 129 attendees.

Around this same time work started on MPEG-1 Part 4 *Compliance Testing*. At least one national standards body had stated in its CD comments that it would veto the DIS if the MPEG group did not provide some form of compliance testing. ISO standards are required to provide testing of normative requirements. Consequently, the normative requirements were collected together for each layer of syntax to form a Part 4 on compliance testing. The MPEG-1 video subgroup chose not to provide bitstreams.

18.4 ITU-T joins MPEG-2 video and systems ○

The MPEG-2 systems and video (but not audio) subgroups became joint efforts with ITU-T Study Group 15 during the summer of 1992. Unlike JPEG and JBIG, the joint effort was not emphasized in the name. To assist in ATM coding, some new video requirements were introduced. The means to meet the new requirements were mostly already in place in MPEG.[4]

Because of the potential applicability of MPEG-2 to cable TV, representatives from cable companies were notably present at the September 1992, MPEG meeting in Tarrytown, New York, USA.

---

[4]MPEG-2 Video (ISO 13818-2) also became ITU-T Recommendation H.262. MPEG-2 Systems (ISO 13818-1) is also ITU-T Recommendation H.222.1, as part of its Broadband-ISDN audio visual telecommunications terminal.

18.5   MPEG-3 dropped ○

The November 1992, MPEG meeting in London, UK, made significant progress on scalability. An ad hoc group produced the first video working draft (WD) for MPEG-2. A consensus was growing that there would not be a need for an MPEG-3 for HDTV because that need could be met with MPEG-2. Profiles and levels were introduced at this meeting, in order to encourage commonality for particular applications. The concept of a "core profile" was defined; this was later renamed "main profile". Core experiments had six parallel groups working on prediction, frequency domain scalability, compatibility with MPEG-1 and H.261, quantization, low delay, and ATM transmission with error resilience. From the experimental results for prediction the video subgroup chose intra slices over "leaky prediction". From the quantization experiments, a new different VLC table for the intra-coded blocks showed significant gains. A nonlinear `quantizer_scale` was discussed. The use of separate quantization matrices for luminance and chrominance was restricted to 4:2:2 and 4:4:4 picture formats. The quantization matrices could be downloaded at the picture layer instead of just at the sequence layer. The video test model 3 (TM 3.0) was produced at this meeting. The MPEG-1 DIS ballot was closed on Oct. 30, 1992, and at the end of this meeting WG11 approved the text for the MPEG-1 IS. MPEG-4 was started; its original purpose was to work at very low bitrates.

18.6   MPEG-1 software simulation ○

The MPEG-1 video subgroup started a technical report containing source code for a software simulation. Its purpose was to correct a perception of poor MPEG quality, perhaps derived from some early MPEG video products that did not implement the full MPEG system. The MPEG-1 video Part 2 specification only defined the decoder. The software was intended to include a software encoder example. The simulation software was also called a "tool for compliance testing". Its reference decoder was implemented with everything arithmetically correct. This technical report was expected to be available from national standard bodies as a separate document.[5]

18.7   MPEG-2 main profile frozen ○

At the January 1993 meeting in Rome, Italy, the MPEG-2 video meetings focused on preparation for the upcoming Sydney meeting. Technical details needed to be worked out if the main profile video specification was to be

---

[5]A very small font is used in this document, at least in the versions available to the authors. Some readers will find this a problem.

frozen at the next meeting. The systems group targeted freezing the systems specification at the end of July 1993. Test model 4 (TM4) was produced at this meeting. The implementations subgroup recommended removing macroblock stuffing, setting 15 Mbit/s as a reasonable upper limit to coded bitrate for MP@ML, and restricting vertical motion vectors to $\pm$ 128.

In April 1993, in Sydney, Australia, the MPEG-2 video subgroup was successful in freezing the Working Draft's technical content for the main profile; however, editorial improvements in the documents were still expected and the scalable extensions still needed refinement and convergence to a consensus. At this meeting it was decided that all MPEG-2 decoders had to be able to decode MPEG-1 bitstreams. Verification of MPEG-2 video syntax was begun. A new direction was taken on the digital storage media (DSM) document.[6] Test model 5 (TM5) was defined containing only main profile experiments. Dual prime was kept in main profile but restricted to P-pictures with no intervening B-pictures. Main profile was restricted to only two quantization matrices, and downloading of matrices would be done in traditional zigzag scan order. Quantiser_scale could be selected at the picture layer. Skipped pictures were allowed in the main profile for the low delay condition. A new mismatch control solution replaced the MPEG-1 oddification. The main profile would not allow slice gaps. Macroblock stuffing was removed from the general MPEG-2 syntax.

18.8   MPEG-1 Part 4 Conformance testing CD ○

The WG11 requested registration of MPEG-1 Part 4: *Conformance testing* as a CD at the Sydney, Australia, meeting. Part 4 specified how tests could be designed to confirm that bitstreams and decoders met the requirements of Parts 1, 2, and 3. Encoders were not specifically addressed. The Working Drafts of MPEG-2 Systems, Video, and Audio were also used to request registration as CDs. This would then give them their ISO standards number.

18.9   MPEG-2 balloting ○

The July 1993, meeting in New York, New York, USA, concentrated on cleaning up the scalability aspects of the video working draft document. MP@HL was defined. Picture "objects" were introduced. They are pictures with slice gaps, multiplexed at the system level. An undefined MPEG-2 system PES packet length added for video layer only. A systems WD was completed.

---

[6]It later evolved into Part 6.

Cleaning up the scalability video continued through the September 1993 meeting held in Brussels, Belgium. The VBV was generalized to include variable bitrate. The working draft document was completed for review.

At the next meeting in November 1993, Seoul, Korea, the Committee Draft (CD) 13818 *Systems, Video, and Audio* was published. The participating standards national bodies then had a three-month ballot. The March 1994 meeting in Paris, France responded to the ballot comments and prepared the Draft International Standard (DIS) 13818 document.

18.10   MPEG-2 Part 6: DSM CC ○

At the July 1994 meeting in Norway, MPEG-2 Part 6: *Digital Storage Medium Command and Control* (DSM CC) was spawned out of early projects that were running as subgroups or task groups. This common command and control interface was adopted to encourage interoperability of MPEG applications and equipment. In particular, cable TV servers and CD-ROM integrators were expected to benefit from these generic commands and controls.

18.11   MPEG-2 Part 7: NBC ○

A call for proposals was also issued at the July 1994 meeting for the development of a multichannel audio coding standard that is not backward compatible to MPEG-1 Audio but still part of the MPEG-2 standard. This became MPEG-2 Part 7: *Non-backwards compatible audio* (NBC). The cost of compatibility with MPEG-1 and MPEG-2 audio coding was estimated to be about 2 dB or 20%. The goal was to develop a totally new syntax that was optimized and more efficient for 5.1 coding.

18.12   MPEG-2 Part 8: 10-bit video ○

MPEG-2 Part 8: *10-bit video* was started in recognition of the need to provide support for studio-quality video signals quantized with 10 bits per component. A call for proposals was issued with a deadline for response of March 1995.

18.13   Promotions to IS ○

The November 1994 MPEG meeting in Singapore gave final approval to IS 11172 Part 4: *Conformance Testing* (MPEG-1) and to IS 13818 (MPEG-2) Part 1: *Systems*, Part 2: *Video*, and Part 3: *Audio*. The MPEG-2 Audio

ISO 11172 (MPEG-1)	IS promotion	Pub. year
Part 1: *Systems*	Nov. 1992	1993
Part 2: *Video*	Nov. 1992	1993
Part 3: *Audio*	Nov. 1992	1993
Part 4: *Conformance testing*	Nov. 1994	1995
Part 5: *Simulation software*	Mar. 1995	

Table 18.2: List of MPEG-1 Parts, dates promoted to IS, and publication year.

standard extends the MPEG-1 Audio to multichannels (up to five) and to lower sampling frequencies and lower bitrates.

18.14 Software reference simulations ○

The full software implementation of the three parts of MPEG-1, contained in Part 5 of MPEG-1, was promoted to Draft Technical Report (DTR) status at the Singapore meeting. The software only addressed the arithmetic precision and correct implementation of syntax aspects of compliance. The question of dynamic performance (i.e., can it decode pictures in real time without skipping pictures?) is a function of the specific implementation.

A technical report on MPEG-2 video including reference software for encoding and decoding was approved for balloting as Proposed Draft Technical Report 13818-5. The Working Draft of MPEG-2 Part 4 (Conformance testing) was reviewed and approved as CD 13818-4 and submitted for balloting. A working draft was created of the *Digital Storage Media Command and Control* (DSM-CC) MPEG-2 extension. A Call for Proposals for MPEG-4, the next item of work, was issued [Chi94].

18.15 MPEG-2 Part 9: RTI ○

MPEG-2 Part 9: *Real Time Interface* (RTI) extends the MPEG-2 Systems standard to specify a delivery model for the bytes of an MPEG-2 Systems stream to a real decoder rather than the idealized model used previously. It was promoted to Working Draft status at the Singapore meeting.

18.16 Summary of MPEG standards ○

Table 18.2 lists the five Parts of MPEG-1, the dates in which they were promoted to international standard status, and the publication (Pub.) year.

ISO 13818 (MPEG-2)	IS promotion	Pub. year
Part 1: *Systems*	Nov. 1994	1996
Part 2: *Video*	Nov. 1994	1996
Part 3: *Audio*	Nov. 1994	1995
Part 4: *Compliance testing*	July 1995	
Part 5: *Simulation software*	July 1995	
Part 6: *Digital storage media command and control*	Sept. 1995	
Part 7: *Non-backwards compatible audio*	Mar. 1997	
Part 8: *10 bit video extension*	July 1996	
Part 9: *Real-time interface*	Nov. 1995	

Table 18.3: List of MPEG-2 Parts, dates promoted or expected to be promoted to IS, and publication (Pub.) year.

ISO 14496 (MPEG-4)	Expected IS promotion
Part 1: *Systems*	Nov. 1998
Part 2: *Video*	Nov. 1998
Part 3: *Audio*	Nov. 1998

Table 18.4: List of MPEG-4 Parts and expected IS promotion dates.

The official publication dates are often a year or more later. Table 18.3 gives the nine parts of MPEG-2 with their IS promotion dates (or expected dates) and their publication year. Table 18.4 shows the three parts of MPEG-4 and the estimated dates for promotion to international standard status.

19

Other Video Standards

This chapter provides a brief overview of several families of video standards. Then it goes into some detail about ITU-T Recommendation H.261, an established teleconferencing video standard, and Draft ITU-T Recommendation H.263, a new teleconferencing video standard. A brief overview of MPEG-4 is found in Section 2.5.

19.1 Digital Video standards ○

Table 19.1 presents an overview of several families of digital video standards. The umbrella name is listed in the left column. Some of the standards in the family are listed by number in the center column and the purpose of each standard is shown in the comments column. This table illustrates how multiple standards are needed to create a complete video application.

19.1.1 p×64 teleconferencing standard ○

The p×64 family is a digital teleconferencing standard that operates in the range of 64 kbits/s to 2 Mbits/s. The ITU-T Rec. H.261 video portion alone is not sufficient to created an entire system. ITU-T Rec. H.221 defines the syntax used to multiplex the audio and video packets and ITU-T Rec. H.230 defines the initial handshake that determines the capabilities of the terminals attempting to establish a telephony session. ITU-T Rec. H.242 (version published in 1993) provides the system control protocol to establish communication between audiovisual terminals. ITU-T Rec. H.320 is analogous to MPEG systems. It describes all of the other recommendations needed to define a complete system. This family of standards is formally known as H.320.

There are several choices of audio standards. ITU-T Rec. G.711 is the ISDN default for coded audio. Sampling is at 8 bits/sample and 8,000 sam-

Umbrella name	Related standards	Comments
p×64	H.261	Video
	H.221	Communications
	H.230	Initial handshake
	H.320	Terminal systems
	H.242:1993	Control protocol
	G.711	Companded audio (64 kbits/s)
	G.722	High quality audio (64 kbits/s)
	G.728	Speech (LD-CELP @ 16 kbits/s)
LBC	H.263	Video
	H.324	Terminal systems
	H.245	Control protocol
	H.223	Multiplexing protocol
	G.723:1995	Speech (5.3 & 6.3 kbits/s)
American TV	A/52	Audio (up to 448 kbits/s)
	A/53	HDTV terrestrial transmission
	A/54	HDTV receivers
	A/55	Program guide
	A/56	HDTV transport layer
	IEEE 802	Cable TV
	CableLabs WG	Satellite & Cable transmission
European DVB	ETR 154	Video (MPEG-2)
	DVB-SI	Transport
	DVB-C	Cable transmission
	DVB-S	Satellite transmission
	DVB-CS	SMATV distribution system
	DVB-T	Terrestrial transmission
	DVB-TXT	Teletext in DVB bitstreams
	DVB-CI	Common Interface
DVC	ISO	Digital video cassette (6 mm)
MJPEG	JPEG	Still image compression
	ODC	Wrapper

Table 19.1: Digital video standards and related standards.

ples/s and uses μ-law and A-law compression. For higher fidelity audio, ITU-T G.722 offers 7 kHz audio-coding within 64 kbits/s. For even better coding efficiency the ITU-T Rec. G.728 provides low-delay code-excited linear prediction (LD-CELP) at 16 kbits/s. Some other specialized standards are also allowed.

19.1.2 Low bitrate teleconferencing ○

The low-bitrate communication (LBC) teleconferencing family is newer than the p×64 family. ITU-T Draft Rec. H.263 is a low bitrate video standard for teleconferencing applications that has both MPEG-1 and MPEG-2 features. It operates at 64 kbits/s or below. ITU-T Rec. H.324 provides the low-bitrate-terminal system capability. ITU-T Rec. H.245 establishes the control protocol and ITU-T Rec. H.223 defines the multiplexing protocol for interleaving the video and speech data streams. ITU-T Rec. G.723 (published in 1995) provides low bit rate speech at 5.3 kbits/s and 6.3 kbits/s using algebraic code excited linear prediction (ACELP). Note that the earlier G.723 standard associated with H.261 is unrelated to this later recommendation.

19.1.3 American TV ○

The Advanced Television Systems Committee (ATSC) was chartered in 1987 to create the American terrestrial television transmission standard(s) that would be the successor(s) to the 1953 NTSC standard. [1] This standards effort is still in progress, but had settled many of the significant technical details. Current draft documents can be obtained from their web site. [2]

The U.S. Advanced Television (ATV) system characteristics are described in the 1995 Digital Television Standard [DTV95a] and in a guide to its use[DTV95b]. It is based on MPEG-2 video and systems, but not MPEG-2 audio. Instead, Dolby AC-3 audio compression is used [ATS94].

The Joint Committee on InterSociety Coordination (JCIC) was formed by several U.S. standards organizations that were interested in ATV. At the time of the approval of the standard, the member organizations were the Electronic Industries Association (EIA), the Institute of Electrical and Electronics Engineers (IEEE), the National Association of Broadcasters (NAB), the National Cable Television Association (NCTA), and the Society of Motion Picture and Television Engineers (SMPTE).

The United States Advanced Television Systems Committee (ATSC) Technology Group on Distribution (T3) prepared the digital television standard for HDTV transmission. The members of the T3 committee approved

---

[1] NTSC was also adopted by Canada and parts of Japan.
[2] The ATSC web site is www.atsc.org.

the document for a letter ballot by the full ATSC committee on February 23, 1995. The document was approved April 12, 1995.

Document A/52 describes the audio representation (Dolby AC-3). The HDTV video is based on the MPEG-2 main profile. Restrictions and levels can be found in documents A/53 and A/54, and the program guide information is described in document A/55. The HDTV transport layer (MPEG-2 transport streams) is contained in document A/56. The cable TV version is being standardized as IEEE 802 standard; satellite and cable transmission is being defined in a CableLabs working group (WG). [3]

19.1.4 European DVB ○

The European Digital Video Broadcast (DVB) consortium was established in 1993. [4] More than 200 private and public organizations (including more than 80 companies) from 25 countries participated as members or associate members. This group produced a set of documents defining video, transport, teletext, cable transmission, satellite transmission, satellite master antenna television (SMATV) distribution systems, and terrestrial transmission for Europe. All of these systems use MPEG-2 video, audio, and systems. These standards are available either from the European Telecommunication Standards Institute (ETSI) or the European Broadcasting Union (EBU).

19.1.5 Digital video cassette consortium ○

The Digital Video Cassette (DVC) Consortium is defining a new 6 mm tape format for professional-quality video recording. The current solution adopts an MPEG-2 intraframe-like coding technique for recording video at 25 Mbits/s. The coding method includes the ability to code individual macroblocks in either field or frame format with adaptive quantization.

19.1.6 Motion JPEG ○

The last entry in Table 19.1 is for motion JPEG (MJPEG). JPEG is the acronymn for the Joint Photographic Experts Group [PM93]. [5] This still image compression standard (i.e., always intra coding) is extensively used in moving-picture editing applications. Since there is no inter coding and high quality image capture in real time has been available since 1989, the original scenes are often digitized as a sequence of independent high-quality JPEG images. Only the final edited version is converted into MPEG, a

---

[3] More information is available at the CableLabs' web site at www.cablelabs.com.

[4] Most of the information about the DVB effort was extracted from information found at the DBV web page at http://www.ebu.ch/dvb_home.html

[5] The JPEG standard is issued as both ITU-T Rec. T.81 and ISO/IEC 10918-1:1993.

step that increases compression by a factor of three. Since only the final MPEG compressed pictures are released, it was not critical that the different MJPEG implementations used proprietary protocols for the sequences. However, in 1995 the Open Doc Consortium (ODC) created a public wrapper for MJPEG.

19.2   ITU-T Recommendation H.261. ○

MPEG-1 and MPEG-2 both assumed that low cost was more important for decoders than for encoders. These standards only specified the syntax and decoder requirements while leaving great flexibility in encoder design. For the DSM market, only a few encoders are needed to generate the master data streams, whereas every player needs a decoder. The encoders do not need to operate in real time. Similarly, for broadcast satellite systems, the encoder can be a software system off-line while every receiver needs a real-time decoder. Tradeoffs were made to enable decoders to run in real time while placing the decision-making burden on the encoder. Encoders have to be tens to hundreds of times more complex than the decoders to fully exploit the video compression performance capabilities.

Teleconferencing, however, is an application in which both the encoder and decoder need to operate in real time. Therefore, a large asymmetry between encoder and decoder complexity is not acceptable. Some loss in compression performance is likely in order to keep the delay between capture of source images and display of reconstructed images low.

In December 1984, several years before the first MPEG committee was formed, CCITT Study Group XV organized a specialist group on coding for visual telephony to develop a video transmission standard for ISDN services [NH95, ITU90, SGV89]. Originally, the group intended the video codec (encoder and/or decoder) to operate at multiples of 384 kbit/s channels (from 1 to 5). However, the video codec they developed has more aggressive specifications, operating at p×64 kbits/s for $p = 1, \ldots, 30$ and the H series became known as p×64 (pronounced p times 64).[6] This section will focus exclusively on the H.261 video compression and ignore the other the H series and G series (audio) recommendations needed for a complete system. [7]

19.2.1   Comparision of H.261 and MPEG ◑

H.261 has many elements in common with MPEG-1 compression. This is not an accident. Members from the CCITT committee participated in the

---

[6]Their world wide web page describes the family as "known informally as p * 64 (and pronounced 'p star 64'), and formally as standard H.320".

[7]For purposes of brevity the official names of the video standards have been shortened from ITU-T Recommendation H.26x to just H.26x.

MPEG effort and, although the first call for MPEG proposals gave points if the intra coding was JPEG-like, H.261 compatibility quickly became more important.

Only the H.261 data stream and decoder are defined, and the encoder is given much flexibility in how valid data streams are generated. The syntax is organized in layers. The macroblocks are the same size and shape as MPEG-1 macroblocks (i.e.,four Y blocks, one Cb block and one Cr block). Also, the chroma components have the same spatial relationship to the luminance. The 8×8 DCT is used and the coefficients are coded in the same zigzag scan order. The intra DC term is quantized with a fixed value of eight and has no dead zone, whereas the inter DC and all AC coefficients are quantized with a dead zone. (See Figure 3.10(b).)

Motion compensation is used in the prediction and motion vectors are coded differentially. Quantization changes are used to control buffer fullness.

Blocks can be skipped within a macroblock. The skipped blocks are organized into a pattern that is identical to the MPEG-1 coded_block_pattern. Some of the VLC tables are identical (macroblock_address_increment and motion vector VLC tables) or fairly close (run/level coefficient and coded_block_pattern) to MPEG-1 tables. Macroblock type tells the decoder if the quantizer is changed, whether the macroblock is motion compensated, and if the blocks are intra or inter coded. Each macroblock must be coded as intra at least every 132 pictures.

However, H.261 is not exactly like MPEG-1. The quantization is a single variable instead of a matrix of 64 terms and can only be changed every 11 macroblocks. Intra-coded AC coefficients have a dead zone. The syntax is simpler with only four layers. The motion vectors are always full pel and have a range of only ± 15 pels. To minimize the delay, only the previous picture is used for motion compensation. Consequently, there are no B-pictures. The macroblock type also indicates if motion-compensated blocks are to be filtered to reduce motion compensation and blocking artifacts.

19.2.2   H.261 source picture formats. ◑

In H.261 the image dimensions are restricted to two sizes, common intermediate format(CIF) and quarter CIF (QCIF), as shown in Figure 19.1. Note that the extra pels on the left and right edges are not coded. Only the significant (also called active) pel area is coded. It is a multiple of 16 for luminance and a multiple of 8 for chrominance. QCIF is required for all decoders, whereas CIF is optional. This common intermediate format allows interworking between 625-(i.e., PAL and SECAM) and 525-line television standards (i.e., NTSC).

The macroblocks are grouped into a fixed format called a group of blocks (GOB). Figure 19.2 shows the numbering of the 33 macroblocks in a GOB.

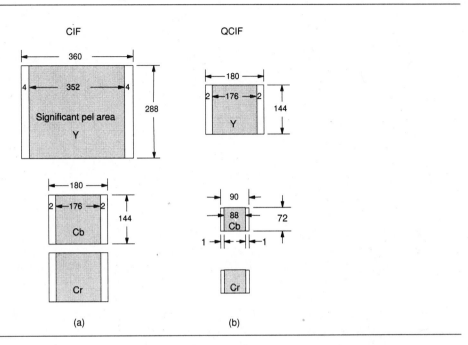

Figure 19.1: H.261 image formats: (a) CIF and (b) QCIF.

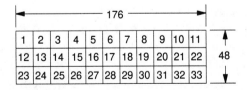

Figure 19.2: H.261 numbering of macroblocks in a GOB.

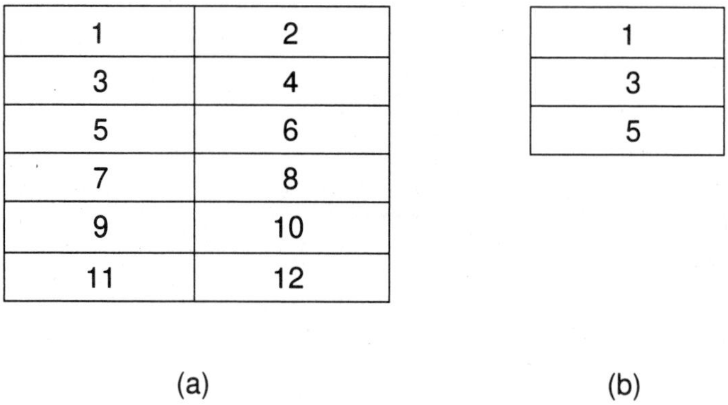

1	2
3	4
5	6
7	8
9	10
11	12

1
3
5

(a)   (b)

Figure 19.3: H.261 numbering of GOBs: (a) CIF and (b) QCIF pictures.

The macroblocks are organized in three rows of 11 macroblocks each. Figure 19.3 shows how GOBs cover the CIF and QCIF images.

19.2.3 H.261 video sequence overview ◑

The four layers of the H.261 video sequence are shown in Figure 19.4. The picture layer has a picture header followed by three or 12 GOBs depending upon whether the required QCIF or the optional CIF picture size is chosen. As many pictures as are desired can be coded. Each GOB has a header followed by information about 33 macroblocks. Macroblocks can be skipped if the motion-compensated prediction is good enough. Macroblocks with some coded blocks start with a macroblock header that may contain a motion vector. The four luminance blocks and two chrominance blocks are skipped or coded according to the block pattern. If a block is coded, then nonzero coefficients are coded with run-level VLC codes. A two-bit end of block terminates each coded block.

The codec operates on noninterlaced pictures occurring approximately 29.97 times per second. The maximum rate of encoders (selected outside of H.261 by another standard such as H.221) is restricted by requiring a minimum of zero, one, two, or three nontransmitted pictures between coded pictures. For the case of three nontransmitted pictures between coded pictures, the pictures actually arrive at a rate of about 7.5 pictures/s. Another standard in the family is used to determine when the sequence stops. The

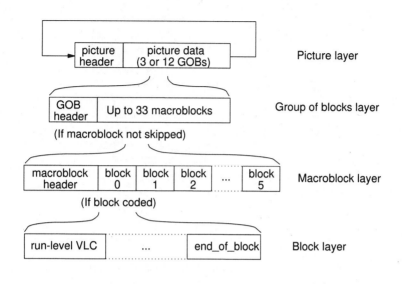

Figure 19.4: H.261 video sequence.

input or output television signals may be analog or digital, component or composite. The conversion of the source input into CIF or QCIF pictures and the conversion back into output television signals is not standardized. However, the number of pels per line was chosen so as to sample the active area of luminance and color difference signals from 525- and 625-line sources at 6.75 and 3.375 MHz, respectively.

The next sections provide details about the H.261 syntax.

19.2.4   H.261 picture layer ●

The previous section gave an overview of the H.261 standard. Now the standard will be explored in more depth using the same pseudocode notation developed in earlier chapters. Uppercase data element names come from the H.261 standard. Implementers are reminded that there is no substitute for the official standard. This pseudocode has not been tested and is presented only to make comparisons with MPEG syntax easier. [8]

The pseudocode in Figure 19.5 is for the picture layer. The pseudocode maintains, as much as possible, an MPEG-like structure, but uses the H.261 uppercase names in order to simplify comparison with MPEG-1 pseudocode.

---

[8]Our copy of the ITU-T Recommendation H.261 was obtained from Global Engineering Documents.

```
picture_layer() {                 /* for H.261                    */
  do {                            /* do pictures                  */
    PSC(20);                      /* r/w 0x00010                  */
    TR(5);                        /* r/w temporal reference       */
    PTYPE(6);                     /* r/w picture type             */
    while (nextbits(1)=='1'){     /* while '1', picture spare info */
      PEI(1);                     /*   r/w '1'                    */
      PSPARE(8);                  /*   r/w byte of pict. spare.   */
    }                             /* end picture spare info.      */
    PEI(1);                       /* r/w '0' to stop p. spare info */
    do {                          /* do group of blocks (GOBs)    */
      group_of_blocks();          /*   r/w GOBs                   */
    } while ((nextbits(16)==GBSC) /* while 0x0001                 */
      && (nextbits(20)!=PSC));    /*    and not 0x00010           */
  } while (nextbits(20)==PSC);    /* while 0x00010                */
}                                 /* end picture_layer() function */
```

Figure 19.5: H.261 `picture_layer()` function.

PTYPE bits	bit='0'	bit='1'
1	Split screen off	Split screen on
2	Document camera off	Document camera on
3	Freeze picture release off	Freeze picture release on
4	Source format QCIF	Source format CIF
5 (HI_RES)	Still image mode on	Still image mode off
6	Reserved	Spare

Table 19.2: H.261 picture information in PTYPE bits.

An H.261 picture starts with a 20-bit picture start code (PSC=0x00010).[9] Note that MPEG start codes are 32-bit byte-aligned codes with 23 leading zeros.

As mentioned at the end of Section 19.2, external controls may require dropping zero to three pictures between transmitted pictures. No picture header is transmitted for these dropped pictures.

The picture layer starts with a picture header. The 20-bit picture start code (PSC) is the first code in the picture header. After the PSC comes a 5-bit temporal reference (TR) number. Its value is calculated by adding one to the last transmitted TR plus a count of the nontransmitted pictures (at 29.97 Hz). Only the five least significant bits are saved in TR (i.e., modulo

---

[9]Since the standard does not indicate if the PSC is byte-aligned, byte-alignment has been left out of this pseudocode.

```
group_of_blocks(){          /* for H.261                        */
  GBSC(16);                 /* r/w 0x0001                       */
  GN(4);                    /* r/w group number                 */
  GQUANT(5);                /* r/w group quantizer information  */
  while (nextbits(1)=='1'){ /* while '1', extra spare info.     */
    GEI(1);                 /*   r/w '1'                        */
    GSPARE(8);              /*   r/w byte of group spare info.  */
  }                         /* end - extra spare info.          */
  GEI(1);                   /* r/w '0' to end extra spare info. */
  do {                      /* do macroblock(s)                 */
    macroblock();           /*   process a macroblock           */
  } while (nextbits(16)!=GBSC)/*  do while not 0x0001           */
}                           /* end - group_of_blocks() function */
```

Figure 19.6: H.261 `group_of_blocks()` function.

32 operation).

The picture type (PTYPE) uses six bits to describe the options chosen for this picture, as shown in Table 19.2. The most significant bit indicates whether the split screen option is off or on. The second bit indicates whether the document camera is off or on. The third bit relates to the freeze-picture release. The fourth bit indicates the source-image format. The fifth bit determines if the optional still-image mode is on or off. The last bit is a spare bit and is set to '1' until redefined by a later version of H.261.

The picture extra information (PEI) is a bit that, when set, signals that a byte of picture spare information (PSPARE) is next. A PEI bit of zero stops the sequence of picture spare bytes. Since PSPARE is reserved for future standardization, PEI is always set to zero. Any spare or unused bits are set to '1' until defined in later revisions of the recommendation. Current decoders are expected to skip over each PSPARE following PEI set to '1'. This allows for backward compatibility with future revisions.

After the picture header comes either three or 12 group of blocks, depending on picture resolution. Each GOB starts with a 16-bit group of blocks start code (GBSC=0x0001), followed immediately by a four-bit a group number (GN) ranging from 1 to 31. The PSC uses group number 0.

19.2.5   H.261 group of blocks layer ●

The pseudocode in Figure 19.6 shows the GOB layer syntax. Each GOB starts with a group of blocks start code (GBSC) and the group number (GN) codes the position of the GOB. Figure 19.3 shows the relationship between the group number and the GOB position. A 5-bit group quantizer (GQUANT) comes next in the GOB header. The actual quantizer is twice this value. Fi-

```
macroblock(){                          /* for H.261                */
  while (nextbits(11)=='0000 0001 111')/* while MBA_stuffing        */
    MBA_stuffing(11);                  /* r/w '00000001111'         */
  MBA(1-11);                           /* r/w VLC for MB address     */
  MTYPE(1-10);                         /* r/w VLC for MB type        */
  if (mb_MQUANT)                       /* if quantizer change        */
    MQUANT(5);                         /*   r/w MB quantizer         */
  if (mb_MVD) {                        /* if motion vector data      */
    MVD_horizontal(1-11);              /* r/w VLC for horz. MVD      */
    NVD_vertical(1-11);                /* r/w VLC for vert. MVD      */
  }                                    /* end-if motion vector data  */
  if (mb_CBP)                          /* if blocks may be skipped   */
    CBP(3-9);                          /* r/w VLC for CBP            */
  if (mb_TCOEFF){                      /* if any coefficients coded  */
    for (i=1; i<7; i++)                /* for the 6 blocks           */
      block(i);                        /*   r/w block data           */
  }                                    /* end - if any blocks coded  */
}                                      /* end macroblock() function  */
```

Figure 19.7: H.261 `macroblock()` function.

nally, a 1-bit data element **GEI** (group extra information) indicates whether there is a byte of group spare information (**GSPARE**) appended. For now **GEI** is 0 because **GSPARE** is reserved for future standards. Then, the 33 macroblocks in the GOB are processed. The next **GBSC** terminates the macroblock processing for the GOB, and signals that any macroblocks not yet processed are to be skipped.[10]

19.2.6   H.261 macroblock layer ●

The macroblock header provides information about the position of the macroblock relative to the position of the macroblock just coded. It also codes the motion vectors for the macroblock and identifies the blocks in the macroblock that are coded. The pseudocode for the macroblock layer is found in Figure 19.7.

The `macroblock()` procedure starts with an optional 11-bit macroblock address stuffing code (**MBA_stuffing**) that may be repeated as many times as needed. It is the same VLC code used in MPEG-1. The macroblock address (**MBA**) is then sent. For the first transmitted macroblock **MBA** in each GOBs, **MBA** is the macroblock numbering shown in Figure 19.2 (i.e previous address is considered to be zero). It is sent coded according to Table 8.6. For

---

[10]Note that MPEG-1 and MPEG-2 do not allow the final macroblock in a slice to be skipped.

Prediction	mb_ INTRA	mb_ MQUANT	mb_ MVD	mb_ CBP	mb_ TCOEFF	VLC code
Inter .	0	0	0	1	1	1
Inter+MC+ FILTER	0	0	1	1	1	01
Inter+MC+ FILTER	0	0	1	0	0	001
Intra	1	0	0	0	1	0001
Inter	0	1	0	1	1	0000 1
Inter+MC+ FILTER	0	1	1	1	1	0000 01
Intra	1	1	0	0	1	0000 001
Inter+MC	0	0	1	1	1	0000 0001
Inter+MC	0	0	1	0	0	0000 0000 1
Inter+MC	0	1	1	1	1	0000 0000 01

Table 19.3: VLC for MTYPE in H.261 pictures.

later macroblocks in each GOB the difference between the full macroblock address and the address of the last transmitted macroblock is sent using the codes in Table 8.6, too. Since the maximum address is 33, the MPEG macroblock_escape is not needed and is not part of the H.261 version of this code table.

The macroblock type (MTYPE) is sent with the variable length codes given in Table 19.3. MTYPE determines the type of prediction used (intra or inter), whether motion compensation (MC) is present without or with filtering (FILTER), whether data elements are present, and whether any transform coefficients are coded. Using MPEG-1-like conventions, mb_INTRA indicates whether the macroblock is intra-coded. Intra-coded macroblocks always send transform coefficients (mb_TCOEFF) for all blocks and so do not need to transmit the coded block pattern (CBP) to indicate skipped blocks. mb_MQUANT signals when MQUANT, a new quantizer, is to be transmitted. mb_MVD indicates if the horizontal component and then the vertical component of the motion vector data are to be transmitted. mb_CBP indicates if the coded block pattern (CBP) is to be transmitted. mb_TCOEFF shows when there are coded transform coefficients. For two out of the ten types of MTYPE only motion vectors are needed. No further data is transmitted per macroblock.

Table 19.4 gives the variable length codes for motion vector data. The motion vector data are sent as horizontal and vertical differential changes in the vectors. Since both horizontal and vertical components have integer values less than ± 16, the differential values could be in the range from -30 to $+30$; however, only 31 out of the 61 values are valid (i.e., within ± 15).

motion vector data	VLC code	bits
15, −17	0000 0011 010	11
14, −18	0000 0011 100	11
13, −19	0000 0011 110	11
12, −20	0000 0100 000	11
11, −21	0000 0100 010	11
10, −22	0000 0100 10	10
9, −23	0000 0101 00	10
8, −24	0000 0101 10	10
7, −25	0000 0110	8
6, −26	0000 1000	8
5, −27	0000 1010	8
4, −28	0000 110	7
3, −29	0001 0	5
2, −30	0010	4
1	010	3
0	1	1
−1	011	3
−2, 30	0011	4
−3, 29	0001 1	5
−4, 28	0000 111	7
−5, 27	0000 1011	8
−6, 26	0000 1001	8
−7, 25	0000 0111	8
−8, 24	0000 0101 11	10
−9, 23	0000 0101 01	10
−10, 22	0000 0100 11	10
−11, 21	0000 0100 011	11
−12, 20	0000 0100 001	11
−13, 19	0000 0011 111	11
−14, 18	0000 0011 101	11
−15, 17	0000 0011 011	11
−16, 16	0000 0011 001	11

Table 19.4: H.261 variable length codes for MVD. Only one of the pair of values gives a motion vector within ± 15.

Consequently, if the motion vector is out of range on the positive side, 32 is subtracted from it. Similarly, if it is out of range on the negative side, 32 is added to it. Both possible values are shown as a pair in Table 19.4, but only one will be valid for each component of the motion vector.

Table 19.4 can be compared to Table 8.9 for MPEG-1 where the final bit s of the VLC is 0 for positive values and 1 for negative values. Except for the separate +16 entry in Table 8.9 the variable length codes match. However, the rules for when the previous motion vectors are considered zero differ. For H.261 the motion vector is sent directly (i.e., the prediction from the previous motion vector is zero) for macroblocks 1, 12, and 23, after any skipped macroblocks, and following any macroblocks that did not include MC in MTYPE.

The same motion vector is used for all luminance blocks. For the chrominance blocks the magnitude of the vector components are divided by two and truncated towards zero to create integer values, just as is done in MPEG-1 for full pel precision. Predictions from pels in the preceding picture that are spatially above or to the left give negative motion vector components. Motion vectors are not allowed to reference pels outside the coded picture.

The CBP VLC codes are the same as for MPEG-1 (See Table 8.8) except that CBP=0 (a forbidden value) is missing. The table is repeated here because it uses a different numbering convention for the blocks. This version of the table is organized so that the codes are in ascending numeric value order, a form particularly useful for decoding.

If mb_TCOEFF (derived from MTYPE) indicates that transform coefficients are coded, then the blocks are processed.

19.2.7 H.261 block layer ●

The block layer is the lowest H.261 layer and is quite similar to the block layer of MPEG-1. The pseudocode for coding the ith block of a macroblock is given in Figure 19.8.

Table 19.6 shows the variable length codes for the run/level coding of the transform coefficients (TCOEFF). Note that this table is identical to the MPEG-1 VLC Table 5.5 if MPEG-1 codes longer than 14 bits are ignored.

If the run/level is not found in Table 19.6, then the run is coded as a six-bit binary number (fixed length code) ranging from 0 to 63 (identical to MPEG-1), followed by the level coded with an eight-bit binary number (fixed length code) ranging from −127 to +127 (see Table 19.7). The values of zero and −128 are forbidden levels. This limits the range of levels such that, for small quantizer values, the full dynamic range of coefficient levels may not be able to coded. MPEG uses the forbidden binary number for −128 ('1000 0000') to extend the levels from −128 to −255 with 16-bit

CBP		block #	CBP	bits
decimal	binary	1234 5 6 YYYY CbCr	VLC code	
39	100111	c..c c c	0000 0001 0	9
27	011011	.cc. c c	0000 0001 1	9
59	111011	ccc. c c	0000 0010 0	9
55	110111	cc.c c c	0000 0010 1	9
47	101111	c.cc c c	0000 0011 0	9
31	011111	.ccc c c	0000 0011 1	9
58	111010	ccc. c .	0000 0100	8
54	110110	cc.c c .	0000 0101	8
46	101110	c.cc c .	0000 0110	8
30	011110	.ccc c .	0000 0111	8
57	111001	ccc. . c	0000 1000	8
53	110101	cc.c . c	0000 1001	8
45	101101	c.cc . c	0000 1010	8
29	011101	.ccc . c	0000 1011	8
38	100110	c..c c .	0000 1100	8
26	011010	.cc. c .	0000 1101	8
37	100101	c..c . c	0000 1110	8
25	011001	.cc. . c	0000 1111	8
43	101011	c.c. c c	0001 0000	8
23	010111	.c.c c c	0001 0001	8
51	110011	cc.. c c	0001 0010	8
15	001111	..cc c c	0001 0011	8
42	101010	c.c. c .	0001 0100	8
22	010110	.c.c c .	0001 0101	8
50	110010	cc.. c .	0001 0110	8
14	001110	..cc c .	0001 0111	8
41	101001	c.c. . c	0001 1000	8
21	010101	.c.c . c	0001 1001	8
49	110001	cc.. . c	0001 1010	8
13	001101	..cc . c	0001 1011	8
35	100011	c... c c	0001 1100	8
19	010011	.c.. c c	0001 1101	8

Table 19.5: (a) H.261 coded block pattern (CBP) VLC codes. Blocks labeled
"." (bit=0) are skipped, whereas blocks labeled "c" (bit=1) are coded.

| CBP | | block # | CBP | bits |
| decimal | binary | 1234 5 6 | VLC code | |
		YYYY CbCr		
11	001011	..c. c c	0001 1110	8
7	000111	...c c c	0001 1111	8
34	100010	c... c .	0010 000	7
18	010010	.c.. c .	0010 001	7
10	001010	..c. c .	0010 010	7
6	000110	...c c .	0010 011	7
33	100001	c... . c	0010 100	7
17	010001	.c.. . c	0010 101	7
9	001001	..c. . c	0010 110	7
5	000101	...c . c	0010 111	7
63	111111	cccc c c	0011 00	6
3	000011	.... c c	0011 01	6
36	100100	c..c . .	0011 10	6
24	011000	.cc. . .	0011 11	6
62	111110	cccc c .	0100 0	5
2	000010	.... c .	0100 1	5
61	111101	cccc . c	0101 0	5
1	000001	 c	0101 1	5
56	111000	ccc. . .	0110 0	5
52	110100	cc.c . .	0110 1	5
44	101100	c.cc . .	0111 0	5
28	011100	.ccc . .	0111 1	5
40	101000	c.c. . .	1000 0	5
20	010100	.c.c . .	1000 1	5
48	110000	cc.. . .	1001 0	5
12	001100	..cc . .	1001 1	5
32	100000	c... . .	1010	4
16	010000	.c.. . .	1011	4
8	001000	..c. . .	1100	4
4	000100	...c . .	1101	4
60	111100	cccc . .	111	3

Table 19.5: (b) Continuation of H.261 coded block pattern (CPB) VLC codes. Blocks labeled "." (bit=0) are skipped, whereas blocks labeled "c" (bit=1) are coded.

run/level	VLC	bits
0/1	1s (first)	2
0/1	11s (next)	3
0/2	0100 s	5
0/3	0010 1s	6
0/4	0000 110s	8
0/5	0010 0110 s	9
0/6	0010 0001 s	9
0/7	0000 0010 10s	11
0/8	0000 0001 1101 s	13
0/9	0000 0001 1000 s	13
0/10	0000 0001 0011 s	13
0/11	0000 0001 0000 s	13
0/12	0000 0000 1101 0s	14
0/13	0000 0000 1100 1s	14
0/14	0000 0000 1100 0s	14
0/15	0000 0000 1011 1s	14
1/1	011s	4
1/2	0001 10s	7
1/3	0010 0101 s	9
1/4	0000 0011 00s	11
1/5	0000 0001 1011 s	13
1/6	0000 0000 1011 0s	14
1/7	0000 0000 1010 1s	14
2/1	0101 s	5
2/2	0000 100s	8
2/3	0000 0010 11s	11
2/4	0000 0001 0100 s	13
2/5	0000 0000 1010 0s	14
3/1	0011 1s	6
3/2	0010 0100 s	9
3/3	0000 0001 1100 s	13
3/4	0000 0000 1001 1s	14
4/1	0011 0s	6
4/2	0000 0011 11s	11
4/3	0000 0001 0010 s	13

Table 19.6: (a) Variable length codes for TCOEFF. The sign bit s is '0' for positive and '1' for negative.

run/level	VLC	bits
5/1	0001 11s	7
5/2	0000 0010 01s	11
5/3	0000 0000 1001 0s	14
6/1	0001 01s	7
6/2	0000 0001 1110 s	13
7/1	0001 00s	7
7/2	0000 0001 0101 s	13
8/1	0000 111s	8
8/2	0000 0001 0001 s	13
9/1	0000 101s	8
9/2	0000 0000 1000 1s	14
10/1	0010 0111 s	9
10/2	0000 0000 1000 0s	14
11/1	0010 0011 s	9
12/1	0010 0010 s	9
13/1	0010 0000 s	9
14/1	0000 0011 10s	11
15/1	0000 0011 01s	11
16/1	0000 0010 00s	11
17/1	0000 0001 1111 s	13
18/1	0000 0001 1010 s	13
19/1	0000 0001 1001 s	13
20/1	0000 0001 0111 s	13
21/1	0000 0001 0110 s	13
22/1	0000 0000 1111 1s	14
23/1	0000 0000 1111 0s	14
24/1	0000 0000 1110 1s	14
25/1	0000 0000 1110 0s	14
26/1	0000 0000 1101 1s	14
EOB	10	2
Escape	0000 01	6

Table 19.6: (b) Continuation of variable length codes for TCOEFF.

```
block(i){                      /* for H.261                       */
  if (pattern_code[i]){        /* if ith block coded              */
    if (mb_INTRA) {            /* if intra-coded macroblock        */
      INTRADC(8);              /* r/w 8 bits for intra DC          */
    else {                     /* else not intra-coded macroblock  */
      dct_coeff_first(2-20);   /* r/w VLC 1st run-level            */
    }                          /* end else not intra-coded mb      */
    while (nextbits(2)!='10')  /* while not end-of-block           */
      TCOEFF(3-20);            /* r/w VLC next run-level           */
    EOB(2);                    /* r/w '01'                         */
  }                            /* end if ith block coded           */
}                              /* end block(i) function            */
```

Figure 19.8: H.261 block() function.

LEVEL	VLC code
−128	Illegal
−127	1000 0001
−126	1000 0010
...	...
−3	1111 1101
−2	1111 1110
−1	1111 1111
0	Illegal
1	0000 0001
2	0000 0010
3	0000 0011
...	...
126	0111 1110
127	0111 1111

Table 19.7: In H.261 the LEVEL is coded by its fixed length eight-bit binary representation following the run code.

codes and the forbidden level of zero ('0000 0000') to extend its levels from
+128 to +255 with another set of 16-bit codes.

The dequantization of the transform coefficients has already been covered in Chapter 12 in Section 12.6. Except for the intra DC coefficient,
all coefficients are reconstructed with a dead zone around zero and with
oddification. (MPEG-1 got its oddification technique from this standard.)
The intra DC coefficient is reconstructed as a multiple of eight. There is no
prediction of the DC coefficients.

The accuracy requirement on the IDCT matches the MPEG-1 requirements given in [IEE91] and discussed in Section 12.4.1. Samples from intra
macroblocks are clipped to values from 0 to 255 after the IDCT, whereas
for inter macroblocks, the 9-bit result is clamped to values from -256 to 255.
The motion vector prediction (always full pel forward prediction within $\pm$
15) is summed and the final result clipped to 0 to 255.

The prediction process may optionally be modified at the macroblock
layer by a two-dimensional spatial filter (FILTER) selected with MTYPE. This
2-D filter is separable into one-dimensional horizontal and vertical filters
with taps of 1/4, 1/2, 1/4. At block edges the filter taps are changed to 0,
1, 0.[11] Full arithmetic precision is maintained until final rounding to 8-bit
integers. Values at 0.5 are rounded up.

19.2.8   H.261 summary of syntax data elements ●

Table 19.8 is a summary chart listing in alphabetical order all of the video
data elements set in the four H.261 layers. Parameters derived from these
data elements are indented under their main parent, thereby helping to
explain how the information is used in the syntax pseudocode. The four
columns labeled "set" have a column each for picture (p), GOB (g), macroblock (m), and block (b) layers. The columns labeled "used" have the
same four layers, and a black dot in a particular column indicates that the
data element is used in the pseudocode defined for that layer. The "number
of bits" column identifies the range of bits the given data element can have.
The final column shows the value range allowed. Upper case data elements
names are used in the standard. Lower case names were created for the
pseudocode representation in order to keep it similar to MPEG's.

A video data element is usually one of three basic data types (dt): unsigned integer (U), variable length coded (V), or bit string (B) (see Table 8.11. Unsigned integers are the only integer format embedded in the
bitstream; if signs are needed, they are coded as separate bits. Variable
length codes are always of the unique prefix class, such that the completeness of the code can be determined from the value at any point during the

---

[11]This is equivalent to no filtering

video data element name	set p	g	m	b	used p	g	m	b	dt	# of bits	value range
CBP			●						V	3-9	1...63
pattern_code[1]			○					●			0,1
...						○		●			0,1
pattern_code[6]			○					●			0,1
dct_coeff_first				●					V	2-20	
EOB				●				○	B	2	'10'
GBSC		●			●	●			B	16	0x0001
GEI		●				○			U	1	'0'
GN		●							U	4	1...12
GQUANT		●							U	5	1...31
GSPARE		●							U	8	0xFF
INTRADC				●					U	8	0...254
MBA				●					V	1-11	1...33
MBA_stuffing				●			○		B	11	'00000001111'
MQUANT				●					U	5	1...31
MTYPE				●					V	1-10	
mb_CBP			○					●	-	1	0,1
mb_INTRA			○					●	-	1	0,1
mb_MVD			○					●	-	1	0,1
mb_MQUANT			○					●	-	1	0,1
mb_TCOEFF			○					●	-	1	0,1
MVD_horizontal				●					V	1-11	-15...15
MVD_vertical				●					V	1-11	-15...15
PEI	●								U	1	'0'
PSC	●				●				B	20	0x00010
PSPARE	●								U	8	0xFF
PTYPE	●								U	6	
Split screen	○								U	1	0,1
Document camera	○								U	1	0,1
Freeze picture release	○								U	1	0,1
Source format	○								U	1	0,1
HI_RES	○								U	1	0,1
Spare	○								U	1	'1'
TCOEFF				●					V	2-20	
TR	●								U	5	0...31

Table 19.8: H.261 video syntax data element summary.

reading of the bitstream. Bit strings are bit sequences with predetermined patterns and fixed size.

19.3 Draft ITU-T Recommendation H.263. ○

Another MPEG-like teleconferencing video standard is H.263 [ITU95]. It was developed for video coding for low bitrate communication (LBC). [12] The scope of H.263 explicitly acknowledges that its video coding is based on H.261. However, ITU-T Rec. H.262 (MPEG-2 video) is also one of the normative references.

H.263 has MPEG-like blocks and macroblocks with prediction and motion compensation. The zigzagged quantized coefficients are coded using the MPEG run-level methods although with different tables.

Four optional modes enhance the functionality of the H.263 standard. First, the unrestricted motion vector mode allows references that are outside the picture, edge samples being duplicated to create the missing samples. This mode also allows larger motion vectors. Second, the arithmetic coding mode replaces the variable length codes. The greater coding efficiency of arithmetic coding gives the same picture quality with fewer coded bits. Third, the advanced prediction mode sents a motion vector for each luminance 8×8 block for P-pictures. Fourth, PB-frames mode codes two pictures as one unit. A P-picture is coded from the previous reference frame and then an inbetween B-picture is coded using both P-pictures. The macroblock in the P-picture is immediately followed by the corresponding macroblock from the B-picture.

19.3.1 Comparison of H.263 with H.261 and MPEG ◗

H.263 is based on H.261 and so continues most of the H.261 commonalities with MPEG-1. The luminance/chrominance color space (4:2:0 YCbCr) is identical to that used in MPEG-1 and H.261. The chrominance is always half the number of samples in both the horizontal and vertical dimensions, and chrominance samples are located in the center of four luminance samples as in H.261 (see Figure 9.1). Only noninterlaced pictures are coded. The macroblock and block definitions agree with MPEG-1 and H.261, and the IDCT equations are the same. The inverse quantized coefficients are oddified following the MPEG-1 (and H.261) method. The run-level coding of the zigzagged coefficients is used. Decoders must have motion compensation capability.

---

[12]Note that this section is based on the Draft ITU-T Recommendation H.263 dated Dec. 5, 1995, and draft standards can change.

In addition to the H.261 CIF and QCIF image formats (see Figure 19.1), H.263 supports one smaller source image format and two larger formats. There is no ability to specify arbitrary horizontal and vertical sizes as can be done in MPEG. The GOBs contain complete macroblock rows and for the two largest picture formats contain multiple macroblock rows. The quantization continues the H.261 method of a single variable instead of the MPEG matrix of 64 terms. This variable is doubled to create the quantization step size. Except for the intra DC coefficient that is uniformly quantized with a step size of 8, the intra AC and all inter reconstructed coefficients have a dead zone around zero. The run-level coefficient coding has new tables in which the end of block is replaced with run-level combinations having an extra designation as to whether they are the "last" coefficient in the block. This replaces the separate end of block code used in the other two standards.

The syntax has four layers with the same names as the H.261 syntax (picture, GOB, macroblock, and block). Many data elements have common names with H.261 data elements, but often, the number of choices is expanded and the number of bits increased. For example, the picture start code (PSC) is increased from 20 bits in H.261 to 22 bits in H.263. The PSC code is explicitly required to be byte aligned. The group start code and end of sequence codes may optionally be byte aligned.

A major departure from H.261 is the GOB format. The group of blocks always contains complete macroblock rows. For the larger pictures sizes, two or four macroblocks rows are contained in the GOB. Unlike H.261, the quantization is not restricted to being changed only every 11 macroblocks. It can be changed in the picture, GOB, and macroblock layers.

Another change from H.261 and MPEG is that decisions specifying Cb and Cr block coding are folded into the macroblock type VLC codes. The decisions for the four luminance blocks are sent separately with VLC codes.

The motion vectors are always in half-pel units (like MPEG-2). The prediction motion vector calculation is based on three neighboring macroblock motion vectors. More details are found in the standard.

19.3.2   H.263 source-picture formats. ◑

Figure 19.1 showed the H.261 image formats of CIF and quarter CIF (QCIF). The image formats for H.263 are listed in Table 19.9. In addition to CIF and QCIF, sub-QCIF, 4CIF, and 16CIF are allowed. Note that the number of chrominance samples is always half the number of luminance samples in both directions. The maximum number of bits per coded picture BPPmaxKb is given in the final column of Table 19.9. A larger number can be negotiated between the encoder and decoder by external means (outside of H.263).

Figure 19.9 shows the numbering of the GOBs for H.263. The three largest image formats have 18 GOBs numbered from 0 to 17. The QCIF

Format	luminance		chrominance		BPPmaxKb
	pels/line	lines	pels/line	lines	
sub-QCIF	128	96	64	48	64
QCIF	176	144	88	72	64
CIF	352	288	176	144	256
4CIF	704	576	352	288	512
16CIF	1408	1152	704	576	1024

Table 19.9: H.263 source picture formats and the maximum number of coded bits per picture (BPPmaxKb).

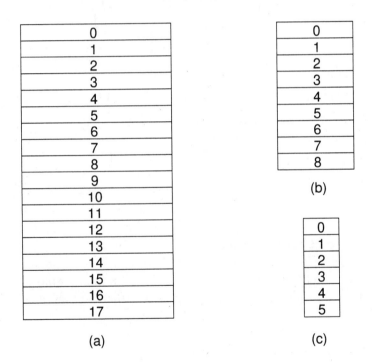

Figure 19.9: H.263 numbering of GOBs in (a) CIF, 4CIF, and 16CIF, (b) QCIF, and (c) sub-QCIF.

Format	GOBs	MB rows/GOBs	MB/row	MB in GOBs
sub-QCIF	6	1	8	8
QCIF	9	1	11	11
CIF	18	1	22	22
4CIF	18	2	44	88
16CIF	18	4	88	352

Table 19.10: H.263 group of blocks for source picture formats.

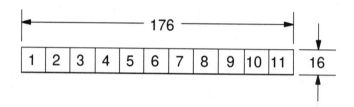

Figure 19.10: H.263 numbering of macroblocks in a QCIF GOB.

format has nine GOBs numbered from 0 to 8. The sub-QCIF pictures contain six GOBs numbered from 0 to 5. Each GOB covers the complete width of the picture. The 4CIF and 16CIF formats contain two and four macroblock rows respectively in each GOB.

Table 19.10 tabulates for each source picture format the total number of GOBs, the number of macroblock rows per GOB, the number of macroblocks per row, and the total number of macroblocks in a GOB.

H.263 numbers the macroblocks in a GOB from left to right and from top to bottom for multiple macroblock rows. The first macroblock in the upper left corner is always numbered 1. Figure 19.10 shows the numbering for a QCIF GOB.

19.3.3   H.263 video sequence overview ◑

In Figure 19.11 an H.263 video sequence is shown. The picture layer consists of a picture header followed by the data contained in the GOBs. Multiple pictures can be coded and optionally the whole sequence can be terminated with an end of sequence (EOS) code. The H.263 video sequence is multiplexed with audio or teletext bitstreams through outside standards (see Table 19.1).

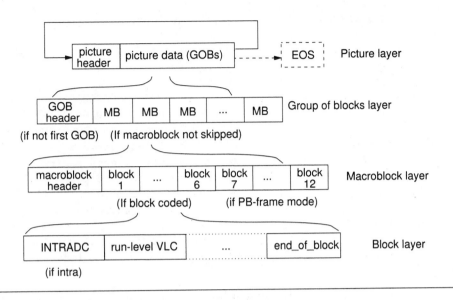

Figure 19.11: H.263 video layers.

The group of blocks layer starts each group of blocks except the first one with a GOB header followed by the macroblocks in numeric order. The macroblock layer codes a macroblock header unless the macroblock is completely skipped. Up to six blocks normally follow the header. The first four are the luminance blocks and the next two are for chrominance blocks. Some of these blocks may be skipped. If the optional PB-frame mode is on, up to another six blocks for the B-picture are also coded.

If an intra block is coded, the DC coefficient is transmitted first. It is followed by as many run-level VLC codes as are needed to code the nonzero coefficients. For inter blocks, the DC coefficient is coded with the AC coefficients. The last nonzero coefficient is coded as the *last* for the block.

The following subsections will go into more detail about the H.263 standard.

19.3.4   H.263 picture layer ●

The `picture_layer()` function for H.263 has byte-aligned picture start codes (PSC). This function assumes that the optional byte-alignment of the group of blocks start code (GBSC) and the end of sequence code (EOS) has not been selected. As mentioned for the H.261 functions, uppercase data elements are the names used in the H.263 standard. The picture layer first writes the picture header, then, the GOB data. If the 17-bit GBSC follows

```
picture_layer(){                  /* for H.263                       */
  next_picture_start_code();      /* find next byte-aligned PSC      */
  do {                            /* do pictures                     */
    picture_header();             /* r/w picture header              */
    do {                          /* do Group of Blocks layer        */
      group_of_blocks();          /*  r/w GOBs                       */
    } while ((nextbits(17)==GBSC)/* while '00000000000000001'        */
      && (nextbits(22)!=EOS)      /*  and !'0000000000000000111111'*/
      && (nextbits(22)!=PSC))     /*  and !'0000000000000000100000'*/
    if (eos_included) {           /* if end of sequence desired      */
      if (estuf_included)         /* if EOS byte aligned desired     */
        ESTUF(0-7);               /*  r/w zeros to byte boundary     */
      EOS(22);                    /*  r/w '0000000000000000111111'   */
    }                             /* end - if eos_included           */
    PSTUF(0-7);                   /* r/w zeros to byte boundary       */
  }while (nextbits(22)==PSC);     /* while '0000000000000000100000'*/
}                                 /* end picture_layer() function    */
```

Figure 19.12: H.263 picture_layer() function.

a GOB and the next five bits do not signal a EOS or PSC code, the next group of blocks is processed. After the last picture in the sequence, the optional EOS can be apppended with optional byte-alignment. Picture stuffing (PSTUF) is always used to pad the final byte with zeros.

19.3.5   H.263 picture header ●

Pseudocode for the picture_header() function is shown in Figure 19.13. The first data element is the 22-bit PSC. The PSC code looks like the GBSC for the first block, but there is no conflict because the first GOB is sent without a header. The next data element is the temporal reference TR. This eight-bit data element starts with zero for the first picture and is incremented for each picture, whether skipped or coded.

Table 19.11 gives the meaning of the picture information PTYPE bits. The first bit is always 1 to avoid start code emulation. The second bit is always 0 to distinguish the header from H.261. The third bit (split screen indicator) means that the top and bottom halves of the picture are to be displayed side by side. It does not effect the picture coding. The fourth bit indicates that the document camera is off or on. The fifth bit is a freeze picture release (external control is used to freeze the picture). This release tells the decoder to exit from the freeze picture mode and display the picture as usual.

The next three bits of PTYPE specify the source picture format. Ta-

```
picture_header(){        /* for H.263                      */
  PSC(22);               /* r/w '0000 0000 0000 0000 1000 00'*/
  TR(8);                 /* r/w temporal reference          */
  PTYPE(13);             /* r/w picture type information    */
  PQUANT(5);             /* r/w picture quantizer information*/
  CPM(1);                /* r/w continuous presence multipt. */
  if (CPM==1)            /* if cont. presence multipoint on */
    PSBI(2);             /* r/w pict. sub-bitstream indicator*/
  if (PTYPE_13) {        /* if PB-frame                     */
    TRB(3);              /* r/w B-picture temporal reference */
    DBQUANT(2);          /* r/w B-picture quantization delta */
  }                      /* end - if PB-frame               */
  while (nextbits(1)=='1'){ /* while '1', picture spare info.  */
    PEI(1);              /*   r/w '1'                       */
    PSPARE(8);           /*   r/w byte of picture spare info.*/
  }                      /* end - picture spare info.       */
  PEI(1);                /* r/w '0' to stop spare info.     */
}                        /* end - picture_header() function */
```

Figure 19.13: H.263 `picture_header()` function.

PTYPE bits	bit='0'	bit='1'
1	Forbidden	Avoids start code emulation
2	Distinguishes from H.261	Forbidden
3	Split screen off	Split screen on
4	Document camera off	Document camera on
5	Freeze picture release off	Freeze picture release on
6-8	Source format (Table 19.12)	Source format
9	Picture coding type INTRA	Picture coding type INTER
10	Optional unrestricted MV off	Optional unrestricted MV on
11	Optional arith. coding off	Optional arith. coding on
12	Normal Prediction	Optional advanced prediction
13	Normal I- or P-picture	PB-frame

Table 19.11: Picture information in H.263 PTYPE bits.

Source format	PTYPE bits 6-8
forbidden	000
sub-QCIF	001
QCIF	010
CIF	011
4CIF	100
16CIF	101
reserved	110
reserved	111

Table 19.12: H.263 source format as indicated by **PTYPE** bits 6-8.

DBQUANT	$BQUANT$
00	$(5 \times QUANT)/4$
01	$(6 \times QUANT)/4$
10	$(7 \times QUANT)/4$
11	$(8 \times QUANT)/4$

Table 19.13: H.263 **DBQUANT** fixed length codes which select the relationship between $BQUANT$ and $QUANT$.

ble 19.12 shows the coding of the five allowed source picture formats. The picture format is only allowed to change at I-pictures. The ninth bit gives the picture coding type as INTRA or INTER. This information is used to select the correct table for decoding of macroblock type. The final four bits specify the four optional modes. These modes can be mixed.

The picture quantizer information **PQUANT** sets up a default quantizer scale until overruled by the GOB quantizer scale **GQUANT** or the macroblock difference quantizer **DQUANT**. The continuous presence multipoint mode **CPM** data element determines if up to four sub-bitstreams are present. The picture sub-bitstream indicator **PSBI** identifies which bit stream when continuous presence mode is active. Bit 13 of the **PTYPE** determines if the optional PB-frame mode is on. If so, then the temporal reference for the B-picture **TRB** and the quantization delta for the B-picture **DBQUANT** are sent next. **TRB** gives the number of nontransmitted pictures plus one between the last reference I- or P-picture and the B-picture.

Table 19.13 shows the fixed length codes for **DBQUANT**. The quantization $BQUANT$ for the B-picture macroblock is scaled up from the quantization of the P-picture macroblock $QUANT$. The "/" in the table indicates truncation. If $BQUANT$ is greater than 31, it is clamped to 31.

If the picture spare information **PEI** is 1, then eight more bits of picture

```
group_of_blocks();                /* for H.263                          */
  if (GN!=0) {                    /* skip GOBs header for first GOBs    */
    if (gstuf_included)           /* if GBSC byte aligned desired       */
      GSTUF(0-7);                 /* stuff zeros to byte boundary       */
    GBSC(17);                     /* r/w '00000000000000001'            */
    GN(5);                        /* r/w group number                  */
    if (CPM==1)                   /* if cont. presence multipoint on    */
      GSBI(2);                    /* r/w group sub-bitstream indicator  */
    GFID(2);                      /* r/w group frame ID                 */
    GQUANT(5);                    /* r/w group quantizer information    */
  }
  do {                            /* do macroblock(s)                   */
    macroblock();                 /*    process a macroblock            */
  } while (nextbits(17)!=GBSC)    /*    do while not'00000000000000001' */
}                                 /* end group_of_blocks() function     */
```

Figure 19.14: H.263 group_of_blocks() function.

spare information PSPARE follow. This may be repeated as often as needed. When PEI is 0, the loop stops. For now, PEI must be set to 0, but future versions of the standard may specify allowed values of PSPARE. Consequently, current decoders are expected to skip over any such PSPARE. It is already specified that the last PSPARE (i.e, the one followed by PEI equals 0) is not allowed to end in six trailing zeros, to avoid accidental start code emulation.

19.3.6   H.263 group of blocks layer ●

Figure 19.14 gives the pseudocode for the group_of_blocks() function. The first GOB header is skipped. If the GOB start code is to be byte-aligned, up to seven zeros are stuffed (GSTUF) to complete the previous byte. The 17-bit GOB start code is followed immediately by the group number GN. (GN values of 0 and 31 are reserved for the PSC and EOS start codes.) If continuous presence multipoint operation is on, the group sub-bitstream indicator GSBI (same purpose as PSBI) is transmitted. The group frame ID GFID in the group_of_blocks() function is fixed for a picture. If PTYPE doesn't change, then GFID is the same as in the previous picturep. If PYTPE does change, then GFID also changes.

The quantizer information GQUANT is used for the GOB until updated by a DQUANT in the macroblock header. Then macroblocks are processed up to the next GBSC.

```
macroblock(){                           /* for H.263                      */
  if (PTYPE_9==1)                       /* if PTYPE bit 9 means INTER     */
    COD(1);                             /*   r/w coded MB indication      */
  if (PTYPE_9==0)||(COD==0) {           /* if INTRA or MB coded           */
    MCBPC(1-9);                         /* r/w MB type & CBP for CbCr     */
    if (MB_type!=stuffing) {            /* if MCBPC !='000000001'         */
      if (PTYPE_13==1) {                /* if PB-frame                    */
        MODB(1-2);                      /* r/w MB mode for B-blocks       */
        if (mb_CBPB)                    /* if CBPB included               */
          CBPB(6);                      /* r/w CBP for B-blocks           */
      }                                 /* end - if PB-frame              */
      CBPY(2-6);                        /* r/w CBP for luminance          */
      if (mb_DQUANT)                    /* if DQUANT included             */
        DQUANT(2);                      /* r/w diff. quantizer info.      */
      if (mb_MVD) {                     /* if MVD included                */
        MVD_horizontal(1-13);          /* r/w VLC for horz. MVD          */
        MVD_vertical(1-13);            /* r/w VLC for vert. MVD          */
      }                                 /* end if MVD included            */
      if (mb_MVD_2-4==1) {              /* if 4 MVD                       */
        MVD_2_horizontal(1-13);        /* r/w VLC for horz. MVD_2        */
        MVD_2_vertical(1-13);          /* r/w VLC for vert. MVD_2        */
        MVD_3_horizontal(1-13);        /* r/w VLC for horz. MVD_3        */
        MVD_3_vertical(1-13);          /* r/w VLC for vert. MVD_3        */
        MVD_4_horizontal(1-13);        /* r/w VLC for horz. MVD_4        */
        MVD_4_vertical(1-13);          /* r/w VLC for vert. MVD_4        */
      }                                 /* end - if 4 MVD                 */
      if (mb_MVDB==1) {                 /* if MVD for B-blocks            */
        MVDB_horizontal(1-13);         /* r/w VLC for horz. MVDB         */
        MVDB_vertical(1-13);           /* r/w VLC for vert. MVDB         */
      }                                 /* end if MVD for B-blocks        */
      for (i=1; i<7; i++)               /* for the 6 blocks               */
        block(i);                       /*   r/w block data               */
      if ((PTYPE_13==1)                 /* if PB-frame                    */
        &&(mb_CBPB==1)) {               /*   and B-blocks coded           */
        for (i=7; i<13; i++)            /* for the 6 B-blocks             */
          block(i);                     /*   r/w block data               */
      }                                 /* end if PB-frame and B-block    */
    }                                   /* end if MCBPC != '000000001'    */
  }                                     /* end if INTRA or MB coded       */
}                                       /* end macroblock() function      */
```

Figure 19.15: H.263 `macroblock()` function.

CBPC		INTRA	MB_type	MCBPC	bits
decimal	binary	5 6		VLC code	
		CbCr			
0	00	..	3	1	1
1	01	.c	3	001	3
2	10	c.	3	010	3
3	11	cc	3	011	3
0	00	..	4	0001	4
1	01	.c	4	0000 01	6
2	10	c.	4	0000 10	6
3	11	cc	4	0000 11	6
-	--	--	stuffing	0000 0000 1	9

Table 19.14: H.263 VLC for I-picture MB_type and CBP for chrominance (MCBPC).

CBPC		INTER	MB_type	MCBPC	bits
decimal	binary	5 6		VLC code	
		CbCr			
0	00	..	0	1	1
1	01	.c	0	0011	4
2	10	c.	0	0010	4
3	11	cc	0	0001 01	6
0	00	..	1	011	3
1	01	.c	1	0000 111	7
2	10	c.	1	0000 110	7
3	11	cc	1	0000 0010 1	9
0	00	..	2	010	3
1	01	.c	2	0000 101	7
2	10	c.	2	0000 100	7
3	11	cc	2	0000 0101	8
0	00	..	3	0001 1	5
1	01	.c	3	0000 0100	8
2	10	c.	3	0000 0011	8
3	11	cc	3	0000 011	7
0	00	..	4	0001 00	6
1	01	.c	4	0000 0010 0	9
2	10	c.	4	0000 0001 1	9
3	11	cc	4	0000 0001 0	9
-	--	--	stuffing	0000 0000 1	9

Table 19.15: H.263 VLC codes for P-picture MB_type and CBP for chrominance (MCBPC).

MB_type	3	4	stuffing
Name	Intra	Intra+Q	−
mb_CBPY	1	1	0
mb_COD	0	0	0
mb_DQUANT	0	1	0
mb_INTRA	1	1	0
mb_MCPBC	1	1	1
mb_MVD	0	0	0
mb_MVD_2-4	0	0	0

Table 19.16: Macroblock type and included data elements in intra-coded H.263 pictures (PTYPE bit 9=0).

MB_type	not coded	0	1	2	3	4	stuf-fing
Name	−	Inter	Inter+Q	Inter4V	Intra	Intra+Q	−
mb_CBPY	0	1	1	1	1	1	0
mb_COD	1	1	1	1	1	1	1
mb_DQUANT	0	0	1	0	0	1	0
mb_INTRA	0	0	0	0	1	1	0
mb_MCBPC	0	1	1	1	1	1	1
mb_MVD	0	1	1	1	0	0	0
mb_MVD_2-4	0	0	0	1	0	0	0

Table 19.17: Macroblock type and included data elements in inter-coded H.263 pictures (PTYPE bit 9=1).

19.3.7   H.263 macroblock layer ●

The macroblock layer pseudocode is given in Figure 19.15. If the picture type is INTER (PTYPE bit 9 = 1), then the coded macroblock indicator COD is the first data element in the macroblock header. The COD is 0 when the macroblock (MB) is coded and 1 when nothing further is transmitted for the macroblock. If nothing further is transmitted, then the macroblock motion vector is zero and there is no coefficient data.

If the picture is an I-picture (PTYPE bit 9 = 0) or a P-picture and the data element COD signaled that the macroblock was coded, the macroblock type and coded block pattern (CBP) for chrominance MCBPC are next in the macroblock header. The VLC codes for MCPBC are given in Table 19.14 for I-pictures and Table 19.15 for P-pictures. For each macroblock type MB_type except stuffing there are four combinations of how the Cb and Cr blocks are

MB_type	not coded	0	1	2	3	4	stuf-fing
Name	—	Inter	Inter+Q	Inter4V	Intra	Intra+Q	—
mb_CBPB	0	c	c	c	c	c	0
mb_CBPY	0	1	1	1	1	1	0
mb_COD	1	1	1	1	1	1	1
mb_DQUANT	0	0	1	0	0	1	0
mb_INTRA	0	0	0	0	1	1	0
mb_MCBPC	0	1	1	1	1	1	1
mb_MODB	0	1	1	1	1	1	0
mb_MVD	0	1	1	1	1	1	0
mb_MVD_2-4	0	0	0	1	0	0	0
mb_MVDB	0	c	c	c	c	c	0

Table 19.18: Macroblock type and included data elements in PB-frame H.263 pictures (PTYPE bits 9 and 13 are 1). "c" means that the presence of the data elements is conditional, depending on the value of MODB.

mb_MVDB	mb_CBPB	MODB VLC code	bits
0	0	0	1
1	0	10	2
1	1	11	2

Table 19.19: H.263 VLC codes for MODB. A '1' means the data element is included in the macroblock.

to be coded or skipped.

Tables 19.16, 19.17, and 19.18 show how the I-, P-, and PB-frame pictures use the picture type and MB_type to determine which data elements are present. If mb_XXX is 1 then the XXX data element is included; a 0 means it is excluded. The "c" in the third table means that the inclusion is conditional, depending on MODB. If the MCBPC equals '0000 0000 1', then stuffing was intended and the rest of the macroblock layer is skipped. The macroblock count is not incremented when stuffing is being done.

If the PTYPE bit 13 indicates the PB-frame mode, then the macroblock mode for the B-blocks MODB is included. Table 19.19 shows how MODB specifies whether the motion vector difference for the B-blocks MVDB and the CBP for the B-blocks CBPB are to be included (mb_MVDB=1 and (mb_CBPB=1 respectively). If mb_CBPB is 1, then a fixed six bit code shows which B-blocks are coded (see Table 8.8). The CBP for luminance CBPY variable-length codes

CBPY		INTRA	INTER	CBPY	bits
decimal	binary	1234 YYYY	1234 YYYY	VLC code	
0	0000	....	cccc	0011	4
1	0001	...c	ccc.	0010 1	5
2	0010	..c.	cc.c	0010 0	5
3	0011	..cc	cc..	1001	4
4	0100	.c..	c.cc	0001 1	5
5	0101	.c.c	c.c.	0111	4
6	0110	.cc.	c..c	0000 10	6
7	0111	.ccc	c...	1011	4
8	1000	c...	.ccc	0001 0	5
9	1001	c..c	.cc.	0000 11	6
10	1010	c.c.	.c.c	0101	4
11	1011	c.cc	.c..	1010	4
12	1100	cc..	..cc	0100	4
13	1101	cc.c	..c.	1000	4
14	1110	ccc.	...c	0110	4
15	1111	cccc	....	11	2

Table 19.20: H.263 coded block pattern for luminance CBPY VLC codes. Blocks labeled "." are skipped, whereas blocks labeled "c" are coded. The intra-coded macroblocks VLC codes mean exactly the opposite of the inter-coded macroblocks VLC codes in terms of which blocks are skipped and coded.

Change to QUANT	DQUANT
−2	01
−1	00
1	10
2	11

Table 19.21: H.263 fixed length codes for DQUANT. The differential values are ±1 or ±2.

```
block(i){            /* for H.263                      */
  if (mb_INTRA)      /* if intra-coded macroblock       */
    INTRADC(8);      /* r/w 8 bits for intra DC         */
  if (pattern_code[i]) {  /* if block coded             */
    do {             /* do TCOEFF                       */
      TCOEFF(3-22);  /*   r/w VLC for trans. coefficients*/
    } while (!last); /* while not last coefficient      */
  }                  /* end - if block coded            */
}                    /* end block(i) function           */
```

Figure 19.16: H.263 block() function.

are given in Table 19.20. If the quantization is changed, DQUANT selects one out of four differential changes as shown in Table 19.21.

If mb_MVD is 1, then normal motion vectors are included. The horizontal differential motion vector MVD_horizontal component and then the vertical differential motion vector MVD_vertical component are included using the VLC codes in Table 19.22. The table has been organized to emphasize the symmetry of the two halves of the table. The last bit determines which column contains a pair of differential motion vectors. Only one of the pair will create a motion vector within the allowed range. Note that all motion vectors are on half-pel units. The first 16 entries match the H.261 table. If mb_MVD_2-4 is set, three more sets of motion vectors are needed. These are coded using Table 19.22, alternating between horizontal and vertical components. If motion vectors are needed for B-blocks, then another pair is sent using the same tables.

The macroblock header is followed by coded block data. Normally only six blocks are coded. If PTYPE bit 13 indicates that the PB-frame mode is on and the MODB has indicated that some blocks are coded (mb_CBPB), then up to another six B-blocks are coded.

19.3.8   H.263 block layer ●

Figure 19.16 shows the pseudocode for the H.263 block layer. Since all of the intra-coded blocks are coded, the first test determines if the macroblock is intra-coded. If so, then the intra DC coefficient is coded by an eight-bit fixed length code. Table 19.23 shows the reconstruction levels for the fixed length codes (FLC). Note that the codes '0000 0000' and '1000 0000' are not used. The code '1111 1111' is used for INTRADC level 128.

If pattern_code[i] indicates that the ith block is coded, then variable length codes for the coefficients are sent. Table 19.24 gives the more frequently used run-level combinations. Each code is also identified as the

MVD s = 0	MVD s = 1	VLC code	bits
	0	1	1
0.5, −31.5	−0.5, 31.5	01s	3
1.0, −31.0	−1.0, 31.0	001s	4
1.5, −30.5	−1.5, 30.5	0001 s	5
2.0, −30.0	−2.0, 30.0	0000 11s	7
2.5, −29.5	−2.5, 29.5	0000 101s	8
3.0, −29.0	−3.0, 29.0	0000 100s	8
3.5, −28.5	−3.5, 28.5	0000 011s	8
4.0, −28.0	−4.0, 28.0	0000 0101 1s	10
4.5, −27.5	−4.5, 27.5	0000 0101 0s	10
5.0, −27.0	−5.0, 27.0	0000 0100 1s	10
5.5, −26.5	−5.5, 26.5	0000 0100 01s	11
6.0, −26.0	−6.0, 26.0	0000 0100 00s	11
6.5, −25.5	−6.5, 25.5	0000 0011 11s	11
7.0, −25.0	−7.0, 25.0	0000 0011 10s	11
7.5, −24.5	−7.5, 24.5	0000 0011 01s	11
8.0, −24.0	−8.0, 24.0	0000 0011 00s	11
8.5, −23.5	−8.5, 23.5	0000 0010 11s	11
9.0, −23.0	−9.0, 23.0	0000 0010 10s	11
9.5, −22.5	−9.5, 22.5	0000 0010 01s	11
10.0, −22.0	−10.0, 22.0	0000 0010 00s	11
10.5, −21.5	−10.5, 21.5	0000 0001 11s	11
11.0, −21.0	−11.0, 21.0	0000 0001 10s	11
11.5, −20.5	−11.5, 20.5	0000 0001 01s	11
12.0, −20.0	−12.0, 20.0	0000 0001 00s	11
12.5, −19.5	−12.5, 19.5	0000 0000 111s	12
13.0, −19.0	−13.0, 19.0	0000 0000 110s	12
13.5, −18.5	−13.5, 18.5	0000 0000 101s	12
14.0, −18.0	−14.0, 18.0	0000 0000 100s	12
14.5, −17.5	−14.5, 17.5	0000 0000 011s	12
15.0, −17.0	−15.0, 17.0	0000 0000 010s	12
15.5, −16.5	−15.5, 16.5	0000 0000 0011 s	13
	−16.0, 16.0	0000 0000 0010 1	13

Table 19.22: H.263 variable length codes for MVD. Only one of the pair of values gives a motion vector within ± 15.

Intra DC value	Reconstruction level	FLC code	
		binary	decimal
1	8	0000 0001	1
2	16	0000 0010	2
3	24	0000 0011	3
4	32	0000 0100	4
...	...	...	...
125	1000	0111 1101	125
126	1008	0111 1110	126
127	1016	0111 1111	127
128	1024	1111 1111	255
129	1032	1000 0001	129
130	1040	1000 0010	130
131	1048	1000 0011	131
...	...	...	...
251	2008	1111 1011	251
252	2016	1111 1100	252
253	2024	1111 1101	253
254	2032	1111 1110	254

Table 19.23: H.263 fixed length codes (FLC) for intra DC coefficients (INTRADC).

last or not (i.e, more nonzero coefficients follow (last=0) or this is the last nonzero coefficient (last=1)). Not having the separate end-of-block code is a significant difference relative to MPEG-1 and H.261.

Less frequently encountered run-level-last combinations are constructed as a 22-bit code from the seven-bit escape, 1-bit last, 6-bit run, and 8-bit level. The longer 22-bit codes may be substituted for the VLC codes in Table 19.24 too. After the 7-bit escape code comes the 1-bit last indicator. It is '0' if there are more nonzero coefficients and '1' if this is the last nonzero coefficient. It is followed by run, coded as a six-bit fixed length binary number. Then level is coded as a fixed eight-bit binary number in exactly the same manner as in H.261 (see Table 19.7). The levels of 0 and −128 are forbidden. The 22-bit codes may be used in place of the shorter VLC codes in Table 19.24.

The pel reconstruction matches H.261 and is described in Section 12.6. The IDCT accuracy requirements follow that given in [IEE91] (see Section 12.4.1). For intra-coded blocks the pels are taken from the output of the IDCT. For inter-coded blocks the pels are summed with the prediction. The reconstructed pels are clipped to the range 0 to 255.

run/level	last	TCOEFF VLC code	bits
0/1	0	10s	3
0/2	0	1111 s	5
0/3	0	0101 01s	7
0/4	0	0010 111s	8
0/5	0	0001 1111 s	9
0/6	0	0001 0010 1s	10
0/7	0	0001 0010 0s	10
0/8	0	0000 1000 01s	11
0/9	0	0000 1000 00s	11
0/10	0	0000 0000 111s	12
0/11	0	0000 0000 110s	12
0/12	0	0000 0100 000s	12
1/1	0	110s	4
1/2	0	0101 00s	7
1/3	0	0001 1110 s	9
1/4	0	0000 0011 11s	11
1/5	0	0000 0100 001s	12
1/6	0	0000 0101 0000 s	13
2/1	0	1110 s	5
2/2	0	0001 1101 s	9
2/3	0	0000 0011 10s	11
2/4	0	0000 0101 0001 s	13
3/1	0	0110 1s	6
3/2	0	0001 0001 1s	10
3/3	0	0000 0011 01s	11
4/1	0	0110 0s	6
4/2	0	0001 0001 0s	10
4/3	0	0000 0101 0010 s	13
5/1	0	0101 1s	6
5/2	0	0000 0011 00s	11
5/3	0	0000 0101 0011 s	13
6/1	0	0100 11s	7
6/2	0	0000 0010 11s	11
6/3	0	0000 0101 0100 s	13
7/1	0	0100 10s	7
7/2	0	0000 0010 10s	11

Table 19.24: (a) H.263 variable length codes for TCOEFF. The sign bit 's' is '0' for postive and '1' for negative.

run/level	last	TCOEFF VLC code	bits
8/1	0	0100 01s	7
8/2	0	0000 0010 01s	11
9/1	0	0100 00s	7
9/2	0	0000 0010 00s	11
10/1	0	0010 110s	8
10/2	0	0000 0101 0101 s	13
11/1	0	0010 101s	8
12/1	0	0010 100s	8
13/1	0	0001 1100 s	9
14/1	0	0001 1011 s	9
15/1	0	0001 0000 1s	10
16/1	0	0001 0000 0s	10
17/1	0	0000 1111 1s	10
18/1	0	0000 1111 0s	10
19/1	0	0000 1110 1s	10
20/1	0	0000 1110 0s	10
21/1	0	0000 1101 1s	10
22/1	0	0000 1101 0s	10
23/1	0	0000 0100 010s	12
24/1	0	0000 0100 011s	12
25/1	0	0000 0101 0110 s	13
26/1	0	0000 0101 0111 s	13
0/1	1	0111 s	5
0/2	1	0000 1100 1s	10
0/3	1	0000 0000 101s	12
1/1	1	0011 11s	7
1/2	1	0000 0000 100s	12
2/1	1	0011 10s	7
3/1	1	0011 01s	7
4/1	1	0011 00s	7
5/1	1	0010 011s	8
6/1	1	0010 010s	8
7/1	1	0010 001s	8
8/1	1	0010 000s	8

Table 19.24: (b) Continuation of H.263 variable length codes for TCOEFF.

run/level	last	TCOEFF VLC code	bits
9/1	1	0001 1010 s	9
10/1	1	0001 1001 s	9
11/1	1	0001 1000 s	9
12/1	1	0001 0111 s	9
13/1	1	0001 0110 s	9
14/1	1	0001 0101 s	9
15/1	1	0001 0100 s	9
16/1	1	0001 0011 s	9
17/1	1	0000 1100 0s	10
18/1	1	0000 1011 1s	10
19/1	1	0000 1011 0s	10
20/1	1	0000 1010 1s	10
21/1	1	0000 1010 0s	10
22/1	1	0000 1001 1s	10
23/1	1	0000 1001 0s	10
24/1	1	0000 1000 1s	10
25/1	1	0000 0001 11s	11
26/1	1	0000 0001 10s	11
27/1	1	0000 0001 01s	11
28/1	1	0000 0001 00s	11
29/1	1	0000 0100 100s	12
30/1	1	0000 0100 101s	12
31/1	1	0000 0100 110s	12
32/1	1	0000 0100 111s	12
33/1	1	0000 0101 1000 s	13
34/1	1	0000 0101 1001 s	13
35/1	1	0000 0101 1010 s	13
36/1	1	0000 0101 1011 s	13
37/1	1	0000 0101 1100 s	13
38/1	1	0000 0101 1101 s	13
39/1	1	0000 0101 1110 s	13
40/1	1	0000 0101 1111 s	13
Escape	0	0000 011	7

Table 19.24: (c) Continuation of H.263 variable length codes for TCOEFF.

video data element name	set p g m b	used p g m b	dt	# of bits	value range
CBPB	··●·	····	U	6	1...63
pattern_code[7]	··○·	···●	-	1	0,1
...	··○·	···●	-	1	0,1
pattern_code[12]	··○·	···●	-	1	0,1
CBPY	··●·	····	V	2-6	
pattern_code[1]	··○·	···●	-	1	0,1
...	··○·	···●	-	1	0,1
pattern_code[4]	··○·	···●	-	1	0,1
COD	··●·	··●·	U	1	0,1
CPM	●···	●●··	U	1	0,1
DBQUANT	●···	····	U	2	0...3
BQUANT	○···	····	U	5	1...31
DQUANT	··●·	····	U	2	0...3
EOS	●···	●···	B	22	63
ESTUF	●···	····	V	0-7	0
GBSC	·●··	●●··	B	17	1
GFID	·●··	····	U	2	0...3
GN	·●··	·●··	U	5	1...17
GQUANT	·●··	····	U	5	1...31
GSBI	·●··	····	U	2	0...3
GSTUF	·●··	····	V	0-7	0
INTRADC	···●	····	U	8	0...253
MCBPC	··●·	····	V	1-9	
CBPC	··○·	··○·	-	2	0...3
pattern_code[5]	··○·	··●·	-	1	0,1
pattern_code[6]	··○·	··●·	-	1	0,1
MB_type	··○·	··○·	-	3	0...4
mb_CBPY	··○·	··●·	-	1	0,1
mb_COD	··○·	··●·	-	1	0,1
mb_DQUANT	··○·	··●·	-	1	0,1
mb_INTRA	··○·	··●·	-	1	0,1
mb_MCPBC	··○·	··●·	-	1	0,1
mb_MODB	··○·	··●·	-	1	0,1
mb_MVD	··○·	··●·	-	1	0,1
mb_MVD_2-4	··○·	··●·	-	1	0,1
stuffing	··●·	··○·	B	9	'000000001'

Table 19.25: (a) H.263 video syntax data element summary.

video data element name	set p	g	m	b	used p	g	m	b	dt	# of bits	value range
MODB			●						V	1-2	
mb_CBPB			○				●		-	1	0,1
mb_MVDB			○				●		-	1	0,1
MVD_horizontal			●						V	1-13	-31.5 ... 31.5
MVD_vertical			●						V	1-13	-31.5 ... 31.5
MVDB_horizontal			●						V	1-13	-31.5 ... 31.5
MVDB_vertical			●						V	1-13	-31.5 ... 31.5
MVD_2_horizontal			●						V	1-13	-31.5 ... 31.5
MVD_2_vertical			●						V	1-13	-31.5 ... 31.5
MVD_3_horizontal			●						V	1-13	-31.5 ... 31.5
MVD_3_vertical			●						V	1-13	-31.5 ... 31.5
MVD_4_horizontal			●						V	1-13	-31.5 ... 31.5
MVD_4_vertical			●						V	1-13	-31.5 ... 31.5
PEI	●				○				U	1	'0'
PQUANT	●								U	5	1 ... 31
PSBI	●								U	2	0 ... 3
PSC	●				●				B	22	32
PSPARE	●								U	8	1 ... 255
PSTUF	●								V	0-7	0
PTYPE	●								U	13	
avoid start codes	○								U	1	1
distinguish H.261	○								U	1	0
split screen	○								U	1	0,1
document camera	○								U	1	0,1
freeze picture release .	○								U	1	0,1
source format	○								U	3	1 ... 5
picture coding type ..	○						●		U	1	0,1
unrestricted MV	○								U	1	0,1
arithmetic coding	○								U	1	0,1
advanced pred. mode .	○								U	1	0,1
PB-frames mode	○				●		●		U	1	0,1
TCOEFF				●					V	3-22	
last				○				●	-	1	0,1
TR	●								U	8	0 ... 255
TRB	●								U	3	1 ... 7

Table 19.25: (b) Continuation of H.263 video syntax data element summary.

19.3.9 H.263 summary of syntax data elements ●

Table 19.25 is a summary chart listing in alphabetical order all of the video data elements that are set in the four video layers. The explanation given in Section 19.2.8 also applies to this H.263 summary.

Appendix A

MPEG-1 Part 2 Notes

This appendix contains comments and errata pertaining to ISO 11172 Part 2: Video. To the best of our knowledge there are no technical corrigenda for MPEG-1 video.

These notes assume that the reader has a copy of the standard available. We have arranged suspected errors and our comments under the respective sections of the video standard. Typographical errors are expressed as "change" lists. We used ISO/IEC 11172-2 First edition 1993-08-01 Part 2: Video, purchased from ANSI in February 1994. Please remember that this appendix has no official significance.

A.1 Section 1: General ◑

1.2: Identical to subclause 1.2 in Part 1 except for the reference to Part 1 instead of Part 2 in the first normative reference.

1.2: Change "IEEE Draft Standard P1180/D2 1990" to "IEEE Standard 1180-1990". Change "*Specification*" to "*IEEE Standard Specifications*". Change "*implementation*" to "*implementations*".

A.2 Section 2: Technical elements ◑

2.1: Identically the same phrases are defined in subclause 2.1 in Part 1. The definitions are essentially the same except for nomenclature that refers to parts of the standard.

2.2: Identical to subclause 2.2 in Part 1.

2.4.3.4: In vbv_delay 90 kHz was chosen as the greatest common multiple of various video and audio sample rates. Note that the R used in the equation is the full precision bit rate. Note also that the 90 kHz means 90,000 and not 90×1024.

2.4.3.6: Change (under `coded_block_pattern`) "If `macroblock_intra` is zero, `cbp=0`." to "If `macroblock_pattern` is zero, `cbp=0`." This sentence did not appear in the DIS or the balloted IS.

2.4.4.2: Change "`pattern[i]`" to "`pattern_code[i]`".

2.4.4.3: Change "`pattern[i]`" to "`pattern_code[i]`".

A.3   Annex A: 8 by 8 Inverse discrete cosine transform ◑

The IEEE Std 1180-1990 "IEEE Standard Specifications for the Implementations of 8×8 Inverse Discrete Cosine Transform" was approved December 6, 1990. It can be obtained from the Institute of Electrical and Electronics Engineers, Inc., 345 East 74th Street, New York, NY 10017, USA.

A.4   Annex B: Variable length code tables ◑

Table B.4: Change column label "motion VLC code" to "VLC code". Change column label "code" to "motion value".

Table B.5a: Change the column label "`dct_dc_size_luminance`" to "`dc_size_luminance`". Otherwise, `dc_size_luminance` is never defined. Subclause 2.4.3.7 explicitly states that `dc_size_luminance` is the number of bits, whereas the data element `dct_dc_size_luminance` is the variable length code.

Table B.5b: Change column label from "`dct_dc_size_chrominance`" to "`dc_size_chrominance`". (See note on Table B.5a.)

A.5   Annex C: Video buffering verifier ◑

C.1.4 4th paragraph: Be aware that "the entire video bitstream" refers only to the video data from the first `sequence_start_code` and the `sequence_end_code` inclusive. It does not cross sequence boundaries.

C.1.4 equations: MPEG-1 assumes constant picture rate. These equations only work for the case where the pictures are removed at constant time intervals between picture extractions.

Figure C.1: The horizontal axis should have the t_n lined up under the B_n and the t_{n+1} lined up under the B_{n+1}. The occupancy line should have had a jog at t_2.

A.6   Annex D: Guide to encoding video ◑

Annex D is the first informative annex in Part 2. Informative annexes can not contain any requirements and are present for informational purposes only; consequently, errors in this annex are not catastrophic. In those cases where the informative annexes are in conflict or inconsistent with the "normative" portions of the document, the standards conventions are clear that the normative portions govern. Annex D is known to contain technical errors, even in the IS publication.

D.5.1.9 second paragraph: Change "3 712" to "4 640". 1.856 Mbits/s divided by 400 equals 4,640. 3,712 times 400 equals 1.4848 Mbits/s. At one time, approximately 1.5 Mbits/s was going to be the constrained parameter limit.

Table D.9: Change "0000 11 -4" to "0000 111 -4". (See Table B.4.)

Table D.13: Change "3 to -2" to "-3 to -2".

Figure D.30: The numbers in the figure and explanation for the zigzag scanning order run from 1 to 64. Do not confuse this figure with the zigzag scanning matrix given in subclause 2.4.4.1 in which the numbers run from 0 to 63. Coefficient 1 refers to the DC term and not the first AC coefficient in this subclause.

A.7   Annex E: Bibliography ◑

Note that the references are all prior to 1992.

Reference 1: The 1995 second edition is available and includes significant discussion of MPEG-1 and MPEG-2 [NH95].

Reference 6: This reference is available as an appendix in [PM93]. The final ISO 10918-1 International Standard was published February 15, 1994, and the CCITT version was official more than a year earlier.

A.8   Annex F: List of patent holders ◑

The Committee Draft version of Part 2 did not mention patents. The Draft International Standard version of Part 2 had a table of company names with "X"s under columns for Systems, Audio, and Video indicating that the patent holders claimed patents. That format was dropped in the final International Standard. It included names not listed in the final version of Annex F.

Chapter 15 describes the Cable Laboratory consortium efforts to answer questions about licensing MPEG video patents.

Bibliography

[AB85]   E. H. Adelson and J. R. Bergen. Spatiotemporal models and the perception of motion. *J. Opt. Soc. Am. A*, 2(2):284–95, Feb 1985.

[ANR74]   N. Ahmed, T. Natarajan, and R. K. Rao. Discrete Cosine Transform. *IEEE Trans. on Computers*, C-23:90–3, 1974.

[AP92]   A.H Ahumada, Jr. and H. A. Peterson. Luminance-Model-Based DCT Quantization for Color Image Compression. In *SPIE Vol. 1666 Human Vision, Visual Processing and Digital Display*, pages 365–74, 1992.

[ATS94]   Digital Audio Compression (AC-3). ATSC Standard A/52, 1994.

[BA83]   P. Burt and E. Adelson. The Laplacian pyramid as a compact image code. *IEEE Trans. Comm.*, COM-31(4):532–40, 1983.

[BCW90]   Timothy C. Bell, John G. Cleary, and Ian H. Witten. *Text Compression*. Prentice Hall, Englewood Cliffs, New Jersey, 1990.

[Bea94]   G. Beakeley. MPEG vendor worksheet. StellaCom, Inc., March 1994.

[Bec95]   J. Beck. MPEG vendor worksheet. Digital Equipment Corp., May 1995.

[Ben85]   K. Blair Benson, editor. *Television Engineering Handbook*. McGraw-Hill, New York, 1985. ISBN 0-07-004779-0.

[BK95]   V. Bhaskaran and K. Konstantinides. *Image and Video Compression Standards: Algorithms and Architectures*. Kluwer Academic Publishers, Boston, 1995.

[Cal85]   R. J. Caldwell. *Electronic Editing*, chapter 20. In Benson [Ben85], 1985. ISBN 0-07-004779-0.

[CCJC91] Liang-Gee Chen, Wai-Ting Chen, Yeu-Shen Jehng, and Tzi-Dar Chiueh. An Efficient Parallel Motion Estimation Algorithm for Digital Image Processing. *IEEE Trans. Circuits Syst. Video Tech.*, 1(4):378–85, Dec 1991.

[Chi91] L. Chiariglione, Convenor. Fifteenth meeting attendance list, ISO/IEC JTC1/SC2/WG11 — MPEG91/180, Aug 1991.

[Chi94] L. Chiariglione, Convenor. Press release, ISO/IEC JTC1/SC29/WG11 MPEG94/N0822, Nov 1994.

[CJK+93] Cheng-Tie Chen, Fure-Ching Jeng, Masahisa Kawashima, Sharad Singhal, and Andria H. Wong. Hybrid extended MPEG video coding algorithm for general video applications. *Signal Proc.: Image Comm.*, 5:21–37, 1993.

[CJOS94] Tae-Yun Chung, Kyu-Hwan Jung, Young-Nam Oh, and Dong-Ho Shin. Quantization control for improvement of image quality compatible with MPEG2. *IEEE Trans. Consumer Electronics*, 40(4):821–5, Nov 1994.

[CL93] Keith Hung-Kei Chow and Ming L. Liou. Genetic Motion Search Algorithm for Video Compression. *IEEE Trans. Circuits Syst. Vid. Tech.*, 3(6):440–5, Dec 1993.

[CLCR93] K. W. Chun, K. W. Lim, H. D. Cho, and J. B. Ra. An Adaptive Perceptual Quantization Algorithm for Video Coding. *IEEE Trans. Consumer Electr.*, 39(3):555–8, Aug 1993.

[Cor70] Tom N. Cornsweet. *Visual Perception*. Harcourt Brace Jovanovich, Inc., Orlando, Florida, 1970. ISBN 0-15-594936-5.

[CP84] Wen-Hsiung Chen and W. K. Pratt. Scene Adaptive Coder. *IEEE Trans. Comm.*, COM-32(3):225–32, Mar 1984.

[CR94] K. W. Chun and J. B. Ra. An improved block matching algorithm based on successive refinement of motion vector candidates. *Signal Proc: Image Communications*, 6:115–22, 1994.

[CWN87] J. G. Cleary, I. H. Witten, and R. M. Neal. Arithmetic Coding for Data Compression. *Commun. of the ACM*, 30(6):520–40, Jun 1987.

[DeL58] H. DeLange. Research into the Dynamic Nature of the Human Fovea-Cortex Systems with Intermittent and Modulated Light. *J. Opt. Soc. Am.*, 48:777–84, 1958.

[dHB94]   Gerard de Haan and Paul W. A. C. Biezen. Sub-pixel motion estimation with 3-D recursive search block-matching. *Signal Proc: Image Comm.*, 6:229–239, 1994.

[DL95]   W. Ding and B. Liu. Rate-Quantization Modeling for Rate Control of MPEG Video Coding and Recording. In *SPIE Digital Video Compression: Algorithms and Technologies '95*, volume 2419, pages 139–50, 1995.

[DTV95a]   Digital Television Standard for HDTV Transmission. ATSC Document A/53, 1995. April 12.

[DTV95b]   Guide to the Use of Digital Television Standard for HDTV Transmission. ATSC Document A/54, 1995.

[EF95]   S. Eckart and C. Fogg. ISO-IEC MPEG-2 software video codec. In *SPIE Digital Video Compression: Algorithms and Technologies '95*, volume 2419, pages 100–9, 1995.

[Eng94]   B. Engel. MPEG vendor worksheet. General Instruments Corp., November 1994.

[FAE⁺94]   C. E. Fogg, P. Au, S. Eckart, T. Hanamura, K. Oosa, B. Quandt, and H. Watanabe. ISO/IEC software implementation of MPEG-1 video. In *SPIE. Digital Video Compression on Personal Computers: Algorithms and Technologies*, volume 2187, pages 249–52, 1994.

[Fil94]   L. Filippini. MPEG vendor worksheet. Center for Advanced Studies, Research, and Development in Sardinia, October 1994.

[Fin85]   D. G. Fink. *Reference Data and Equations*, chapter 22. In Benson [Ben85], 1985. ISBN 0-07-004779-0.

[FL90]   E. Feig and E. Linzer. Discrete Cosine Transform Algorithms for Image Data Compression. In *Proc. Electronic Imaging '90 East*, pages 84–7, Boston, MA, 1990.

[Ger88]   B. Gerod. Eye movements and coding of video sequences. *SPIE Visual Communications and Image Processing*, 1001:398–405, 1988.

[Gha90]   M. Ghanbari. The Cross-Search Algorithm for Motion Estimation. *IEEE Trans. Comm.*, 38(7):950–3, Jul 1990.

[Gil88]   Michael Gilge. A High Quality Videophone Coder using Hierarchical Motion Estimation and Structure Coding of the Prediction Error. pages 864–74, 1988.

[Gil90]   Michael Gilge. Motion estimation by scene adaptive block matching (SABM) and illumination correction. pages 355–66, 1990.

[Gir93]   Bernd Girod. Motion-Compensating Prediction with Fractional-Pel Accuracy. *IEEE Trans. Comm.*, 41(4):604–12, Apr 1993.

[Gon91]   C. A. Gonzales, 1991. Private communication.

[Gon96]   C. A. Gonzales. Private Communication, Jan 1996.

[GV91]   C. A. Gonzales and E. Viscito. Motion Video Adaptive Quantization in the Transform Domain. *IEEE Trans. Circuits Syst. Video Tech.*, 1(4):374–8, Dec 1991.

[Haa94]   J. Haass. MPEG vendor worksheet. Sun Microsystems, Inc., October 1994.

[Hab94]   G. T. Haber. MPEG vendor worksheet. CompCore Multimedia, Inc., November 1994.

[Har94]   S. Hartmann. MPEG vendor worksheet. Hartmann Multimedia Service, October 1994.

[Has72]   B. G. Haskell. Buffer and channel sharing by several interframe Picturephone® coders. *Bell Syst. Tech. J*, 51(1):261–89, Jan 1972.

[HMC72]   B. G. Haskell, F. W. Mounts, and J. C. Candy. Interframe Coding of Videotelephone Pictures. *Proc. IEEE*, 60(7):792–800, Jul 1972.

[Hof97]   Roy Hoffman. *Data Compression in Systems*. Chapman & Hall, New York, NY, 1997.

[HPN97]   Barry G. Haskell, Atul Puri, and Arun N. Netravali. *Digital Video: An Introduction to MPEG-2*. Chapman & Hall, New York, NY, 1997.

[Huf52]   D. A. Huffman. A Method for the Construction of Minimum-Redundancy Codes. *Proc. IRE*, 40(9):1098–101, Sep 1952.

[IEC86]   Time and control code for video tape recorders. IEC Standard Publication 461, Second Edition, 1986.

[IEE91]   IEEE Standard Specifications for the Implementations of 8x8 Inverse Discrete Cosine Transform. IEEE Std 1180-1990, Mar 1991.

[ISO93a]   Information technology — Coding of moving pictures and asso-
            ciated audio for digital storage media at up to about 1,5 Mbits/s
            —. International Standard ISO/IEC 11172-2, Part 2, 1993.

[ISO93b]   Test Model 5. Test Model Editing Committee. Standards docu-
            ment MPEG93/457, 1993.

[ISO95]   MPEG-4 Proposal Package Description (PPD) - Revision
            2. ISO/IEC JTC1/SC29/WG11. AOE Sub Group. MPEG
            95/N0937, Mar 1995.

[ITU90]   Recommendation H.261 – Video Codec for Audiovisual Services
            at p × 64 kbit/s. ITU-T (CCITT), Aug 1990. Specialist Group
            on Coding for Visual Telephony.

[ITU93]   Recommendation H.261: Video Codec for Audiovisual Services
            at $p \times 64$ kbits. ITU-T (CCITT), Mar 1993.

[ITU95]   Draft Recommendation H.263: Video Coding for Low Bitrate
            Communication. ITU-T (CCITT), Dec 1995.

[JCC94]   Her-Ming Jong, Liang-Gee Chen, and Tzi-Dar Chiueh. Paral-
            lel Architectures for 3-Step Hierarchical Search Block-Matching
            Algorithm. *IEEE Trans. Circuits Syst. Vid. Tech.*, 4(4):407–16,
            Aug 1994.

[JJ81]   Jaswant R. Jain and Anil K. Jain. Displacement Measurement
            and Its Application in Interframe Image Coding. *IEEE Trans.
            Comm.*, COM-29(12):1799–808, Dec 1981.

[Kat93]   Y. Katayama. Protection from IDCT Mismatch. Technical Re-
            port MPEG 93/283, ISO/IEC JTC1/SC2/WG11, Mar 1993.

[Kel61]   D. H. Kelly. Visual responses to time-dependent stimuli. I. am-
            plitude sensitivity measurements. *J. Opt. Soc. Am.*, 51:917–8,
            1961.

[KLH$^+$81] T. Koga, K. Linuma, A. Hirano, Y. Iijima, and T. Ishiguro.
            Motion-compensated interframe coding for video conferencing.
            In *NTC 81 Proc.*, pages G5.3.1–5, New Orleans, LA, Dec 1981.

[KN90]   Tero Koivunen and Ari Nieminen. Motion field restoration using
            vector median filtering on high definition television sequences.
            pages 736–42, 1990.

[Knu73]   D. E. Knuth. *Searching and Sorting, Vol 3, The Art of Computer
            Programming.* Addison-Wesley, Reading, MA, 1973.

[Kop94a]   T. Kopet. MPEG vendor worksheet. Array Microsystems, Inc., October 1994.

[Kop94b]   T. Kopet. MPEG vendor worksheet addendum. Array Microsystems, Inc., November 1994.

[Kor94]   R. Korosec. MPEG vendor worksheet. IBM Microelectronics, April 1994.

[KP89]   J. S. Kim and R. H. Park. Feature-based block matching algorithm using integral projections. *Electr. Lett.*, 25(1):29–30, Jan 1989.

[KR83]   S. Kappagantula and K. R. Rao. Motion compensated predictive coding. In *Proc. Int. Tech. Symp. SPIE*, San Diego, CA, Aug 1983.

[Kro94]   J. Krog. MPEG vendor worksheet. Texas Instruments, Inc., November 1994.

[LD94]   J. Lee and B. W. Dickinson. Temporally Adaptive Motion Interpolation Exploiting Temporal Masking in Visual Perception. *IEEE Trans. Image Proc.*, 3(5):513–26, Sep 1994.

[LDD+92]   Nan Li, Stefaan Desmet, Albert Deknuydt, Luc Van Eycken, and Andre Oosterlinck. Motion adaptive Quantization in Transform Coding for Exploiting Motion Masking Effect. pages 1116–23, 1992.

[Lee86]   Jong Soo Lee. Image Block Motion Estimation Using Fit by Generalized Discrete Orthogonal Polynomials. pages 188–92, 1986.

[LG96]   D. Le Gall. MPEG vendor worksheet. C-Cube Microsystems, Inc., August 1996.

[LM75]   J. O. Limb and J. A. Murphy. Measuring the Speed of Moving Objects from Television Signals. *IEEE Trans. Comm.*, 23(4):474–8, Apr 1975.

[Loh84]   H. Lohscheller. Subjectively Adapted Image Communication System. *IEEE Trans. Comm.*, COM-32(12):1316–22, Dec 1984.

[Loo94]   T. Lookabaugh. MPEG vendor worksheet. DiviCom, Inc., October 1994.

[Lou85]   B. D. Loughlin. *Monochrome and Color Visual Information and Transmission*, chapter 4. In Benson [Ben85], 1985. ISBN 0-07-004779-0.

[LR78]   J. O. Limb and C. B. Rubinstein. On the Design of Quantizers for DPCM Coders: A Functional Relationship Between Visibility, Probability and Masking. *IEEE Trans. Comm.*, COM-26(5):573–8, May 1978.

[LRR92]   H. Levkowitz, P. K. Robertson, and B. E. Rogowitz. Color Theory and Models for Computer Graphics and Visualization. Technical Report RC 18192, IBM Research Division, 1992.

[LWLS93] Liang-Wei Lee, Jhing-Fa Wang, Jau-Yien Lee, and Jung-Dar Shie. Dynamic Search-Window Adjustment and Interlaced Search for Block-Matching Algorithm. *IEEE Trans. Circuits Syst. Vid. Tech.*, 3(1):85–7, Feb 1993.

[LZ93]   Bede Liu and Andre Zaccarin. New Fast Algorithms for the Estimation of Block Motion Vectors. *IEEE Trans. Circuits Syst. Vid. Tech.*, 3(2):148–57, Apr 1993.

[Mac94]   A. G. MacInnis. MPEG-2 Systems. In *SPIE. Digital Video Compression on Personal Computers: Algorithms and Technologies*, volume 2187, pages 274–8, 1994.

[Mak94]   B. Makley. MPEG vendor worksheet. FutureTel, Inc., November 1994.

[Man94]   M. Mansson. MPEG vendor worksheet. LSI Logic Corp., October 1994.

[Mit95]   J. L. Mitchell. Private Communication, Nov 1995.

[MN89]   D. L. McLaren and D. T. Nguyen. Activity function for DCT coded images. *Elect. Lett.*, 25(25):1704–5, Dec 1989.

[MPG85]   Hans Georg Musmann, Peter Pirsch, and Hans-Joachim Grallert. Advances in Picture Coding. *Proc. IEEE*, 73(4):523–48, Apr 1985.

[Mul85]   K. T. Mullen. The Contrast Sensitivity of Human Colour Vision to Red-Green and Blue-Yellow Chromatic Gratings. *J. Physiol.*, 359:381–400, 1985.

[NB67]   F. I. Van Ness and M. A. Bouman. Spatial Modulation Transfer in the Human Eye. *J. Opt. Soc. Am.*, 57(3):401–6, Mar 1967.

[NH95]   A. N. Netravali and B. G. Haskell. *Digital Pictures*. Plenum Press, New York and London, 1995. Second Edition.

[Nis88] B. Niss. Prediction of AC Coefficients from the DC values. Technical Report N745, ISO/IEC JTC1/SC2/WG8, May 1988.

[NP77] A. N. Netravali and B. Prasada. Adaptive Quantization of Picture Signals Using Spatial Masking. *Proc. IEEE*, 65(4):536–48, Apr 1977.

[NR79] A. N. Netravali and J. D. Robbins. Motion-Compensated Television Coding: Part I. *BSTJ*, 58(3):631–70, Mar 1979.

[NS79] A. N. Netravali and J. A. Stuller. Motion-Compensated Transform Coding. *BSTJ*, 58(7):1703–18, Sep 1979.

[OR94] A. Ortega and K. Ramchandran. Forward-adaptive quantization with optimal overhead cost for image and video coding with applications to MPEG video coders. volume 2419, pages 129–37, 1994.

[Owe72] W. G. Owen. Spatio-temporal integration in the human peripheral retina. *Vision Res.*, 12:1011–26, 1972.

[PA90] A. Puri and R. Aravind. On comparing motion-interpolation structures for video coding. pages 1560–71, 1990.

[PA91] Atul Puri and R. Aravind. Motion-Compensated Video Coding with Adaptive Perceptual Quantization. *IEEE Trans. Circuits Syst. Video Tech.*, 1(4):351–61, Dec 1991.

[PA94] M. R. Pickering and J. F. Arnold. A Perceptually Efficient VBR Rate Control Algorithm. *IEEE Trans. Image Proc.*, 3(5):527–32, Sep 1994.

[PAC92] M. Pickering, J. Arnold, and M. Cavenor. VBR Rate Control with a Human Visual System Based Distortion Measure. In *Australian Broadband Switching and Services Symposium*, Jul 1992.

[Pan94] D. Pan. An Overview of the MPEG/Audio Compression Algorithm. In *SPIE. Digital Video Compression on Personal Computers: Algorithms and Technologies*, volume 2187, pages 260–73, 1994.

[Pea90] William A. Pearlman. Adaptive Cosine Transform Image Coding with Constant Block Distortion. *IEEE Trans. Comm.*, 38(5):698–703, May 1990.

[Pen96a] W. Pennebaker. MPEG vendor worksheet. Encoding Science Concepts, Inc., July 1996.

[Pen96b]   W. B. Pennebaker. Technical report, in preparation, 1996.

[Pen96c]   W. B. Pennebaker. Technical report, in preparation, 1996.

[Pen96d]   W. B. Pennebaker. Bit-rate Reduction and Rate Control for
            MPEG. Technical Report ESC96-001, Encoding Science Con-
            cepts, Inc., Feb. 1996.

[Pen96e]   W. B. Pennebaker. Motion Vector Search Strategies for MPEG-1.
            Technical Report ESC96-002, Encoding Science Concepts, Inc.,
            Feb. 1996.

[PHS87]   A. Puri, H. M. Hang, and D. L. Shilling. An efficient block-
            matching algorithm for motion compensated coding. In *Proc.
            IEEE ICASSP*, pages 25.4.1–4, 1987.

[Pin94]   J. Pineda. MPEG vendor worksheet. Logician, Inc., October
            1994.

[Pir81]   Peter Pirsch. Design of DPCM Quantizers for Video Signals
            Using Subjective Tests. *IEEE Trans. Comm.*, COM-29(7):990–
            1000, July 1981.

[PM93]   William B. Pennebaker and Joan L. Mitchell. *JPEG Still Image
            Data Compression Standard.* Van Nostrand Reinhold, New York,
            1993. ISBN 0-442-01272-1.

[Poy95a]   C. A. Poynton. Frequently Asked Questions about Colour. *In-
            ternet: www.inforamp.net/ poynton*, pages 1–24, 1995.

[Poy95b]   C. A. Poynton. Frequently Asked Questions about Gamma. *In-
            ternet: www.inforamp.net/ poynton*, pages 1–10, 1995.

[Qua94]   B. Quandt. MPEG vendor worksheet. HEURIS Logic Inc., March
            1994.

[Raz94]   A. Razavi. MPEG vendor worksheet. Zoran Corp., March 1994.

[Rea73]   C. Reader. *Orthogonal Transform Coding of Still and Moving
            Pictures.* PhD thesis, University of Sussex, 1973.

[Rea96]   C. Reader. MPEG4: Coding for content, interactivity, and uni-
            versal accessibility. *Optical engineering*, 35(1):104–8, Jan 1996.

[RF85]   A. R. Robertson and J. F. Fisher. *Color Vision, Representation,
            and Reproduction*, chapter 2. In Benson [Ben85], 1985. ISBN
            0-07-004779-0.

[Ris76] J. J. Rissanen. Generalized Kraft Inequality and Arithmetic Coding. *IBM J. Res. Develop.*, 20:198–203, May 1976.

[RJ91] M. Rabbani and P. W. Jones. *Digital Image Compression Techniques.* SPIE Optical Engineering Press, Bellingham, WA, 1991.

[Rob66] J. G. Robson. Spatial and temporal contrast-sensitivity functions of the visual system. *J. Opt. Soc. Am.*, 56:1141–2, Aug 1966.

[Roc72] F. Rocca. Television Bandwidth Compression Utilizing Frame-to-Frame Correlation and Movement Compensation. In *Symposium on Picture Bandwidth Compression, 1969*, MIT, Cambridge, Mass., 1972. Gorden and Breach.

[Rog83a] B. E. Rogowitz. Spatial/temporal interactions: Backward and forward metacontrast masking with sine wave gratings. *Visual Research*, 23(10):1057–73, 1983.

[Rog83b] B. E. Rogowitz. The Human Visual System: A Guide for the Display Technologist. *Proc. SID*, 24(3):235–52, 1983.

[RY90] K. R. Rao and P. Yip. *Discrete Cosine Transform.* Academic Press, New York, 1990.

[RZ72] Fabio Rocca and Silvio Zanoletti. Bandwidth Reduction Via Movement Compensation on a Model of the Random Video Process. *IEEE Trans. Comm*, pages 960–5, Oct 1972.

[Say96] K. Sayood. *Introduction to Data Compression.* Morgan Kaufmann Publishers, Inc., San Francisco, 1996.

[SB65] A. J. Seylor and Z. L. Budrikis. Detailed perception after scene changes in television image presentation. *IEEE Trans. Inf. Theory*, IT-11(1):31–43, Jan 1965.

[Sey62] A. J. Seyler. The Coding of Visual Signals to Reduce Channel-Capacity Requirements. Technical Report Monograph No. 535E, The Institution of Electrical Engineers, July 1962.

[Sey94] J. Seymour. MPEG vendor worksheet. AuraVision Corp., 1994.

[SGV89] Description of Ref. Model 8 (RM8). CCITT Study Group XV. Specialist Group on Coding for Visual Telephony. Doc. No. 525, Jun 1989.

[Sha49] C. E. Shannon. *The Mathematical Theory of Communication.* The University of Illinois Press, Illinois, 1949.

[Ste94] P. Stevens. MPEG vendor worksheet. Siemens Ltd., Australia, October 1994.

[Tan92] Craig K. Tanner. The Cable TV Industry Looks for Imaging Solutions at Cable Labs. *Advanced Imaging*, pages 50–65, Oct 1992.

[Tay94] M. Tayer. MPEG vendor worksheet. General Instruments Corp., March 1994.

[Tom94] A. S. Tom. MPEG vendor worksheet. Imedia Corp., October 1994.

[TR96] Jeffrey A. Trachtenberg and Mark Robichaux. Crooks Crack Digital Codes of Satellite TV. Wall Street Journal, page B1, January 12, 1996.

[Vas73] A. Vassilev. Contrast sensitivity near boarders: Significance of test stimulus form, size, and duration. *Vision Res.*, 13(4):719–30, Apr 1973.

[VCD94] Video CD Specification Version 2.0: Compact Disc Digital Video. JVC, Matsushita, Philips, and Sony, July 1994.

[WA85] A. B. Watson and A. J. Ahumada, Jr. Model of human visual-motion sensing. *J. Opt. Soc. Am. A*, 2(2):322–41, Feb 1985.

[Wan95] Brian A. Wandell. *Foundations of Vision*. Sinauer Associates, Inc, Sunderland, Massachusetts, 1995.

[WC90] Q. Wang and R. J. Clarke. Motion compensated sequence coding using image pyramids. *Elect. Lett.*, 26(9):575–6, Apr 1990.

[Wil91] R. N. Williams. *Adaptive Data Compression*. Kluwer Academic Publishers, Boston, 1991.

[WMB94] Ian H. Witten, Alistair Moffat, and Timothy C. Bell. *Managing Gigabytes*. Van Nostrand Reinhold, New York, 1994.

[XEO90] Kan Xie, Luc Van Eycken, and Andre Oosterlinck. Motion-Compensated Interframe Prediction. pages 165–75, 1990.

[XEO92] Kan Xie, Luc Van Eycken, and Andre Oosterlinck. A new block-based motion estimation algorithm. *Signal Proc.: Image Communications*, 4:507–17, 1992.

[Yag93] Yagasaki. Oddification problem for iDCT mismatch. Technical Report MPEG 93/278, ISO/IEC JTC1/SC2/WG11, Mar 1993.

[Zur94]   F. Zurla. MPEG vendor worksheet. IBM Microelectronics, March
           1994.

[ZZB91]   Sohail Zafar, Ya-Qin Zhang, and John S. Baras. Predictive Block-
           Matching Motion Estimation for TV Coding – Part I: Inter-Block
           Prediction. *IEEE Trans. Broadcasting*, 37(3):97–101, Sep 1991.

Index

458

464

466

470